I0066270

A DICTIONARY OF
WEIGHTS AND MEASURES
FOR THE BRITISH ISLES:
THE MIDDLE AGES
TO THE TWENTIETH CENTU

A DICTIONARY OF WEIGHTS AND MEASURES FOR THE BRITISH ISLES: THE MIDDLE AGES TO THE TWENTIETH CENTURY

RONALD EDWARD ZUPKO

AMERICAN PHILOSOPHICAL SOCIETY
Independence Square   Philadelphia
1985

Memoirs of the
AMERICAN PHILOSOPHICAL SOCIETY
Held at Philadelphia
For Promoting Useful Knowledge
Volume 168

Copyright 1985 by the American Philosophical Society

Library of Congress Catalog Card No. 68-14038
International Standard Book Number 0-87169-168-X
US ISSN 0065-9738

To

Ramon and Michael

CONTENTS

INTRODUCTION

The complexity of medieval and modern pre-metric weights and measures throughout the British Isles has long presented an obstacle to scholarly research on western European economic history. The problem is really two-fold: first, the approximate dimensions of many nonstandardized measuring units, used by both the Crown and the regional and local markets, varied from time to time and from place to place; second, the specific dimensions even of standard weights and measures used in any given period are often poorly understood. Too many times the researcher, investigating certain facets of economic and social development, has not taken these ambiguities into consideration, or has not even been aware of them, and has automatically assumed that a particular measuring unit contained or equaled a fixed amount. Such assumptions have led to inaccuracies in many textbooks and monographs. Hence, this book is directed toward clarifying some of the confusion and bringing a new focus to the field of metrology in general and a new understanding of the units in particular.

The tables which follow will aid the reader in using the dictionary. Since it would be impractical to give the year and reign for every citation (e.g., 25 Edward III), I have, in most instances, provided only the year in which a manuscript or law was written. Table 1 has been compiled for rapid identification of the ruling English, Scottish, Irish, or Welsh sovereign for any given year in the dictionary. Table 2 contains all of the abbreviations used throughout the dictionary; they are alphabetically arranged for quick reference. Tables 3 and 4 list the current English Imperial, American Customary, and metric units; Table 5 contains the basic equivalents for these weights and measures. The latter table will enable the reader to make further correlations between metric and nonmetric units that are beyond the scope of this book. Table 6 provides definitions for the terms used to describe the weights and measures in the entries.

The dictionary uses a number of textual devices to help the reader gain rapid and accurate access to the material. All entry headings are printed in boldface, and a dash separates them from their variant spellings (e.g., **acre**—1 aecer (OED), æcyr (OED); 1-2...). The variants are arranged according to the centuries in which they were most commonly used; the numbers preceding them identify the centuries:

1 = pre-12th century	6 = 16th century
2 = 12th century	7 = 17th century
3 = 13th century	8 = 18th century
4 = 14th century	9 = 19th and 20th centuries
5 = 15th century	? = no century given in source

If there is no citation for a certain variant spelling within an entry, its source reference is indicated in parentheses (e.g., 1 æcer (OED)...1-2 acr). The abbreviation L preceding a variant indicates that that variant was a Latin form used in scholarly treatises in England.

The etymologies, always in square brackets, immediately follow the variant spellings. Generally when an etymology is well known and can be found easily in the Oxford English Dictionary (OED) or Webster's New International Dictionary, 3rd edition (WNID3), only a shortened form is given in the entry, and the reader should refer to one of these standard etymological dictionaries for further information (e.g., **acre**—1 æcer (OED)...6-7 aiker [ME aker fr OE æcer; see WNID3]). If no etymology is given, an asterisk (*) indicates that the derivation of the word is unknown.

Following the etymological comments either a general explanation for the unit is given or, if there are variations within the unit, each major variation or group of variations is discussed in a separate paragraph or subsection. Every time the name of a unit other than the entry unit appears in the explanation it will be found in capital letters the first time it is used, and readers may refer to entries for these other units to gather additional information. In addition, wherever possible, metric equivalents are included in parentheses; the equivalents have been carried out to two decimal places for the approximate units and usually to three decimal places for the exact. But, if the unit's measurement or description is identical to that of another more commonly known unit, the words "equivalent to" follow the etymological comments. If the unit were different by definition from another unit, but commonly associated with it due to identical physical properties or dimensions, the terms "synonymous with" or "used interchangeably with" are employed.

After each major metrological variation or group of variations there are citations from medieval and modern sources:

The date in boldface type at the beginning of these citations always represents the year in which the manuscript or book was written and never the publication date.

The code name and numbers after the date identify the source (e.g., **1198** Feet 3.8: De vij...Ridon´).

The code name always refers to a corresponding title in the bibliography.

A Roman numeral following the code name, but preceding the period before the page number, supplies the volume (e.g., **1443** Brokage II.7).

An Arabic number in such a position refers to one of several books listed under that particular code name in the bibliography (e.g., Feet 3.8 refers to the third book under the code name Feet).

The number after the period is always the page number. If there is no volume number and the bibliographical code name has only one title listed under it, the page number immediately follows the source reference (e.g., Caernarvon 242).

Whenever a measuring unit has several variations which do not fit into any of the other major sections, or for which there is no explanation in the documents as to their relative value, they are placed at the end of the entry in a separate paragraph.

It should be noted that in the illustrative quotations all manuscript abbreviations have been expanded and underlined (e.g., "Et xl ptice" is changed to "Et xl pertice"). Also, letters superscripted in the source have been placed on the same line as the rest of the word, with brackets indicating the change (e.g., gr^{a}na is amended to gr[a]na). Similarly, whenever Roman numerals in manuscripts were elevated to the right of some number (e.g., V^{xx}), they have been placed on the text line, with brackets again indicating the change. If multiplication or addition is involved, the appropriate arithmetical sign has been placed between the numbers (e.g., V^{xx}XII = 112 is changed to V [X] xx [+] XII = 112). Other abbreviations, such as l., li., and lib. for *liber*, *libra* (pound) and the apothecary symbols ℈ for scruple, ʒ for dram, ℥ for ounce, and ℔ for apothecary pound, have been retained as in the original source.

The bibliography includes only those sources which provide information on individual weights and measures and which discuss some of the problems characteristic of metrology in general. No fictional sources were used in the data compilation and illustrative quotations. Works on the metric system generally are omitted except for those which discuss various aspects of the pre-metric systems. Finally, the bibliography includes the names of several reference books on weights and measures in which the interested reader may find leads to literature on other decimal and duodecimal systems.

ACKNOWLEDGMENTS

For the prompt transmission of the photographic reproductions of the many manuscripts, documents, monographs, and other sources used in the compilation of this book, I am indebted to the Department of Printed Books of the British Museum, the Science Museum Library of South Kensington, London, the University of London Library, the London School of Economics Library, the Service Photographique of the Bibliothèque Nationale, the Library Business Office of the University of Toronto, the Marquette University Memorial Library, the Photographic Section of the University of Pennsylvania Library, the Department of Photography-Cinema of the University of Wisconsin, and the Books and Series Department of University Microfilms. Special mention must be made of the Microfilm Department of the University of Wisconsin-Milwaukee who helped me in numerous ways throughout the months devoted to working among the thousands of books in the Short Title Catalogue Collections.

Furthermore, I am beholden to the National Science Foundation, the American Philosophical Society, and to Marquette University for awarding me fellowship and scholarship funds on a number of occasions. Without their generosity, I would not have been able to conduct my research both in the United States and in Europe over the course of the last ten years.

For their continuous and unstinting intellectual encouragement spanning more than a decade, I must express by deepest gratitude to Dr. Jon B. Eklund, Curator of Chemistry and Metrology in the Division of Physical Sciences at the Smithsonian Institution, Dr. Harald Witthöft, Professor of Medieval History at the University of Siegen, West Germany, and Dr.

Marshall Clagett, Professor of History at the School of Historical Studies, the Institute for Advanced Study, Princeton.

Finally, to Mr. Reginald W. Sprecher, a former student of mine in medieval history and currently my invaluable computer consultant and advisor, I owe my sanity.

Milwaukee, Wisconsin Ronald Edward Zupko

1985

TABLE 1 MONARCHS FOR YEARS CITED IN TEXT

(A) ENGLISH

Saxon Dynasty		House of Normandy	
Egbert	(802-39)	William I	(1066-87)
Ethelwulf	(839-58)	William II	(1087-1100)
Ethelbald	(858-60)	Henry I	(1100-35)
Ethelbert	(860-65)	Stephen	(1135-54)
Ethelred I	(865-71)	House of Plantagenet (Angevin)	
Alfred "the Great"	(871-99)	Henry II	(1154-89)
Edward "the Elder"	(899-924)	Richard I	(1189-99)
Athelstan	(924-39)	John	(1199-1216)
Edmund I	(939-46)	Henry III	(1216-72)
Edred	(946-55)	Edward I	(1272-1307)
Edwy	(955-59)	Edward II	(1307-27)
Edgar "the Peaceable"	(959-75)	Edward III	(1327-77)
Edward "the Martyr"	(975-78)	Richard II	(1377-99)
Ethelred II	(978-1016)	House of Lancaster	
Edmund II	(1016)	Henry IV	(1399-1413)
Danish Dynasty		Henry V	(1413-22)
Canute	(1016-35)	Henry VI	(1422-61)
Harold I	(1035-40)	House of York	
Harthacanute	(1040-42)	Edward IV	(1461-83)
Wessex Dynasty		Edward V	(1483)
Edward "the Confessor"	(1042-66)	Richard III	(1483-85)
Harold II	(1066)		

House of Tudor	
Henry VII	(1485-1509)
Henry VIII	(1509-47)
Edward VI	(1547-53)
Mary I	(1553-58)
Elizabeth I	(1558-1603)
House of Stuart	
James I	(1603-25)
Charles I	(1625-49)
The Commonwealth	
Oliver Cromwell	(1653-58)
Richard Cromwell	(1658-59)
Stuart Restoration	
Charles II	(1660-85)
James II	(1685-88)
William and Mary	(1689-1702)
Anne	(1702-14)
House of Hanover	
George I	(1714-27)
George II	(1727-60)
George III	(1760-1820)
George IV	(1820-30)
William IV	(1830-37)
Victoria	(1837-1901)
House of Saxe-Coburg	
Edward VII	(1901-10)
House of Windsor	
George V	(1910-36)
Edward VIII	(1936)
George VI	(1936-52)
Elizabeth II	(1952-)

(B) SCOTTISH

Pre-Stuart

Kenneth MacAlpin	(843-58)
Donald I	(858-62)
Constantine I	(862-77)
Aed	(877-78)
Eochaid, with Giric	(878-89)
Donald II	(889-900)
Constantine II	(900-43)
Malcolm I	(943-54)
Indulf	(954-62)
Dub	(962-66)
Culen	(966-71)
Kenneth II	(971-95)
Constantine III	(995-97)
Kenneth III	(997-1005)
Malcolm II	(1005-34)
Duncan I	(1034-40)
Macbeth	(1040-57)
Lulach	(1057-58)
Malcolm III	(1058-93)
Donald Bane	(1093-94)
Duncan II	(1094)
Donald Bane	(1094-97)
Edgar	(1097-1107)
Alexander I	(1107-24)
David I	(1124-53)
Malcolm IV	(1153-65)
William	(1165-1214)
Alexander II	(1214-49)
Alexander III	(1249-86)
Margaret	(1286-90)
"13 Claimants"	(1290-92)
John Balliol	(1292-96)
"War of Independence"	(1296-1306)
Robert I	(1306-29)
David II	(1329-71)

House of Stuart

Robert II	(1371-90)
Robert III	(1390-1406)
James I	(1406-37)
James II	(1437-60)
James III	(1460-88)
James IV	(1488-1513)
James V	(1513-42)
Mary	(1542-67)
James VI	(1567-1625)

(C) IRISH

Connor	(817-31)	Brian Boru	(1002-14)
Niall of Callan	(831-43)	Malachy II	(1014-22)
Malachy I	(843-60)	Interregnum	(1022-42)
Hugh Finly	(860-77)	Dermott	(1042-72)
Flann of the Shannon	(877-916)	Turlough O´Brian	(1072-86)
Niall (Black-Knee)	(916-19)	Murtaugh O´Brian	(1086-1119)
Donough	(919-49)	Donald MacLaughlin	(1119-21)
Connell	(949-59)	Interregnum	(1121-36)
Donald O´Niall	(959-80)	Turlough O´Connor	(1136-56)
Malachy II	(980-1002)	Murty O´Neill	(1156-66)
		Roderic O´Connor	(1166-82)

(D) WELSH

Territorial Divisions

Monarchs	Gwynedd	Powys	Seisyllwg	Dyfed
Rhodri the Great	844-78	855-78	872-78	-------
Cadell	-------	-------	878-909	-------
Anarawd	878-916	-------	-------	-------
Hywel the Good	942-50	-------	920-50	904-50
Idwal the Bald	916-42	-------	-------	-------
Owain ap Hywel	-------	-------	954-88	954-88
Maredudd ab Owain	986-99	-------	988-99	988-99
Llywelyn ap Seisyll	1005-23	-------	1018-23	1018-23
Rhydderch ab Iestyn	-------	-------	1023-33	1023-33
Iago ab Idwal	1023-39	-------	-------	-------
Gruffydd ap Llywelyn	1039-63	1039-63	1055-63	1055-63
Bleddyn ap Cynfyn	1063-75	-------	-------	-------
Rhys ap Tewdwr	-------	-------	1081-93	1081-93
Gruffydd ap Cynan	1081-1137	-------	-------	-------
Maredudd ap Bleddyn	-------	1116-32	-------	-------
Madog ap Maredudd	-------	1132-60	-------	-------
Gruffydd ap Rhys	-------	-------	1135-37	1135-37
Owain Gwynedd	1137-70	-------	-------	-------
Rhys ap Gruffydd	-------	-------	1155-97	1155-97
Gruffydd Maelor I	-------	Northern 1160-91	-------	-------

Monarchs	Gwynedd	Powys	Seisyllwg	Dyfed
Owain Cyfeiliog	-------	Southern 1160-95	-------	-------
Dafyddd ab Owain	Eastern 1175-94	-------	-------	-------
Rhodri ab Owain	Western 1175-95	-------	-------	-------
Madog ap Gruffydd	-------	Northern 1191-1236	-------	-------
Gwenwynwyn	-------	Southern 1195-1208	-------	-------
Llywelyn ab Iorwerth	Eastern 1195-1240 Western 1200-1240	Southern 1208-40	1216-40	1216-40
Gruffydd Maelor II	-------	Northern 1236-70	-------	-------
David ap Llywelyn	1240-46	-------	-------	-------
Gruffydd ap Gwenwynwyn	-------	Southern 1240-86	-------	-------
Llywelyn ap Gruffydd	1246 (in part) 1256 (in whole)-1282	Southern 1257-82	1258-82	1258-82

TABLE 2 ABBREVIATIONS

a	=	are	der	=	derivative
acc	=	accusative	dg	=	decigram
adj	=	adjective	dial	=	dialect, dialectal
AF	=	Anglo-French	dim	=	diminutive
AL	=	Anglo-Latin	dkg	=	dekagram
alter	=	alteration	dkl	=	dekaliter
ap	=	apothecary	dkm	=	dekameter
Ar	=	Arabic	dl	=	deciliter
avdp	=	avoirdupois	dm	=	decimeter
bbl	=	barrel	dr	=	dram
BI	=	British Imperial	dst	=	decistere
bu	=	bushel	Du	=	Dutch
c	=	about, around	dwt	=	pennyweight
C	=	hundred	E	=	English
Celt	=	Celtic	F	=	French
cent	=	century	fem	=	feminine
cf	=	compare	fr	=	from
cg	=	centigram	ft	=	foot
cl	=	centiliter	g	=	gram
cm	=	centimeter	G	=	German
coll	=	collective	Gael	=	Gaelic
cu	=	cubic	gal	=	gallon
Cwt	=	hundredweight	Gmc	=	Germanic
Dan	=	Danish	Goth	=	Gothic

gr = grain
Gr = Greek
ha = hektare
hg = hektogram
Hind = Hindustani
hl = hektoliter
hm = hektometer
Icel = Icelandic
Ir = Irish
It = Italian
kg = kilogram
kl = kiloliter
km = kilometer
l = liter
L = Latin
lb = pound
LB = Low Breton
LG = Low German
LL = Late Latin
m = meter
M = thousand or thousandweight
m-a = measure of area
m-c = measure of capacity
MDu = Middle Dutch
ME = Middle English
MedL = Medieval Latin
merc = mercantile
MF = Middle French
mg = milligram
MHG = Middle High German
mi = mile
ml = milliliter
m-l = measure of length
MLG = Middle Low German
mm = millimeter
modif = modification
m-q = measure of quantity
m-v = measure of volume
n = noun
neut = neuter
Nor = Norwegian
Nord = Nordic
OE = Old English
OF = Old French
OHG = Old High German
OIr = Old Irish
OIt = Old Italian
OLG = Old Low German

ON	=	Old Norse	Scand	=	Scandinavian
ONF	=	Old North French	Sem	=	Semitic
OPort	=	Old Portuguese	SI	=	Système International
OPr	=	Old Provençal	Skr	=	Sanskrit
OS	=	Old Saxon	Sp	=	Spanish
OSp	=	Old Spanish	sq	=	square
OSw	=	Old Swedish	st	=	stere
oz	=	ounce	sv	=	sub verbo
part	=	participle	Sw	=	Swedish
perh	=	perhaps	t	=	troy
pk	=	peck	tow	=	tower
pl	=	plural	trans	=	translated, translation
Port	=	Portuguese	ult	=	ultimately
poss	=	possessive	US	=	United States
Pr	=	Provençal	v	=	verso
prob	=	probably	var	=	variant(s)
prop	=	properly	vb	=	verb
pt	=	pint	VL	=	Vulgar Latin
qt	=	quart	W	=	Welsh
s	=	scruple	wt	=	weight
Sc	=	Scottish	yd	=	yard

TABLE 3 CURRENT ENGLISH IMPERIAL AND AMERICAN CUSTOMARY UNITS

Linear Measure

12 inches	=	1 ft
3 ft	=	1 yd or 36 inches
5 1/2 yd	=	1 rod or 16 1/2 ft
40 rods	=	1 furlong or 220 yd
8 furlongs	=	1 statute mi or 5280 ft

Area Measure

144 sq inches	=	1 sq ft
9 sq ft	=	1 sq yd or 1296 sq inches
30 1/4 sq yd	=	1 sq rod or 272 1/4 sq ft
160 sq rods	=	1 acre or 4840 sq yd or 43,560 sq ft
640 acres	=	1 sq mi or 3,097,600 sq yd

Cubic Measure

1728 cu inches	=	1 cu ft
27 cu ft	=	1 cu yd

Liquid or Dry Measure

English Imperial

4 gills	=	1 pt
2 pt	=	1 qt
4 qt	=	1 gal or 277.42 cu inches
2 gal	=	1 pk
8 gal	=	1 bu or 2219.36 cu inches
8 bu	=	1 quarter

Liquid Measure

American Customary

4 gills	=	1 pt
2 pt	=	1 qt
4 qt	=	1 gal or 231 cu inches

Dry Measure

American Customary

2 pt	=	1 qt
8 qt	=	1 pk
4 pk	=	1 bu or 2150.42 cu inches

Troy Weight

24 gr	=	1 dwt
20 dwt	=	1 oz or 480 gr
12 oz	=	1 lb or 5760 gr

Apothecaries Weight

20 gr	=	1 s
3 s	=	1 dr or 60 gr
8 dr	=	1 oz or 480 gr
12 oz	=	1 lb or 5760 gr

Avoirdupois Weight

English Imperial

27 11/32 gr	=	1 dr
16 dr	=	1 oz or 437 1/2 gr
16 oz	=	1 lb or 7000 gr

English Imperial (continued)

14 lb	=	1 stone
2 stone	=	1 quartern or 28 lb
4 quartern	=	1 Cwt or 112 lb
20 Cwt	=	1 ton or 2240 lb

American Customary

27 11/32 gr	=	1 dr
16 dr	=	1 oz or 437 1/2 gr
16 oz	=	1 lb or 7000 gr
100 lb	=	1 short Cwt
112 lb	=	1 long Cwt
20 short Cwt	=	1 short ton or 2000 lb
20 long Cwt	=	1 long ton or 2240 lb

TABLE 4 CURRENT METRIC (SI) UNITS

Prefixes

Prefix	Signification
deka-	10
hekto-	100
kilo-	1000
deci-	0.1
centi-	0.01
milli-	0.001

Linear Measure

10 mm	=	1 cm
10 cm	=	1 dm
10 dm	=	1 m
10 m	=	1 dkm
10 dkm	=	1 hm
10 hm	=	1 km

Area Measure

100 sq mm	=	1 sq cm
100 sq cm	=	1 sq dm
100 sq dm	=	1 sq m
100 sq m	=	1 are
100 ares	=	1 ha
100 ha	=	1 sq km

Volume Measure

10 ml	=	1 cl
10 cl	=	1 dl
10 dl	=	1 l
10 l	=	1 dkl
10 dkl	=	1 hl
10 hl	=	1 kl

Cubic Measure

1000 cu mm	=	1 cu cm
1000 cu cm	=	1 cu dm
1000 cu dm	=	1 cu m
1000 cu m	=	1 cu dkm
1000 cu dkm	=	1 cu hm
1000 cu hm	=	1 cu km

Weight

10 mg	=	1 cg
10 cg	=	1 dg
10 dg	=	1 g
10 g	=	1 dkg
10 dkg	=	1 hg
10 hg	=	1 kg
1000 kg	=	1 metric ton

TABLE 5 BASIC EQUIVALENTS

Linear Measure

centimeter =
0.01094 yd
0.03281 ft
0.3937 inch

decimeter =
0.3281 ft
3.9370 inches

dekameter =
10.9361 yd
393.70 inches

foot =
0.3048 m
30.480 cm

furlong =
201.168 m

hektometer =
19.8838 rods
109.361 yd

inch =
2.5400 cm
25.400 mm

kilometer =
0.6214 mi
1093.6 yd
3280.8 ft

meter =
1.09361 yd
3.2808 ft
39.370 inches

mile (statute) =
1.6093 km
1609.3 m

millimeter =
0.03937 inch

rod =
5.0292 m

yard =
0.9144 m
91.440 cm

Area Measure

acre =
0.4047 ha
4046.8 sq m

are =
0.0247 acre
119.60 sq yd
1076.4 sq ft

hektare =
2.4710 acres
395.367 sq rods

square centimeter =
0.00108 sq ft
0.1550 sq inch

square foot =
0.09290 sq m
929.03 sq cm

square inch =
6.4516 sq cm
645.16 sq mm

square kilometer =
0.3861 sq mi
247.10 acres

square meter =
0.0395 sq rods
1.1960 sq yd
10.764 sq ft
1550.0 sq inches

Area Measure (continued)

square mile =
2.5900 sq km
259.000 ha

square millimeter =
0.00155 sq inch

square rod =
0.00253 ha
25.293 sq m

square yard =
0.8361 sq m
8361.3 sq cm

Capacity or Volume Measure

bushel (English) =
0.36368 hl
3.6368 dkl
36.3677 l

bushel (American) =
0.35238 hl
3.5238 dkl
35.238 l

centiliter =
0.61025 cu inch

cubic centimeter =
0.06102 cu inch

cubic decimeter =
0.00131 cu yd
0.0353 cu ft
61.023 cu inches

cubic foot =
0.02832 cu m
28.316 l

cubic inch =
0.01639 l
1.6387 cl
16.387 ml
16.387 cu cm

cubic meter =
1.3079 cu yd
35.314 cu ft

cubic millimeter =
0.00006 cu inch

cubic yard =
0.7646 cu m
764.54 l

deciliter =
6.1025 cu inches

dekaliter =
0.27497 BI bu
0.28378 US bu

gallon (English) =
4.54596 l
4546.1 cu cm

gallon (American) =
3.7853 l
3785.4 cu cm

gill (English) =
0.14206 l
142.07 cu cm

gill (American) =
0.11829 l
118.295 cu cm

hektoliter =
2.7497 BI bu
2.8378 US bu

Capacity or Volume Measure (continued)

kiloliter =
1.3080 cu yd
35.316 cu ft

liter =
0.00131 cu yd
0.03532 cu ft
61.025 cu inches
0.02750 BI bu
0.02838 US bu
0.21998 BI gal
0.26418 US gal
0.87990 BI qt
1.05671 US liquid qt
0.90810 US dry qt

milliliter =
0.0610 cu inch

peck (English) =
9.0919 l

peck (American) =
8.8096 l

pint (English) =
0.56825 l
568.25 ml
568.26 cu cm

pint (American liquid) =
0.47317 l
473.167 ml
473.17 cu cm

pint (American dry) =
0.5506 l
550.599 ml
550.61 cu cm

quart (English) =
1.1365 l
1136.52 cu cm

quart (American liquid) =
0.9463 l
946.358 cu cm

quart (American dry) =
1.1012 l
1101.23 cu cm

quarter =
2.909 hl

Weight

centigram =
0.15432 gr

decigram =
1.54324 gr

dekagram =
0.35274 avdp oz
5.64383 avdp dr

dram (apothecaries) =
3.88794 g

dram (avoirdupois) =
1.77184 g

grain =
0.0648 g
64.7989 mg
0.00268 ap or t lb
0.00220 avdp lb
0.03215 ap or t oz
0.03527 avdp oz
0.25721 ap or t dr
0.56438 avdp dr
0.64301 dwt
0.77162 s
15.4324 gr

hektogram =
3.52739 avdp oz

Weight (continued)

hundredweight (short) =
45.3592 kg

hundredweight (long) =
50.8023 kg

kilogram =
2.67923 ap or t lb
2.20462 avdp lb

milligram =
0.01543 gr

ounce (apothecaries or troy) =
31.1035 g

ounce (avoirdupois) =
28.3495 g

pennyweight =
1.55517 g
1555.17 mg

pound (apothecaries or troy) =
0.37324 kg
373.242 g

pound (avoirdupois) =
0.45359 kg
453.592 g

quartern =
12.70 kg

scruple =
1.29598 g
1295.98 mg

stone =
6.350 kg

ton (short) =
0.90718 metric ton
907.185 kg

ton (long) =
1.01605 metric ton
1016.05 kg

ton (metric) =
2204.62 avdp lb
1.10231 short ton
0.98421 long ton

TABLE 6 TERMINOLOGY IN WEIGHTS AND MEASURES

General Terms

UNIT A unit is a value, quantity, or magnitude by which other values, quantities, or magnitudes are expressed. Generally a unit is fixed by definition and is independent of such physical conditions as temperature. The pound, bushel, and mile are examples of units used to express a fixed weight, capacity, and length, respectively.

STANDARD A standard is a physical representation of a unit. Generally it is not independent of physical conditions; it is a genuine or absolute representation of a unit only under certain controlled conditions. For example, a physical standard for the yard would vary slightly in length if it were not kept in a hermetically sealed compartment with a controlled constant atmospheric temperature.

MEASURE OF LENGTH A measure of length or linear measure is the distance between two points established according to some standard. The standard may be simple and primary, such as the pace, the palm, the finger, etc.; or it may be based on an arbitrarily defined unit, such as the medieval English inch that was taken as the length of three medium-sized barleycorns placed end to end. Statutes have furthered the use of the defined standard over the simple primary standard: for example, when multiples of the inch were reconciled with the larger units of length such as the yard, ell, fathom, mile, and league.

MEASURE OF AREA A measure of area or superficial measure is commonly the square of the linear unit and is usually defined in terms of square feet, square yards, or square rods (perches). The acre—the principal superficial measure in medieval England—consisted of 40 linear perches in length and 4 in width or 160 square perches. The actual number of square feet in this acre, however, depended on the size of the linear unit. An acre contained 43,560 square feet only when its linear perch equaled 16 1/2 feet.

MEASURE OF CAPACITY A measure of capacity or volume measure is the cube of the linear unit. In medieval England a capacity measure was usually a vessel that contained a certain mass of liquid or dry substance but it did not necessarily have a definite size or shape. Units such as the bundle, bag, box, cage, chest, and sack had varying dimensions depending on the quality, form, and weight of a particular product.

MEASURE OF QUANTITY A measure of quantity is the number, tale, or count of a certain product. In medieval England any related dimensions of quantity measures were usually unspecific and depended upon the physical qualities of the product (e.g., a hundred of hoops versus a hundred of eels). But measures of quantity normally consisted of a specified number. A dozen, for example, was 12 of any item. A hundred was often 100, 106, 112, 120, 124, 160, or 225 depending on the product. A score was usually 20, while a gross was 12 dozen.

MASS The mass of a body is a measure of its inertial property; the "weight" of a body has been used traditionally to designate its mass or to designate a force that is related to gravitational attraction. Since these two concepts are currently considered incompatible and confusing, the present trend is to avoid using the term "weight" in the context of force. Hence, when the term "weight" is used, as in weights and measures, it is considered to be synonymous with mass. Medieval English systems of weight were based either on the barley or on the wheat grain. The pennyweight, containing 24 barleycorns in the troy system or 32 wheat grains in the tower system, was the unit from which the larger weights, such as the scruple, dram, ounce, and pound, were formed. Hence, the troy pound of 5760 grains consisted of 240 pennyweights of 24 grains each or 12 ounces of 480 grains, each ounce containing 20 pennyweights of 24 grains each.

GROSS WEIGHT Gross weight refers to the weight (mass) of both the container and its contents. The best example of this was the butter barrel, which medieval English documents define as a vessel weighing generally 26 pounds and containing 230 pounds of butter. The total or gross weight, thus, was 256 pounds.

Special Terms

ARE An are is a metric unit of area equal to the area of a square 10 meters long on each side; hence, 100 square meters.

ASSAY An assay was a testing of weights and measures to determine whether they were in conformity with Crown standards. Private citizens, lords of manors, abbots, bailiffs, chancellors of Oxford and Cambridge, bishops and archbishops, mayors, guilds, courts leet, justices of assize and of oyer and terminer, sheriffs, and coroners shared the privilege of performing assays, along with clerks of the market and justices of the peace.

ASSIZE An assize was an enactment that regulated the quality, quantity, weight, measure, and price of articles for sale. An example of this type of assize was the Assisa Panis et Cervisiæ of Henry III, issued in 1266. The assize was also the name for a session at which the examination and authentication of local weights and measures took place. Merchants and producers broke the assize when they adulterated their goods, sold defective merchandise, or employed false weights and measures.

AULNAGE Aulnage was the measuring of cloth to determine whether its length and breadth violated any of the specifications laid down by statute.

AULNAGER An aulnager was an official stationed in a port or town who measured the cloth brought in by merchants and textile manufacturers to determine whether its length and breadth conformed to statutory specifications.

CLERK OF THE MARKET A clerk of the market was an appointed official who verified and enforced statutory weights and measures. He represented the Crown in what could be considered a prescriptive office for he had no other function and usually operated independently of local judges and justices. The clerk of the market for the king's household (clericus mercate hospitu regis) looked after the king's standards and saw to it that weights and measures in every district conformed to them. There were also clerks of the market (clerc del marche, clericus merketi, clericus marescalciæ) assigned to the most important shires to oversee and supervise the local use of weights and measures. The authority of the clerks was not always clearly defined.

GRAM A gram is a metric unit of weight equal to 1/1000 kilogram and nearly equal to one cubic centimeter of water at its maximum density.

IMPERIAL GALLON An imperial gallon is the volume occupied by 10 pounds of distilled water of density 0.998859 gram per milliliter weighed in air of density 0.001217 gram per milliliter against weights of density 8.136 grams per milliliter.

KILOGRAM A kilogram is a metric unit of mass (weight) equal to the mass of a particular platinum-iridium standard, the International Prototype Kilogram, kept at the International Bureau of Weights and Measures (Bureau International des Poids & Mesures) in Sèvres, France, and nearly equal to 1000 cubic centimeters of water at the temperature of its maximum density.

LITER A liter is now a special name given to a cubic decimeter. Prior to 1964 it was described as the volume occupied by one kilogram of distilled water at 4° Centigrade (Celsius) and at the standard atmospheric temperature of 760 millimeters.

METER A meter is a metric unit of length equal to 1,650,763.73 wavelengths in a vacuum of the orange-red radiation of krypton 86. The meter is the unit upon which all metric standards and measurements of length, area, and capacity are based.

METRICATION Metrication is the process of converting any unit to its metric equivalent.

NAUTICAL MILE A nautical mile is the length of one minute of the meridian through Greenwich, that is 1/60th of a degree of latitude.

PONDERATOR A ponderator was a locally appointed weigher of agricultural and nonagricultural goods in a village market or in a town weighing station. His services appear to have encompassed all aspects of commerce and trade. He is also known in medieval English documents as a pensarius, pesarius, poiser, ponderarius, and poynder.

SEAL A seal was a mark affixed to weights and measures by either the Crown or local municipal officials to prevent frauds. The practice probably originated during the reign of William I.

SI SI is the accepted abbreviation for Système International d'Unités (International System of Units), the modern form of the metric system finalized at the Eleventh General Conference of Weights and Measures in October, 1960.

STRIKE A strike was usually a wooden board with a straight edge of greater length than the diameter of the measure to be struck (leveled). It was passed over the rim after the measure had been filled as a prevention against the traditional practice of selling wheat and certain other commodities by heaped measure.

A DICTIONARY OF

BRITISH

WEIGHTS AND MEASURES

A

acar, acer, acr, acra. ACRE

acre—1 æcer (OED), æcyr (OED); 1-2, 6 acr; 1-7 L acra; 2 æker (OED); 2-7 aker; 3 akre (Langtoft); 4-9 acre; 5 akere (OED), akyr, akyre (OED), hakere (OED); 5-6 akir; 6 acar, acer (McCaw); 6-7 aiker [ME aker fr OE æcer; see WNID3]). A m-a for land in England, Wales, Scotland, and Ireland which, in its earliest usage, probably referred to the amount of land which one yoke of oxen could plow in a day. Sometimes it was abbreviated a. or ac.

In England the acre was standardized during the High Middle Ages at 160 sq PERCHES of 16 1/2 ft each, or 4840 sq yd, or 43,560 sq ft (0.405 ha). This statutory acre was 40 perches in length and 4 in breadth and was equal to 4 ROODS of 40 sq perches each.—c**1065** St. Edmunds 25: Goduin Aluuini nepos IX acras. c**1175** Clerkenwell II: Et ix acras in prato in eadem villa de Kingestuna. **1198** Feet 3.8: De vij acris terre cum pertinentiis in Ridon´; ibid 65: iiij acras terre et iij rodas. c**1200** Caernarvon 242: Tres pedes faciunt vlnam quinque ulne & dimidia faciunt perticam. Et xl pertice in longitudinem & iiij in latitudine faciunt vnam acram terre. **1200** Feet 3.108: Et j acram juxta domum Willelmi filii Wictiue. **1206** Feet 2.46: De dimidia acra terre. c**1230** Clerkenwell 134: Vna acra terre in parochia de Sidingeburne. **1283** Battle xliii: Et sunt ibidem in campis qui vocantur Horscroftes lxxviij acræ separales. c**1300** Brit. Mus. 18.135v: Quando acra terre continet x perticas in longitudine tunc xvj perticas in latitudine. c**1310** Malmesbury II.323: Pro una acra terræ quæ vocatur la Guldene acre. c**1400** Hall 41: Nota quod lxviii milia lepores possunt sedere in una acra terre mensurata. c**1440** Promp. Parv. 8: Akyr of londe. Acra. c**1461** Hall 7: Et iiii perches en laeure et xl en longure font l acre de terre.... Et quinque ulne et dimidia faciunt perticam et xl pertice in longitudine et quatuor in latitudine faciunt unam acram. **1494** Fabyan 246: An acre conteyneth xl. perches in length, and iiii. in brede. c**1500** Brit. Mus. 6.7: Una Acra...clx pearches.... Di Acr...lxxx pearches. **1502** Arnold 173: Of what lengith soo euer they be, clx. perches make an akir. **1537** Davenport lxxxv: An aker of whete and an aker of barley. **1558** Gray 235: Also Ayther of theym haith one Rige of medo lying in the este field in one plays called the mire Doyle conteyning by estimation two parts of one acar.... Item two Ingdailes lying in the newe Inge in the same contening by estimation one half Acar. **1567** Acts Scotland 3.38: Exceid not vj aikeris of land. **1589** Bellot 15: You knowe that an acre ought to bee of fortie poles in length, and foure poles in breadth, and the kinges pole is of sixteene foote and an halfe. **1603** Henllys 133: For whereas the statute de terris mensurandis appointeth the pole to be xvi foote and di...and that 4 of these in bredth, and 40 in length make the acre. **1613** Tap 1.62: One Acre containeth...Roods. 4...Square Perches. 160...Yards. 4840...Feete. 43560. **1615** Collect. Stat. 464: And forty pearches, and 4. in bredth make an acre. **1616** Hopton 165: So that an Acre hath 43560 square Feete, 4840 square Yards, and 160 square Pearches. **1616** Salignacus 80: If two oxen are 4 akers of land in 23 1/2 dayes, in how many dayes shall 2 oxen are 3 akers. **1624** Huntar 2: The Aiker of land. **1635** Dalton 150: Forty pole in length, and foure in breadth (or 160 pole) doe make an acre. **1647** Digges 1: Five Yards, 1/2. a Pearch: fortie Pearches in length and foure in breadth, an Acre. **1664** Spelman 8: Est autem Acra, mensurata terræ portio, olim incerta, sed nunc Statuto Anni 31 Edowardi primi, bis octogies perticam continens. **1664** Gouldman sv: An acre. Acra, f. jugerum. **1682** H. Coggeshall 2.63: In Land-measure 160 Sq. Perches, at 16 1/2 F. to the Perch, make an Acre. **1784** Ency. méth. 139: L'acre de terre d'Angleterre est de 4 fardingdeales. **1829** Palethorpe sv acre: ACRE, the universal measure of land in England. It contains 4 square roods, each rood 40 square poles or perches of 16 1/2 feet each. **1883** Simmonds sv: The English standard acre is 4840 square yards. **1883** McConnell 13: Imperial Acre = 4840 sq yds. = 43560 sq ft. = 6,272,640 sq. in. **1907** Hatch 23: 4 roods = 1 acre = 160 sq. rods = 4840 sq. yards = 43560 sq. feet = 10 sq. chains; ibid 35: 1 acre = .404684934 hectare. **1956** Economist 7: 4 roods = 1 acre...640 acres = 1 sq. mile; ibid 8: Acre. Imperial...4,840 sq yards. **1969** And. & Bigg 11: 1 acre = 4046.86 m^2 = 0.404686 ha. See CHAIN; FARTHINGDALE

Since the size of the acre was defined in terms of the linear perch, regional variations arose whenever the length of the perch (16 1/2 ft by statute, or 5.029 m) or the number of sq perches in the acre (160 by statute) differed from the statutory standards. For example, acres larger than the statutory acre were used (c1800-1900) in Cheshire,

10,240 sq yd (c0.86 ha); Cornwall, 5760 sq yd (c0.48 ha); Lincolnshire, 5 roods (c0.51 ha); Staffordshire, nearly 2 1/4 statute acres (c0.911 ha); Westmorland, 6760 sq yd (0.565 ha) or 160 perches of 6 1/4 sq yd each; Ireland, called the Irish plantation acre, 7840 sq yd (0.655 ha) or 160 sq perches, each perch equal to 7 yd; and Scotland, 6150 4/10 sq yd or 55,353.6 sq ft (c0.51 ha) or slightly more than 5/4 of an English statute acre (1.2707438 exactly) (Second Rep. 5, Cyclopædia sv weights, and Skilling Preface). Acres smaller than the statutory acre were used (c1800-1900) in Bedfordshire, sometimes 2 roods (c0.20 ha); Dorsetshire, generally 134 sq perches (c0.34 ha); Herefordshire, 2/3 of a statute acre (c0.27 ha); Leicestershire, 2308 1/4 sq yd (c0.19 ha); Worcestershire, 90 to 141 sq perches (c0.23 to c0.36 ha); and North Wales, 4320 sq yd (0.361 ha) for the ERW or standard acre and 3240 sq yd (0.271 ha) for the STANG or customary acre (Second Rep. 5 and Cyclopædia sv weights). Some regions had acres (c1800-1900) both larger and smaller than the statutory acre: Hampshire, 107 to 180 sq perches (c0.27 to c0.45 ha) and Sussex, 100, 107, 110, 120, 130, 180, or 212 sq perches (c0.25 to c0.54 ha) (Cyclopædia sv weights, Second Rep. 5, and Donisthorpe 204). A "hop acre" in Herefordshire (c1800-1900) was a section of land containing 1000 plants, equal to approximately 1/2 statute acre (Cyclopædia sv weights and Donisthorpe 204). Other variations resulting from diverse perch lengths appeared from time to time.—**c1100** Bello 11: Pertica habet longitudinis sedecim pedes. Acra habet in longitudine quadraginta perticas, et quatuor in latitudine.

1400 Henley 68: E pur ceo ke les acres ne sunt mye touz de une mesure kar en acon pays mesurent il par la verge de xviii peez e...de xx peez e...de xxii peez e...de xxiiij peez. c**1475** Hall 14: And sum of thame [perches] be of xviij fote, sum of xx fote, and sum of xxi fote; but of what lengthe be euer thei be, euermore this is yt serteyn, that viii [X] xx perchys make an aker. **1537** Benese 4: An acre bothe of woodlande, and also of fylde land is always xl. perches in length, and iiii. perches in bredth, although an acre of woodlande be more in quantite...because the perche of woodlande is longer. **1589** Bellot 7: And the acre which is measured by the pole of foure and twentie foote, maketh two acres and a roode of the pole of sixteene foote, and foure acres doe make nine acres. **1654** J. Eyre 182: 986 Irish Acres, at 21 foot to the Pole. **1665** Assize 6: In many Countries [= districts] this Pole or Perch doth vary, as in some places it is 18 foot, and in some other places 21 foot.... Of the which Poles...40 in length, and four of them in breadth, make the Acre of Land or Wood. **1867** C. I. Elton 129: Varying indefinitely in length and breadth, it [the Kentish acre] was always a piece of land containing 160 perches of sixteen feet square, i.e. a fraction over 4,551 square yards. **1888** Taylor 179: We occasionally meet with records of acres which are said to be measured by the perch of 10, 16, 19, 20, 22, and 25 feet...and also of acres at 18, 20, and 22 perches to the acre instead of 40. **1889** Francis 11: The acre of Devonshire and Somersetshire contains 160 perches of 15 ft.... Lancashire, 160 perches of 21 ft., or 70,560 square ft. **1897** Maitland

375: Even if the limits of variation are given by rods of 12 and 24 feet, this will enable one acre to be four times as large as another.

acreme [*; see OED]. A late medieval and early modern law term which designated an area of land containing 10 acres or 48,400 sq yd (4.050 ha). It appears to be synonymous with the FARTHINGDALE.—**1669** Worlidge 321: An Acreme of Land is ten Acres. **1725** Bradley sv: Acreme of Land, ten Acres of Land.

æcer, æcyr, æker. ACRE

aghendole—6 akendoule; 6-7 aghendole; 7 aighendole [perh OE aghtand, an eighth part, + dole, DOLIUM]. A m-c for grain in the counties of Lancaster and York equal to approximately 1/8 COOMB or 1/2 bu (cl.76 dkl).—**1586** Shuttleworths 1095: 2 metts and 3 akendoule...15 s. 1 d. **1605** Ibid: 1 peck, 2 s. 6 d.; 1 aghendole, 7 1/2 d. **1617** Ibid: 4 score and 15 metts and 3 aighendole... £38 3 s.

aighendole. AGHENDOLE

aiker. ACRE

akendoule. AGHENDOLE

aker, akere, akir, akre, akyr, akyre. ACRE

alm, alme. AUME

alna. ELL

ambær, ambar. AMBER

amber—1 ambær (OED), ambar (OED), amber, ambre, L ambrum, ambyr (Prior), awmbyr (Prior), awmyr (Prior), omber (OED), ombor (OED), ombra; 1-3 L ambra [OE amber, vessel, pail, dry measure; akin to OS ēmbar, pail, OHG ambar, borrowed in Gmc fr L amphora, two-handled narrow-necked jar]. A m-c for grain and liquids that varied in size, with 4 bu (c1.41 hl) being the most common.—**c900** Select Doc. 73: XXX ombra gades uuelesces. **c940** Du Cange sv ambra: De duabus meis firmis dent eis singulis mensibus Ambra plena farinæ. **c1000** Ibid: Et reddere debet 120. mensuras, quas Angli dicunt Ambres, de sale. **1086** Sussex 98: Ibi v salinæ de cx ambris salis; ibid 104: Ibi æcclesia et vi salinæ de xx solidis et x ambræ salis. **c1100** Bello 35: Willelmus...dedit et concessit...de dominio suo...unam quoque hidam terræ...et annuatim centum ambras. **1208** Bish. Winch. 4: Idem reddunt compotum de cxxxiiij sextariis dimidio, iiij ambris salis.... In carne salienda, lard [ario], xxviij sextaria j ambra. **c1283** trans in Battle xiii: To carry 2 ambræ, 2 bushels and a half of salt. **1285** trans in Cal. Char. 2.300: And of twenty ambers (ambras) of salt yearly at Leya. **1664** Spelman 29: Ambra & Ambrum. Vas seu mensuræ genus, apud Anglo-Saxones. **1678** Du Cange sv ambra: Ambrum, Amber, Anglo-Saxonibus, Vasis vinarii genus, vel mensura. **1772** Richmond 257: Ambræ salis. Mensuræ genus apud Anglo-Saxones, & Anglo-Normannos, ex Latinorum Amphora.... Dicuntur hic xxiv Ambræ salis facere xii Quarteria secundum mensuram Londini. **1872** Robertson 68: The Amber, which survives apparently in the German Ohm, the Scandinavian Ahm, was a measure of 4 bushels in the thirteenth century, by the London standard. **1886** Sussex 135: The Ambra was a Saxon measure of four bushels, used for salt.

ambra, ambre, ambrum, ambyr. AMBER

ame. AUME

anaphorum. OENOPHORUM

ancel, ancell. AUNCEL

anchor. ANKER

anker—7 ankor (OED); 7-9 anker; 8 anchor (OED) [Du and G anker fr MedL ancheria, small barrel, prob fr OHG hant-kar, hand vessel]. Before 1800 a m-c for wine which in England contained approximately 10 wine gal (3.785 dkl) and in Scotland, 20 Scots pt (3.41 dkl) (Second Rep. 6, Jessop 26, Klimpert 11, and Palethorpe sv). Since the establishment of the Imperial system, the anker has been reckoned at 10 gal (4.546 dkl), the half-anker at 5 gal (2.273 dkl), and the quarter-anker at 2 1/2 gal (1.1365 dkl) throughout the United Kingdom (Waterston 144 and Economist 54).

ankor. ANKER

ansul. AUNCEL

archa. ARK

ark—3 L archa; 4-7 ark; 7 arke [ME ark fr OE arc; akin to OHG arahha, ark, ON örk, Goth arka; all fr a prehistoric Gmc word borrowed fr L arca, chest, box, coffer]. A m-c, a large CHEST, COFFER, or bin of no standard dimensions, for fruit, grain, and similar products.—**1208** Bish. Winch. 67: In j rota de novo facta et archa reparanda, xx d. **1604** Cawdrey 19: Arke, shippe or chest. **1717** Dict. Rus. sv: Ark, a large Chest to put Fruit or Corn in.

arke. ARK

asine [MF asine fr L asinus, an ass]. A m-c (the load or burden of one ass, prob a sack-load) used principally for wine, without standard dimensions.—**1371** York Mem. 1.14: Et que chescun estraunge marchaunt des vins paie, pur chescun asine de vyne Rynois amesne a la citee et mys a la vent, ij s.

auchlet [Sc aucht, eight, + -let, dim, or lot, a part; see OED]. A m-c in Scotland (c1600-1800) for grain and frequently called a half-peck: Kirkcudbrightshire, all grain, 1478.375 cu inches (2.423 dkl): Wigtownshire, wheat, peas, and beans, 1075.21 cu inches (1.762 dkl), oats, barley, and malt, 1537.815 cu inches (2.521 dkl). The auchlet in both shires was reckoned as 4 LIPPIES or FORPITS or 1/16 BOLL (Swinton 94-95, 128-29).

auln, aulne, aum. AUME

aume—5-7 alm (OED), alme (OED); 6-7 awme; 7-8 ame (OED), auln, aulne, aum; 7-9 aume, awm [prob fr MedL āma, wine measure]. A m-c for wine containing 40 gal (1.51 hl) or sometimes equal to a wine TIERCE of 42 gal (1.59 hl).--**1590** Rates 2.39: Wine called Renish wine the Awne. **1607** Cowell sv aulne: Aulne of Renish wine. a. I. Ed. 6. ca. 13. aliàs, Awme of Renish Wine. I. Iaco. ca. 33. is a vessell that conteineth 40. gallons. **1696** Phillips sv auln: Aum of Renish Wine, a measure containing 40 Gallons, and as many pints over and above. **1717** Dict. Rus. sv: Aume, (of Rhenish Wine) a Measure containing 160 Paris-Pints, or 40 English Gallons. **1721** Bailey sv aulne: Of Rhenish Wine, a Vessel that contains 40 Gallons. **1756** Rolt sv scavage:

Rhenish wine, the awn. **1783** Beawes 865: Rhenish, the Auln. **1820** Second Rep. 6: Aume or Awm...A tierce of wine, or 42 gallons. **1895** Donisthorpe 204: AUME or AWM: A tierce of wine, or 42 gallons.

auncel—4-5 auncere, aunsell, aunselle (OED), aunser (OED); 4-7 auncel; 5 hauncere; 5-7 auncell, auncelle (OED); 6, 8 ancell; 6-7 ancel (OED); 7 L ansul, avuncell, awnsel (OED), awnsell; 8 auricel (error for auncel) [ME auncel fr AF auncelle fr OF lancelle fr OIt lancella, small balance, fr lance, balance, fr L lanx, scalepan]. An illegal scale which was similar to a primitive steelyard. It consisted of a rod or beam suspended or supported at a specified point near the end from which the goods to be weighed were hung, while along the graduated longer section of the rod an auncel weight was moved until equilibrium was attained. In the Middle Ages the weigher usually used his forefinger or the edge of his hand as a fulcrum. By the early modern period most auncels were supplied with a handle at the fulcrum for lifting. It was very easy for the weigher to cheat and relatively difficult for the customer to check him, for the former could tilt the scale very slightly or use defective auncel weights.—**1351** Rot. Parl. 2.239: Item, Pur ceo que divers Marchantz usont d´achater & poiser Laines & aultes Marchandises par un Pois que est appelle Auncel, a grant damage & deceit del poeple: Prie la Commune, que cel Pois appelle Auncel soit de tout oustee. c**1430** Salzman 2.60: There was take one branche of disceit away that hurte many men sore, the which was called an Hauncere, whiche greved many a trewe man. **1431** Rot. Parl. 4.381: Serroit poise par le auncell.

c**1435** Amundesham I.53: Et omnia alia pondera, "aunceres" vulgo dicta, adnullentur, penitus ab usibus totius vulgi extirpentur. c**1461** Hall 13: Aunsell weyght is forboden by the Parlement; and also holy Chyrche hath cursyd theym that by or sell by that weyght, for itt is...false. **1470** Year Bk. 158: Les stokks en chescun vil sont ordenew par le statut de anno xxv E. iii cap. ii pur ceux... qe vsent les auncelx weyghts. **1517** Hall 51: The Ancell Beame, which being altogeather prohibited yet are used by many; ibid 53: Which is the Ancell Weight which yarne choppers and others doe buie by. **1587** Stat. 20: It is accorded and stablished, that this weight called Auncell, betwixt buyers and sellers shall be wholie put out. And that euerie sale and buying be by the balance. **1607** Cowell sv auncell weight: It may probably be thought to be called (awnsell weight, quasi hand sale weight) because it was and is performed by the hand, as the otheris by the beame. **1615** Collect. Stat. 465: That this weight called Auncell...shall be wholly take away. **1657** Tower 79: The print forbidding Auncel weights...agreeth with the Record. **1678** Du Cange sv ansul: Genus ponderis apud Anglos, idem forte quod etiamnum Avuncell weight dicunt. **1717** Dict. Rus. sv auricel-weight: Quasi Hand-Sale-Weight...is a kind of Weight with Scales hanging, or Hooks fasten´d at each end of a Beam or Shaft, which a Man us´d to lift up from his Fore-finger or Hand. **1755** Postlethwayt II.186: In the reign of Edward III. an act passed to take away the weight called ancell. **1756** Rolt sv: AUNCEL weight, an ancient kind of balance. **1883** Simmonds sv: Auncel, the old name for weighing by the steelyard. **1964** Breed 13: Up to the time of Edward III, articles of avoirdupois were weighed by the Auncel. <u>See</u> BISMAR and PUNDLAR

auncell, auncelle, auncere, aunsell, aunselle, aunser. AUNCEL

auoirdupois, auoyxdepois. AVOIRDUPOIS

auricel. AUNCEL

aveirdepeis, averdepays, averdepois, averdepoise, averdepoiz, averdepoys, averdupois, averdupoise, averdupoize, avoirdepois, avoirdepoiz, avoirdepoys. AVOIRDUPOIS

avoirdupois--4 avoirdepoys; 4-5 haberdepase (Glazebrook); 4-7 avoirdepois; 5 averdepays (Shuttleworths), habertypoie, haburdepeyse, haburdepoyse, haverdepous; 6 auoirdupois, auoyxdepois (OED), avoirdepoiz, habardepayce, habardepayse, habardepayx (OED), habardepoix, habardipoys, habardypeyse (Nicholson), haberdepoiz, haberdepoyie, haberdepoysse (Hall), haberdipoys, haburdypeyse; 6-7 haberdepois, haberdepoise, haverdupois; 7 averdepoise (OED), averdepoiz, averdepoys (Sheppard), avoyrdepoyce, haberdepoies, haberdepoys, haburdypoyse, hauerdepiz, haverdepoise, haverdupoiz, haverdupoize; 7-8 averdupoize; 7-9 averdepois, averdupois, avoirdupois, avoirdupoise; 8 averdupoise, avoirdupoiz; ? aveirdepeis (Prior), avoirdupoys (Prior), haberdepayes (Prior), haberdupois (Eng. Cyclo.) [ME <u>avoirdepois</u>, <u>averdepeis</u>, goods sold by weight, fr OF <u>avoirdepois</u>, <u>averdepeis</u> fr <u>aver</u>, <u>avoir</u>, goods, property, + <u>de</u>, of, + <u>peis</u>, <u>pois</u>, weight, fr L <u>pensum</u>; <u>see</u> WNID3]. A system of wt (abbreviated <u>av.</u>, <u>avdp.</u>, or <u>auoir.</u>) which originally applied to goods sold by wt rather than by capacity, the piece, or otherwise (see POUND and OUNCE).—c**1350** Swithun 80: Una bala cujuslibet avoir de poys. **1353** Report 1.420: Itempur ces que nous avons entendu que ascuns marchauntz achatent avoir de pois leynz, et autres merchandises per un pois, et vendent per un autre. c**1461** Hall 12: The wegthes of Ynglond be made by nunbyr; for (there) be iij maner of weyghtes, that is to say: Troy and Aunsell, and also lyeng weyghtes odyrwyse callyd Haburdy Poyse; ibid 13: And aftyr be leynge weyght callyd Habur de Poyse. **1474** Cov. Leet 396: The seid xxxij graynes of whete take out of the myddes of the Ere makith a sterling peny & xx sterling makith a Ounce of haburdepeyse; and xvj Ouncez makith a li. **1496** Keith 1.24: The same time ordayned that xvi onces of Troye maketh the Haverdepous a li for to by and sell spice by. **1496** Seventh Rep. 29: The same tyme ordeined that xvi uncs of Troie maketh the Haberty poie. **1517** Hall 48: So makyth the whete afore namyd the Habar de Payse once.... And xvi of that onces the trewe habar de poix lib. c**1525** Ibid 40: Item xvi onces Habar de Payce ys. a lib.... Item xviii onces di. of Troy weyghte makys xvi onces Habar de payse. **1532** Seventh Rep. 31: Beef, pork, mutton, and veal shall be sold by weight called haver-du-pois. **1545** Rates 1.52: Thys Lyinge and Haburdy peyse is all one: the pounde conteinyng. xvi. ounces of troye. **1566** Recorde K iii: But commonly there is used an other waight called Haberdepoise, in which 16 Ounces make a pounde.... But if yt be Haberdepoyie, you must diuide...by 16. **1577** D. Gray 7: 112. lib. haberdepoiz; ibid 47: The whiche beeing haberdipoys waight is 16 ounces.... 12. ounzes to bee 3/4 of the lib. habardipoys. **1588** Hall 45: Avoir de poiz waight is to bee used for other commodities, ffor Merchandize, and for Grocers. **1595** Powell C: There is also an other weight named Avoirdupois weight, whereunto there is xvi. ounces for the pound. **1600** Hill 66: 16. Ounces of hauerdepoise weight maketh 1. Pound of hauerdepoise. **1603** Henllys 138: And all spice, Iron, Rosen, pitche and other drugges uttered by the mercers are sold by the haberdepoies pound; ibid 139: Iron is sold by the stone which consisteth of xvj haberdepoys. **1606** Hall 37: There is onely two sortes of waightes used in England the which are allowed by Statute, the one called Troy waight, the other Haberdepois waight; ibid 38: This waight of Haberdepois is allowed alsoe by Statute being 16 oz. to the pound waight with the which is wayed all phisick drugges, grocery wares, rozen, wax, pitch, tarr, tallowe, sope, hempe, fflaxe, all metalles and mineralles. **1607** Cowell sv avoir de pois: Avoir de pois, is in true French (avoir du poix. i. habere pondus, aut iusti esse ponderis). It signifieth in our common lawe, two things: first, a kinde of weight divers from that, which is called Troy weight conteining but 12. ounces to the pound, where as this conteineth sixteene.... Then also it signifieth such merchandize, as are waied by this weight, and not by Troy weight; ibid sv weigh: 256. pounds of avoyr de poyce. **1616** Bullokar sv haberdepoise: A pound weight conteineth sixteene ounces. **1628** Hunt B3: By Hauerdepiz, Haberdepois, or Auerdepois is weighed all Grocery Wares and Phisicall

Drugs. **1635** Dalton 143: Averdepois weight is by custome...and thereby are weighed all kind of Grocerie wares, Physicall drugs, Butter, Cheese, Flesh, Wax, Pitch, Tarre, Tallow, Wools, Hemp, Flax, Yron, Steele, Lead. **1657** Tower 419: That there may no more be taken for weighing in any place of the Realm for any Aver-depoiz than in London. **1660** Bridges 21: Averdupoize Little weight. This weight is distinguished into Drams, Ounces and Pounds. **1661** Hodder 15: Addition of Haverdupoize weight; ibid 22: Subtraction of Haverdupoiz weight. **1665** Assize 2: There is also another weight named Avoirdupois weight, whereunto there is 16 ounces for the pound. **1682** Hall 29: Aver-du-pois conteynes: every pound, 16 ounces; every ounce, 8 drgmes [sic]; every dragme, 3 scruples; every scruple, 20 graines. **1688** Bernardi 137-38: Libra equidem Avoirdupois qua solent populares mei graviores mercium æstimare quam pretiosiores, 1/112 Hundredi sui sive centenarii crassi, 16 unciæ, 128 = 16 X 8 drachmæ; ibid 138: Habet et libra Avoirdupois scripulos suos 384 = 128 X 3, gravans nobis 1,2169, sed ratione Wybardica 17/14 = 1,2413 libræ de Troy. **1690** Barbon 12: There are Two Sorts of Weights in Common Use, the Troy, and Averdupois. **1699** Hatton 1.153: Troy weight, and the Avoirdupoise. **1701** Hatton 3.7: Add Ounces in Averdupoize-weight. **1708** Chamberlayne 206: But the Avoirdupois Pound is more than the Troy Pound, for 14 Pound Avoirdupois are = to 17 Pound Troy-Weight. **1710** Harris 1. sv weight: And the other is called Averdupois, containing 16 Ounces in the Pound. **1717** Dict. Rus. sv dram: Dram or Drachm, the just Weight of sixty Grains of Wheat; in

Avoir-du-pois Weight, the sixteenth part of an ounce; ibid sv hundred-weight: But ordinarily a Pound is the least Quantity taken notice of in Aver-du-pois Gross Weight; ibid sv pound: A sort of Weight containing 16 Ounces Avoir-du-pois. **1732** J. Owen 126: Avoirdupoiz Weight. The Denominations are Tuns, Hundreds, Quarters, Pounds, Ounces and Drams. **1737** Hall 47: Two solid pounds Averdupoise, all extraordinary well sized and adjusted. **1742** Account 1.553: The single Averdupois Bell Pound, against the flat Averdupois Pound Weight was found...to be heavier by Two Troy Grains and a half. **1750** Reynardson 6: The Pound Avoirdepois at 7000...such Grains. **1778** Diderot XXVI.420: L´avoir-du-pois est de seize onces. **1790** Jefferson 1.986: So that the pound troy contains 5760 grains, of which 7000 are requisite to make the pound avoirdupois. **1793** Leake 30-31: This Avoirdupois originally signified no more than Goods in gross, or by wholesale. **1794** Martin 15: The new proposed pound is equal to 20 ounces Avoirdupoise. **1814** Eliot 4: 210 lbs. avoirdupoise. **1868** Eng. Cyclo. 822: But in buying and selling medicines wholesale, averdupois weight is and always has been used. **1878** Wedgwood 34: Averdepois...goods that sell by weight and not by measurement. **1964** Breed 12: It is evident that, in this statute [1353], the word aver de pois, the old spelling of avoirdupois, refers to the nature of the goods and does not mean a particular kind of pound.

avoirdupoise, avoirdupoys, avoirdupoiz, avoyrdepoyce.

AVOIRDUPOIS

avuncell. AUNCEL

awm. AUME

awmbyr. AMBER

awme. AUME

awmyr. AMBER

awnsel. AUNCEL

B

baele. BALE

bag--3-7 bagge; 4-9 bag; 5 bague (Southampton 1); 6 bage; 6-8 bagg [ME bagge fr ON baggi]. A m-c, generally a large canvas sack, varying in size according to its contents (c1600-1850): almonds, 3 Cwt (152.406 kg); aniseed, 3 to 4 Cwt (152.406 to 203.208 kg); cocoa, 1 Cwt (50.802 kg); coffee, 1 1/4 to 1 1/2 Cwt (63.502 to 76.203 kg); cotton yarn, 2 1/2 to 4 1/4 Cwt (113.397 to 192.776 kg); currants, 4 Cwt (203.208 kg); goats-hair, 2 to 4 Cwt (101.604 to 203.208 kg); lime, 1 heaped bu (c4.50 dkl); pepper, 1 1/4 to 3 Cwt (61.235 to 146.964 kg); pimento, 1 Cwt (45.359 kg); sage, 1 Cwt (50.802 kg); and Spanish wool, 240 lb (108.862 kg) (Dict. Rus. sv, Waterston 147, D. Digges 44, Palethorpe sv, Hatton 3.220, Pasley 114, and Second. Rep. 6). It was sometimes abbreviated b. or bg. See HUNDRED

The bag also had local variations (c1800-1900): Devonshire, wheat, 2 bu totaling 140 lb (63.503 kg); Kent and Surrey, hops, 2 1/2 Cwt (127.005 kg); Shropshire, wheat, 3 bu (c1.06 hl); Staffordshire, wheat, 210 lb (95.254 kg); Westmorland, potatoes, 7 1/2 bu (c2.64 hl); Scotland, flour, 91 English lb (41.277 kg), and barley, 279 or 280 English lb (126.552 or 127.005 kg); and South Wales, oats, 7 heaped MEASURES or 8 1/2 striked or leveled measures, making 170 qt or 5 bu and 10 qt (c2.99 hl) (Second Rep. 6, Cyclopædia sv weights, Eliot 4, and Britten 167).

However, bags of aloes, alum, brush-making materials, fish, ginger, hops, and soap do not appear to have had specific sizes.—**c1420** Gras 20]

1.461: xii bagges de aloe. **1443** Brokage II.174: 1 parvo bagge saponis. **1509** Gras 1.564: xxv bages aluminis; ibid 566: 1 packe cum ii bages ginger continent iii [X] c libras; ibid 567: ii bages spletes; ibid 569: i bage cum hethe pro brusshes. c**1610** Lingelbach 113: Alam by the Bagge. **1704** Mer. Adven. 243: Ffor bearing to the Weighouse a bagg of hops and weighing 2 [d] per C. **1706** Holroyd 15: 11 Barrells of seals...2 Baggs of Each.

bage, bagg, bagge, bague. BAG

bail, bal, bala. BALE

balatt. BALET

bale--3 boillun, boyllum, boylun, buyllon (Cal. Lib. 2); 3-4 L bala; 4-9 bale; 5 baele (Southampton 1); 5-6 bal; 5-7 bayl (OED); 6 balle; 6-8 ball; 7 bayll (Halyburton); 7-8 bail [ME bale fr OF bale, balle, of Gmc origin]. A m-q or m-c, variously defined for different items. Originally it denoted a large bundle of more of less cylindrical shape, but by the late Middle Ages it had come to designate a closely pressed, rectangularly shaped package, wrapped generally in canvas and tightly corded or hooped with copper or iron. It sometimes was abbreviated bl.

The bale was used most often for buckram, 60 pieces; fustian, generally 40 or 45 half-pieces; hay or straw, generally 224 lb (101.604 kg); paper, 10 REAMS; and wool, 180 lb (81.646 kg).—**1502** Arnold 206: A balle bokrom conteyneth lx. pecis...a balle fustian conteyneth xlv. half peces. **1507** Gras 1. 697: Fustyon´ the balle containing xl hallfe peces. c**1590** Hall 25: The bale of paper is 10 reames of paper. **1616**

Hopton 164: A Bale of Paper is 10 Reame, or 200 Quires. **1656** Rawlyns 70: A Bail of Paper containes...Reams 10. **1660** Bridges 31: 1 Quire is 25 Sheets. 20 Quire a Ream. 10 Reams a Bale. **1934** Int. Traders 71: Bale (wood)...United Kingdom...180 pounds. **1956** Economist 50: Bale: Hay and Straw = 224 lb; ibid 69: Paper measures...1 bundle = 2 reams. 1 bale = 5 bundles.

The bale was also used (c1600-1800) for almonds, 3 Cwt (146.964 kg); boultel (bolting cloth), 20 pieces; caraway seeds, 3 Cwt (152.406 kg); cochineal, 1 1/2 Cwt (76.203 kg); coffee, 2 to 2 1/2 Cwt (101.604 to 127.005 kg); cotton yarn, 3 to 4 Cwt (136.077 to 181.436 kg); flaxen yarn, 240 lb (108.862 kg); hemp, 20 Cwt (1016.040 kg); licorice, 2 Cwt (101.604 kg); madder, 8 Cwt (406.416 kg); pipes, 10 gross or 1440 in number; raw silk, 1 to 4 Cwt (50.802 to 203.208 kg); Spanish wool, 2 1/4 Cwt (114.304 kg); and thread, 100 bolts (Rates 2.2ff, Dict. Rus. sv, Palethorpe sv, Second Rep. 6, and Waterston 147). See HUNDRED

Bales used for the following items did not have standard dimensions.—**1239** trans in Cal. Lib.1.367: And a bale (boyllum) of ginger...a bale (boylun) of cinnamon...four bales (boilluns) of dates. **1242** trans in ibid 2.154: For a bale (bala) of ginger. c**1300** Swithun 80: Una bala cujuslibet avoir de poys. **1303** Gras 1.161: Bala de bresil. **1304** Ibid 168: Pro ii bales basane. **1308** Ibid 362: Adduxit ii balas basani. **1323** Ibid 209: De quodlibet balo zucre. **1439** Southampton 2.63: 1 bale panni; ibid 70: 2 balys de streyt. **1443** Brokage II.1: Cum c allei et 1 bale alym; ibid 2: 1x bal´

amigdalorum...1 bale madr´; ibid 3: Cum viii bal´ dates; ibid 15: 1 parvo bal cere. **1509** Gras 1.698: Lycerus the balle. **1545** Rates 1.1: Almondes the bale; ibid 16: Flaxe the balle. c**1550** Welsh 97: 1 balle anmorici; ibid 278: 5 balls flax. **1664** Gouldman sv: A bale of spicery. **1704** Mer. Adven. 243: Ffor a poke or bail of mather.

balet—5 balett, balette (OED); 5-6 balet; 6 balatt, ballett; 6-7 ballet [ME balet fr OF balete, ballete, dim of bale, balle; see BALE]. A m-q or m-c for many products and generally equal to 1/2 bale.--**1439** Southampton 2.12: 4 balett´ de wode; ibid 55: Pro 2 balett´ de wastyng paper; ibid 70: 1 balett panni çontinente 7 pannos sine grano et 18 vergas grany; ibid 72: Pro 2 balett´ pellium vitulinarum continentibus 30 dosyn´; ibid 88: Pro 2 balett´ grani pro panno; ibid 90: Pro i balet granis paradisi; ibid 91: 2 balett´ rys. **1443** Brokage II.1: Cum viii balett wald´; ibid 81: Cum lx balett´ waid. **1509** Gras 1.562: iii balletts annessede. c**1550** Welsh 62: 3 balletts canvas; ibid 73: 1 ballet crassum; ibid 237: 7 balattes...toloss wood. **1628** Hunt B2: A ballet of Canuas.

balett, balette. BALET

ball, ballette. BALE

ballet, ballette. BALET

band—4-5 bande (OED); 6-7 band [ME bande, strip, fr MF bande, strip, edge, side]. A wt for iron, the equivalent, in 1600, of 24 STONE (c152.41 kg) (Shuttleworths 790).

bande. BAND

barayl, barel, barele, barell, barelle, barellus. BARREL

barge-load. KEEL

barile, barill, barillus. BARREL

barleycorn—4-5 L ordeum [ME barly corn, barlye corne; see WNID3, sv barley; see ibid, sv corn]. The artificial standard upon which medieval linear measures and the ap, avdp, and t lb were based. The INCH, for example, was defined by statute as the length of 3 medium-sized barleycorns placed end to end. The foot was then made equal to 12 of these inches; the CUBIT, 18; the yd, 36; the ELL, 45; and the FATHOM, 72. The ap and t lb contained 5760 barleycorns, while the avdp lb contained 7000.—c**1300** Hall 7: Nota quod tria grana ordei de medio spice faciunt pollicem. c**1400** Ibid 9: Sciendum quod tria grana ordei vel quatuor grana frumenti, in medio spice sumpta, in longitudine faciunt pollicem Regis. c**1461** Ibid 14: The lengythe of iij barly cornys make an ynche. **1537** Benese 3: The lengthe of an ynche after some mens opinion, is made by the length of thry barlye cornes, the which rule is not at all tymes true. For the lengthe of a barlye corne of some tyllage is lenger, and of some tyllage is shorter, after the fatnes and leanesse of the lande, where it was sowen upon. Therefore in makynge of an ynche after thys rule, it shulde be sometymes lenger, and sometymes shorter, after the lengthe and shortenes of the barlye cornes. **1606** Hall 38: A graine is deriued from the barlie corne and is the least part proporcionable from an oz. **1616** Hopton 165: Three barley cornes make an Inch. **1682** Hall 28: An Inch is 3 barly cornes dry and round in length. **1717** Dict. Rus. sv: Barley-Corn, is taken for the least of our long Measures, of which three in Length make an Inch. **1850** Alexander 7: Barley-corn; imaginary...0.33 inches. **1964** Breed 8: Measures derived from the barley corn are still in use by cordwainers, as the sizes of footwear in England are based upon it, and shoemakers' tapes and rules are divided not into inches, but into thirds of an inch, which are called sizes. See GRAIN; INCH; POUND

barrall. BARREL

barrel—3-4 L barillus; 4 barayl (OED), L barrellus; 4-5 barele (OED); 4-6 barelle, L barellus, barrelle; 4-7 barel, barell; 5-7 barrell, barylle; 6 barile, barill, baryll, beryll (OED); 6-8 barrall; 7-9 barrel [ME barel, barell fr MF baril, barrel, cask]. A m-c for both wet and dry products. It was a nearly cylindrical wooden vessel generally wider in the middle than at the ends, its length exceeding its breadth. It was often formed of curved staves bound together by hoops. In the early modern period it was commonly abbreviated bar. or brl.

A bbl of ale contained 32 gal (cl.48 hl) and was equal to 4 ale FIRKINS of 8 gal each or 2 ale KILDERKINS of 16 gal each. In 1688 it was changed to 34 gal (cl.57 hl), and in 1803 it was standardized at 36 gal (cl.66 hl). The Irish bbl of ale (c1800) contained 8704.0 cu inches (1.427 hl) or 40 Irish gal of 217.6 cu inches each and was equal to 2 Irish ale kilderkins or 4 Irish ale firkins (Edinburgh XII.572).—**1393** Henry Derby 157: Pro iiij barellis ceruisie. **1517** Hall 49: Be alwayes xxxii galons' to the barell, xvi galons' to the ale kylderkyn, and viii galons to the ale ffyrkyn. **1518** St. Peter´s 304: Item, two ale barels. **1587** Stat. 595: And that euery barrell for ale shall conteine xxxii. gallons, euerie kilderkin...xvi. gallons, and euerie ferkin...viii gallons of the kings standard gallon. c**1600** Brit. Mus. 16.70v: Of Ale the Barell contayneth. 32. gallons. **1635** Dalton 144: 32 gallons maketh the Barrell. **1665** Sheppard 14: Of Ale...The Barrell 32...Gallons. **1682** Hall 29: But Ale hath no more than 32 gallons to the barrell: and therefore but 64 pottles, 128 quarts, and 256 pints. **1701** Hatton 3.10: In a Barrel are...32 Gallons or 9024 Solid Inches. **1707** Acts Scotland 11.407: Thirty four Gallons English Barrel of Beer or Ale amounting to twelve Gallons Scots present measure. **1850** Alexander 7: Barrel...for ale...36.0 gallons.

A bbl of beer contained 36 gal (cl.66 hl) and was equal to 4 beer firkins of 9 gal each or 2 beer kilderkins of 18 gal each. In 1688 it was changed to 34 gal (cl.57 hl), and in 1803 it was fixed once again at 36 gal (cl.66 hl). Since the establishment of the Imperial system the bbl of beer has been reckoned at 36 gal (1.636 hl) everywhere in the United Kingdom except in Ireland, 32 gal (1.455 hl).—**1443** Brokage II.191: ii barellis byre. c**1475** Gras 1.193: Of a barel of bier. c**1500** Brit. Mus. 24.18v: Barrell for beere shall conteyne 36 gallons of the kings standarte gallon. **1502** Arnold 246: The barell of beer, xxxvi galones. **1517** Hall 50: That there shuld´ be no lesse assyse for bere than xxxvi galons to the barelle. **1547** trans in Cal. Pat. 23:397: Licence to the king´s servant Galter de Loenus to export 300 ´tonnes´ of beer in ´buttes, pypes, hoggesheddes, pontions or barrelles´. **1553** Remembrance 47: Yt ys agred the vijth day of October anno 1553 that no bruar that dewllythe within thys towne shall not sell the best bere...above tow s. the baryll. **1587** Stat. 595: And that euerie barrell for beere shall conteine xxxvi. gallons. c**1590** Hall 22: The firkin conteynyth 9 galons: the barill contenith 36 gallons. **1607** Cowell sv barrell: For a barrell of beere conteineth 36. gallons. **1635** Dalton 148: And so Beere measure containeth in the barrell foure gallons more than Wine, or any other vessel. **1675** Mayne 51: Any number of Inches are reduced into the parts of a Beer Barrel, if divided by 10152. **1682** Hall 29: 1 Barrell conteynes: 2 Kilderkins, 4 Firkins, 36 Gallons, 72 Pottles, 144 Quarts, 288 Pints. **1701** Hatton 3.9: In a Barrel are...36 Gallons or 10152 solid Inches. **1707** Forbes 55: Thirty Six Gallons of Beer, and Thirty Two of Ale...go to a Barrel of Beer and Ale in London.... But Thirty Four Gallons are reckoned a Barrel of Beer or Ale in all other places of England. **1883** Simmonds sv: The beer barrel is 36 gallons, or 2 kilderkins. **1956** Economist 54: Barrel = 36 gallons (32 gallons in Ireland).

The capacity of a bbl of butter or soap conformed to the ale bbl capacity of 32 gal (cl.48 hl), but the weights for butter, soap, and their casks were equally important. Generally, the butter bbl weighed 256 lb (116.119 kg) or 26 lb (11.793 kg) for the cask and 230 lb (104.326 kg) for the butter, whereas the soap bbl weighed 280 lb (127.005 kg) or 32 lb (14.515 kg) for the cask and 248 lb (112.490 kg) for the soap.—**1420** Gras 1.506: Pro xxi barellis saponis. **1443** Brokage II.1: Cum iiii barellis saponis; ibid 17: 1 barello saponis nigri. c**1475** Gras 1. 193: Of a barel sope. **1502** Arnold 246: The barell of soep, xxx [sic] galones. **1507** Gras 1. 695: Butter the barelle; ibid 702: Sope called blacke sope the barrelle. **1587** Stat. 595: That all maner of sope makers within this realme of England, which shall put to sale anie sope by barrell. c**1590** Hall 24: The barill of butter waieth, caske and all, 256 poundes waight haberdepoyse; whereof the caske wayeth 26 poundes waight; so ther remaynith in the caske of clean Butter 230 poundes waight haberdepoysse.... The barill of sope, caske and all, wayeth 280.... The barill of soap empty nowe 32 waight. **1595** Powell C: And euerie Sope barrell to holde and containe 32. Gallons. **1635** Dalton 149: Sope, the barrell...shall bee of the same content that ale is.... Butter shall be of the same measure that sope is of. **1665** Assize 4: And every Sope-Barrel to hold and contain 32 gallons...and shall weigh being empty xxvi pounds of Avoirdupois weight; ibid 5: Which is twelve score and sixteen pounds...and the barrel of Butter is of like weight. **1682** Hall 30: The Barrells for herrings, Butter and Soape are the same with Ale measures. **1829** Palethorpe sv: The barrel of soap 256 lbs.

A bbl of herrings or eels usually contained 30 gal fully packed (cl.14 hl). For salmon, and sometimes for eels, the bbl contained 42 gal (cl.59 hl) and was equal to 1/2 salmon PIPE or 1/12 salmon LAST. For most other fish, including occasionally herrings and eels, the bbl conformed to the 32 gal capacity of the ale bbl (c1.48 hl) except in the case of pilchards or salted mackerel where the capacity was standardized in 1800 at 50 gal (c1.89 hl). The Scots salmon bbl contained 14 gal (c1.90 hl) to 1573, 12 gal (c1.63 hl) from 1573 to approximately 1625, and was standardized at 10 gal (c1.36 hl) thereafter; the herring bbl was 9 gal (c1.22 hl) until it was fixed at 8 1/2 gal (c1.16 hl) around 1625. The present Imperial fish bbl used throughout the United Kingdom contains 26.6 gal (1.209 hl).—**c1300** Topham 144: 2 barillos picis. **1324** Gras 1.376: Pro vii barellis sturgonum. **1341** Ibid 174: xv barrellis de pyk´. **c1400** Ibid 216: De quolibet barello de haddok. **1432** Rot. Parl. 4.256: The barrell of Heryng and Eles. xxx Galons full pakked. **1439** Southampton 2.12: 1 barello salmonum continente 2 dosyn´. **1443** Brokage II.41: Cum ii barellis salmon; ibid 105: Cum 1 barello salmon continente xvi salmon. **1478** Stonor II.73: For ij barell herreng, xxij. s. **1482** Rot. Parl. 6.221: The Barell of Salmon XLII Galons. **1487** Acts Scotland 2.178: ALSA It is statut & ordanit be the thre estates in this parlment/that the barell bind of Salmond suld kepe & contene the assise & mesour of xiiij gallonis. **c1500** Brit. Mus. 24.16v: The barrell [of salmon] 42 gallons.... Hearinge barrell to conteyne 32 gallons. **1507** Gras 1.697: Elys called chaffte ellys the barylle...Elys called pymper eles the barelle; ibid 699: Hadockes the barrelle; ibid 702: Sawlte fyche the barell. **1509** Ibid 569: i di. barellus samonis. **1545** Rates 1.15: Elis called stubbe elis the barell. **1559** Remembrance 67: Item more every barrell of hys salmond to be sold at iii li. vi s. viii d. the barrell. **1573** Acts Scotland 3.82-83: That euerie Salmond Barrel to be maid heirefter sall contene twelf gallounis of the Striuiling pynt and that euerie Barrel of Hering & quhite fische contene nyne gallounis of the samin stop. **1587** Stat. 267: Barels of herring and of eeles. c**1590** Hall 23: The barill of heringe and eeles ought to be 30 gallons in content fully packed.... The but of salmone ought to be 84 gallons fully packed.... The last is 6 buttes conteninge 504 gallons; the barill is 42 gallons. **1595** Powell C: The Hearing Barrell must hold and conteine, thirtie two gallons. The Eale barrell fortie two gallons. **1615** Collect Stat. 466: Nor barrell of Herring nor of Eeles, unles they contain 30. gallons fully packed. **1616** Hopton 162: And know that the barrell, and halfe barrell of Herrings, and likewise of butter and sope, are the same measure used for Ale. **1624** Huntar 4: [In Scotland] The Salmon barrell conteines 10 Gallons. The Herring barrell holdes 8 Gallons and a halfe. **1681** Acts Scotland 8.400: By which act The saids Lords Concluded and Ordaines ane constant measure of Salmond.... Every measure [the bbl] containing Ten gallons. **1682** Hall 92: Barrell fish hath 12 Ale barrels to a Last. **1693** Acts Scotland 9.260: And that ilk Barrell for exporting of Herring contain eight Gallons and two pynts, And ilk Barrell for exporting Salmond ten Gallons. **1779** Swinton 29: [In Scotland] The herring-barrel contained 8 1/2 gallons. The salmon-barrel contained 10 gallons. **1829** Palethorpe sv: The barrel of salmon must contain 42 gallons, the barrel of eels the same. **1895** Donisthorpe 204:

BARREL...of cod fish, wet, 32 gallons...of eels, 42 gallons...32 Ed. 4; but 30 by 2 H. 6; ibid 205: BARREL...of pilchards, or mackerel, salted, 50 gallons. **1956** Economist 53: Fish...1 barrel = 26.6 Imperial gallons.

A bbl of gunpowder weighed 1 Cwt of 100 lb (45.359 kg) and was equal to 1/24 last of gunpowder; a bbl of coals contained nearly 4 Winchester bu (cl.40 hl).—**c1590** Hall 22: The hundred waight of gunpowder is but fyve skore poundes waight, haberdepoyse, to the hundrid.... The last...is 24 barills. **1603** Henllys 139: Coles are sold by the barrell w[hich] is of Bristoll band, or neere about foure Wynchester bushells. **1775** Postlethwayt II.191: A last of gunpowder contains twenty-four barrels, and the barrel a hundred pounds. **1882** Jackson 227: Barrel of gunpowder...100 Lbs. av.

A bbl of wine generally contained 31 1/2 gal (cl.19 hl) and was equal to 1/8 wine tun of 252 gal. The oil and honey bbl conformed to the specifications of the wine bbl, as did the tar bbl after 1750. The Scots bbl of wine contained 6617.856 cu inches (cl.09 hl) or 8 gal, or 32 qt, or 64 pt, or 128 CHOPPINS, or 256 MUTCHKINS, or 1024 GILLS. The Irish bbl of wine (c1800) contained 6854.4 cu inches (1.123 hl) or 31 1/2 Irish gal of 217.6 cu inches each (Edinburgh XII.572).—**c1300** Topham 144: Unum barillum mellis. **1341** Gras 1.174: De lxxii barrellis de tarr´. **c1400** Hall 18: There be also...barrells of terre. **1439** Southampton 2.15: Pro 3 barellis de tarr. **1443** Brokage II.8: Cum v barellis tarr´. **1507** Gras 1.701: Oyle called mette oylle or rape oylle the barrelle. **1525** Jacobus 73: j barile mellis continens x laginas [Scots]. c**1590** Hall 21: The [wine] barill which is 1/8 part of a tonne contenyth 31 gallons 1/2. c**1600** Brit. Mus. 16.70: The Barylle holdeth.31.1/2 Gallones. **1615** Collect. Stat. 467: And euerie [wine] Barrell to containe one and thirtie gallons and an halfe. **1635** Dalton 148: Wine, Oyle, and Honey: their measure is all one. **1682** Hall 29: Wyne, Oyle and Hony Measures: 1 Tunne conteynes...8 Barrells. **1704** Mer. Adven. 243: Ffor the carriage of a barrall of oyl. **1756** Rolt sv: The English barrel, wine-measure, contains the eighth part of a tun, the fourth part of a pipe, and the half of a hogshead; that is, thirty-one gallons and a half. **1850** Alexander 7: Barrel...for wine and brandy...31,5 gallons.

The bbl was also used (c1700-1950) in England for anchovies, 30 lb (13.608 kg); apples, 3 bu (c1.06 hl); barilla, 2 Cwt (101.604 kg); barley, 224 lb (101.604 kg); beef, 32 wine gal (c1.21 hl); candles, 120 lb (54.421 kg); coffee, 1 to 1 1/2 Cwt (50.802 to 76.203 kg); flour, 196 lb (88.904 kg); nuts, 3 bu (c1.06 hl); oatmeal, 2 Cwt (101.604 kg); oats, 196 lb (88.904 kg); plates (white or black), 300 in number; potash, 2 Cwt (101.604 kg); raisins, 1 Cwt (50.802 kg); rosin, 2 Cwt (101.604 kg); Spanish tobacco, 2 to 3 Cwt (101.604 to 152.406 kg); vinegar, 34 gal (c1.29 hl); and wheat, 280 lb (127.005 kg) (Hatton 3.17, Second Rep. 6-7, Seventh Rep. 62, Economist 50, and Simmonds sv).

The bbl had a number of different uses (c1800-1900) outside England: Guernsey and Jersey, charcoal and lime, 120 pots or 60 gal (c2.77 hl);

Ireland, grain, generally 4 bu of 10 gal each (c1.80 hl), but barley and rape, 16 STONE of 14 lb each or 224 lb (101.604 kg), beans, peas, and wheat, 20 stone (127.00 kg), bran, 6 stone (38.101 kg), malt, 12 stone (76.20 kg), oats, 14 stone (88.90 kg), oatmeal, 8 stone (50.802 kg), and potatoes, 20 stone (127.00 kg); Isle of Man, lime, 6 Winchester bu (c2.11 hl); Wales, lime in some counties, 3 provincial bu of 10 gal each, equal to 3 1/4 Winchester bu (c1.14 hl), and culms, 4 heaped bu or 40 gal (c1.80 hl) (Second Rep. 8, J. Sheppard 86, and Britten 167).

In Scotland the bbl was used (c1600-1700) for aqua vitae, 10 gal (c1.36 hl); ashes, 2 Cwt (101.604 kg); barilla, 2 Cwt (101.604 kg); brass, 10 STONE (c63.50 kg); butter from England or Holland, 12 stone (c76.20 kg); lead ore, 5 Cwt (254.010 kg); plates (white or black), 300 in number; and powder, 10 stone (c63.50 kg) (Halyburton 288-341 and Acts Scotland 7.251-254).

barrell, barrelle, barrellus. BARREL

barrow [ME barew, barowe fr OE bearwe, basket, handbarrow]. A m-c for salt containing approximately 6 pk (c5.29 dkl), used in Cheshire, Worcester, and other places in the salt region in the 1880s. It was a conical wicker case or basket in which salt was put to drain (Leigh 14).

baryll, barylle. BARREL

baskatt. BASKET

basket—3-9 basket; 4-5 baskett, baskette; 5 baskatt (OED), baskyt (OED); 6 baskete, basquette (OED), baszkett (OED), baszkette (OED); 7 basquet (OED) [ME basket, prob fr ONF baskot fr (assumed) ONF baskou fr L bascauda, dishpan; see WNID3]. A m-c varying in size according to its contents (c1700-1900): medlars, 2 bu (c7.05 dkl); cherries, Kent, 48 lb (21.772 kg); and asafetida, 20 to 50 lb (9.072 to 22.679 kg) (Bradley sv, Britten 167, Dict. Rus. sv, and Hatton 3.220). It was usually a wickerwork container made from plaited osiers, cane rushes, or other similar materials, and was abbreviated occasionally bkt.

Baskets used for the following items did not have standard dimensions.--**1420** Gras 1. 500: Pro i baskette cum xiiii briste-plates nigr´. **1443** Brokage II.67: iiii baskettys orengys. **1509** Gras 1.565: iiii basketts cum xi [X] c galipotts; ibid 568: i parv´ basket cum ii dossenis et di. felts...i basket cum x dossenis mistel bedes; ibid 570: ii basketts cum iiii cases spectakilles...ii basketts continent´ cv pecias et remanenta teli lini Hasburgh. **1545** Rates 1.7: Bokes unbounde the basket; ibid 43: Trenchers the maunde or baskete. **1783** Beawes 866: Figs, the 18 Baskets, 800 lb.

baskete, baskett, baskette, baskyt, basquet, basquette, baszkett, baszkette. BASKET

bat [prob a special use of bat, stick, club]. A m-a in South Wales (c1800-1900) containing 1 perch of 11 sq ft (1.022 sq m) (Second Rep. 8, Donisthorpe 205, and Britten 167).

batten [F bâton, stick, staff]. A m-c for straw in Durham (early 1800s) equal to 1/12 THRAVE (Dinsdale 134). It was probably the amount of packed straw in a bundle whose breadth was equal to the length of a certain measuring stick.

bay [ME bay fr MF baée, an opening, fr OF baee fr vb baer, to be open]. A m-a of slater's work in Derbyshire (c1800-1900) containing 500 sq ft (46.452 sq m) (Second Rep. 8, Donisthorpe 205, and Britten 167).

bayl, bayll. BALE

beatment [perh fr vb beat, in the sense of a "beating," or quantity to be beaten at once, + -ment]. A m-c for grain (c1800-1900) in Durham, Newcastle, and Northumberland, equal to 1/4 pk (c2.20 l) (Brockett 22, Cyclopædia sv weights, and Britten 167).

belet. BILLET

beryll. BARREL

bescia [F bêche, spade, fr MF besche + -ia L ending]. A m-a in Lincolnshire (c1400) for turf-cutting on the fens. It represented the amount of land that could presumably be dug annually by one man with a spade between May 1st and August 1st (Prior 150).

beyschell. BUSHEL

billet—4-6 billette; 5 bylet; 5-6 belet (OED); 6 billett, byllet (OED), byllot; 6-9 billet; 8 billot (OED) [ME billette fr MF billete (F billette, billot), dim of bille, log, round stick]. A m-l of 3 ft 4 inches (1.016 m) for firewood. A single billet had a circumference of 7 1/2 inches (19.050 cm); a cast billet, 10 inches (25.40 cm); and a two cast billet, 14 inches (35.56 cm).—c**1440** Promp. Parv. 36: Bylet, schyde. **1559** Fab. Rolls 353: In byllot or shydes. **1587** Stat. 171: And euerie billet to conteine in length three foot and foure inches. c**1590** Hall 27: The billettes be of like lengthes, but not of like tycknes. Euery billett ought to be in length 3 foott 4 ynches in lenght. The single billet conteyn´ 7 ynches about and 1/2. A billet caled a cast contenith 10 ynches about. Euery billet caled a cast of 2 contenith 14 ynches about. **1616** Hopton 163: All fuell is used by the Statute, of which there be Shids, Billets, Fagots and Coles. **1665** Assize 18: And every Billet named a single, to contain seven inches and a half about...every Billet named...cast, to contain 10 inches about; and every Billet of two Cast, to contain 14 inches about. **1756** Rolt sv measures: Billets are to be 3 feet long, whereof there should be 3 sorts; a single, a cast, and a cast of two. **1880** Britten 167: Billet of firewood, 3 feet 4 inches long; if single, about 7 1/2 inches.

billett, billette, billot. BILLET

bind—3 binde, L bynda; 3-5 L binda; 4, 6 bynd; 4-6 bynde (OED); 4, 7-9 bind [ME binde fr vb binden, to bind]. A m-q for eels, consisting of 10 STICKS, or 250 in number.—c**1353** Hall 12: La binde de anguilles est de x estikes. c**1272** Report 1.414: Item binda anguillarum constat ex decem stiks. c**1275** Hall 10: Bynda vero anguillarum constat ex decem stickes; et quelibet sticke ex viginti et quinque anguillis. **1290** Fleta 120: Item lunda [bynda?] anguillarum constat ex x. stikis. c**1300** Brit. Mus. 13.29: Bind anguillarum constat ex. x. stikes. c**1300** Ibid 1.148v: Bynd anguillarum constat ex .x. stikes. Et quelibet stike ex. xxv. anguillis. **1495** Ibid 28.156v: Binda anguillarum constat ex x stikes. c**1590** Hall 23: A bynd of eeles consistith 10 stikes. **1615** Collect. Stat. 465: A bind of Eeles consisteth of 10. strikes [sic] and euerie strike [sic] 25. eeles. **1665** Sheppard 61: A Bind of Eels. **1707** Justice 7: Eels, 25 to the strike [sic], and 10 strike [sic] to the Bind. **1717** Dict. Rus. sv: Bind of Eels, a quantity consisting of 250. **1883** Simmonds sv: Bind, in the fish trade, a term applied to 250 eels or ten strikes [sic], each containing a quarter of a hundred. **1895** Donisthorpe 205: BIND: of eels, 10 sticks = 250 eels.

binda, binde. BIND

bing [ME bing, of Scand origin; akin to ON bingr, an enclosure, bin, heap, pile]. A wt of 8 Cwt (406.416 kg) for lead ore in Durham and Northumberland (c1800-1900) (Second Rep. 8, Donisthorpe 205, and Pasley 115).

binne—6 byne; 7 binne [var of BIND]. A m-q for skins, numbering 33.—c**1590** Hall 28: The Byne consisteth 33 skynns. **1615** Collect. Stat. 465: The Binne of skinnes consisteth of 33. skins. **1665** Sheppard 57: But the Binne of Skins doth consist of 33 skins.

bismar—7-9 bismar (OED); 8 bysmar [Sc bismar fr Dan bismer, steelyard, or ON bismari, steelyard]. A type of steelyard (see citation) in the Orkney and Shetland Islands, principally used for weighing barley, oats, malt, and meal.—**1779** Swinton 105: The Bysmar is a beam of wood about three feet long, whereof a little more than the half is a cylinder of about an inch in diameter. The remaining part of the beam, or but-end, is also cylindrical, but much thicker than the other, being about three inches in diameter. In the small end there is a hook, from which the goods are suspended. The small end is marked with iron studs, at unequal distances. These studs correspond to and exhibit the weight of the commodities weighed, from 1 mark to 24 marks, which make a Setteen or Lyspund. When the material to be weighed is hung upon the hook, the Bysmar is horizontally suspended in the bight or loop of a cord. The weigher holds this cord in his hand; shifting its place, until the material weighed equiponderates the but-end of the Bysmar, which serves as the counterpoise. When the instrument is thus brought to an equilibrium, the stud nearest the cord shows the weight of the commodity in marks. This instrument bears relation to the Malt-pundlar, that is, the weights on it are multiples of the Malt-pundlar. See AUNCEL, LISPOUND, MARK, PUNDLAR, and SETTEEN,

blanc, blanck, blancke. BLANK

blank—6-7 blanc (OED), blanke (OED), blanck (OED), blancke (OED); 7-9 blank [ME blank fr MF blanc, of Gmc origin; see OED and WNID3]. A moneyer´s unit of wt equal to 1/24 PERIT or 1/230400 t gr (0.000000281 g). It belonged to a series of imaginary wt used to compute exact coin wt by alternate subdivisions of 20 and 24.—**1665** Sheppard 15: 24 Blanks make a Perit. **1707** Justice 4: One Perit into 24 Blanks. **1725** Bradley sv weights: The Moneyers subdivide the grain thus: 24 Blanks make 1 Perrot; 20 Perrots 1 Dwit; 24 Dwits 1 Mite; 20 Mites 1 grain. **1756** Rolt sv weights: The refiners weights are still a part of the troy, the least of which is the blank; whereof 24 make a periot. **1783** Beawes 893: Blanks, of which 24 make a Perit. **1816** Kelly 84: The Grain Troy is divided into 20 Mites, the Mite into 24 Doits, the Doit into 20

Periots, and the Periot into 24 Blanks. These divisions are imaginary. **1840** Ruding I.411: Memorandum, Twelve ounces make a pound weight troy, twenty pennyweights an ounce, twenty-four grains a pennyweight, twenty mites a grain, twenty-four droits a mite, twenty perits a droit, twenty-four blanks a perit. **1868** Eng. Cyclo. 822: A peroite 24 blanks. This division of the grain into 230,400 parts...is said to have been confined to the moneyers.

blanke. BLANK

bodge—6 bogge (OED); 6-7 bodge [perh fr ME vb bodge, var of botch, to patch clumsily, fr F boce, protuberance]. A false m-c used illegally by chandlers and others in place of the POTTLE. Since it resembled the pottle, the buyer was deceived into believing that it contained 1/2 gal or 2 qt (cl.89 1) even though its actual capacity was less.—**1588** Hall 46: Instead of the Pottle, falce measures are used, called Bodges, and some woodden measures, made under the halfe pint, most deceiptfull and unlawfull.... These measures are most used by chaundlers, milke weomen and diuerse others...contrary to the Statute in that case made and provided. c**1634** Ibid 53: In Baskettes called Prickles...and sometimes not soe much, which causeth them to sell by Bodges. See POTTLE

bogge. BODGE

boillun. BALE

boissel. BUSHEL

bole. BOLL

boll--2-4 L bolla; 4-7 bolle; 5-9 boll; 6-7 boule, boull, bow; 7 bowle; 7,

9 bole [ME bolle fr OE bolla, bowl, beaker, measure]. A m-c in northern England and Scotland for grain, coal, and other dry products.

When the Newcastle coal CHALDER weighed 42 Cwt (2133.684 kg), the boll was 1/20 of that amount or 2 1/10 Cwt (106.684 kg); but when this chalder was increased to 52 1/2 Cwt (2667.105 kg) under Charles II, the boll became 1/21 of the chalder or 2 1/2 Cwt (127.005 kg) and contained 22 1/2 gal. Finally, when the Newcastle chalder was standardized in 1695 at 72 heaped bu totaling 53 Cwt or 5936 lb (2692.506 kg), the boll equaled 1/24 of this chalder or 247 1/3 lb (112.187 kg).—**1603** Hostmen 38: Whereas, tyme out of mynde, yt hath been accustomed that all Colewaynes did usuallie cary and bringe Eighte Boulls of Coles to all the Staythes upon the Ryver of Tyne. **1606** Ibid 244: Paid for two boulles for the measuringe of keeles...paide for 4 bowles. **1608** Ibid 245: Paid for foure newe bolles and for the froneinge of them. **1617** Ibid 247: The 28 of March payd to Cuthbert Cutter for B´les [boules] for measuringe of the waynes had of him. See HUNDRED

The grain boll on the St. Paul´s Estate (c1200) contained 1 gal (c4.62 l) (St. Paul´s cxxxiv), but elsewhere in northern England (c1800-1900) it was considerably larger: Durham and Newcastle, 2 bu (c7.05 dkl); Cumberland, at Wooler, 6 bu (c2.11 hl), at Carlisle, 3 bu (c1.06 hl); Northumberland, at Alnwick, barley and oats, 5 bu (c1.76 hl), peas, rye, and wheat, 4 bu (c1.41 hl); Westmorland, rye, 2 triple bu (c2.11 hl) (Second Rep. 8 and Britten 168).

The boll, however, was used principally in Scotland. Under Robert III it was standardized at 12 gal or the capacity of a vessel 9 inches deep and 72 inches in circumference. By 1600, it was fixed at 4 FIRLOTS or 8789.34 cu inches (1.441 hl) and equal to 4.087267 Winchester bu for wheat, peas, beans, rye, and white salt, and 12,822.096 cu inches (2.101 hl) and equal to 5.962601 Winchester bu for oats, barley, and malt (Swinton 32). Both bolls were equal to 16 pk or 64 LIPPIES or FORPITS. The boll of bark was standardized in 1686 at 22 gal (c2.99 hl).—**c1150** Acts Scotland 1.310: Item bolla debet continere in se sextarium viz. xij. lagenas seruicie. Et bolla ex profunditate debet esse. ix. pollicium. In latitudine superiori debet esse. xxiiij. pollices cum spissitudine ligni vtriusque partis. In rotunditate superiori debet esse. lxxij. pollicium in medio ligni vtriusque partis superioris. In rotundine inferiori debet esse. lxxj. pollicium; ibid 311: De qualibet bolla boni brasei ordiacii nisi. xij. lagenas seruisie taberne. **c1390** Du Cange sv: Bolla debet continere sextarium, videlicet 12. lagenas, et debet esse in profunditate 9. pollicum cum spissitudine ligni utriusque partis. Et in rotunditate superiore continebit 72. pollices, in medio ligni superioris. In rotunditate inferiori 71. pollices. **1425** Acts Scotland 2.10: Thare salbe maid certane mesures of boll firlot & half firlot pek ande galone; ibid 12: Four firlotes to contene a boll.... Item the boll sal contene in breid xxix Inchys within the burdes & abufe xxvij Inche & a half euin cure thort ande in deipness xix Inchys.... Ande the boll contenande four firlotes weyis viij [X] xx & iiij punde. **1525** Jacobus 5: j celdra viij bolle frumentj. **1609** Skene 2.57: The boll...salbe in the deipnes nine inches...and in the Roundnes aboue, it sall contein thrie score and twelue inches. **1618** Acts Scotland 4.587: And that four fulles of either of the foresaids Firlots conteine and bee repute to bee ane just BOLL. **1624** Huntar 5: 4. Firlets makes the Bow. **1686** Acts Scotland 8.608: That tuenty tuo gallones shall be the measure of ane boll of unbeaten bark and soe proportionaly for lesser measures. **1813** Cooke 103: Linlithgow Bear Measure.... 4 Firlots = 1 Bole...12822.096 [cu inches]. Linlithgow Wheat Measure.... 4 Firlots...1 Bole...8789.340 [cu inches]. **1816** Kelly 93: 4 Lippies...1 Peck. 4 Pecks...1 Firlot. 4 Firlots...1 Boll. 16 Bolls...1 Chalder; ibid 93-94: The standard firlot for measuring wheat, peas, beans, rye, and white salt, contains 2197.333 English cubic inches...and the boll...8789.34 cubic inches; ibid 94: The standard firlot for measuring barley, oats, and malt, contains 3205.524 English cubic inches, and the boll...12822 cubic inches. **1829** Palethorpe sv: Boll, in Scotland, a dry measure, containing nearly 6 imperial bushels [for oats, barley, and malt]. **1880** Britten 168: (Scotl), of grain, the boll contains four firlots, nearly 6 Winchester bushels, or more accurately 5.9626.

However, there were many exceptions to these standard Scots bolls. Geographically, the variations (c1600-1900) ran as follows (Bald 447-454, Swinton 53-130, Kelly 96-112, Henderson 263-296, J. Sheppard 95-145, Edinburgh XII.571, Second Rep. 8-10, and Britten 168). North—Nairnshire: wheat, peas, beans, rye, ryegrass-seed, oatmeal, and barleymeal, 10,720.884 cu inches (1.757 hl); barley, 14,294.528 cu inches (2.343 hl); oats, 17,868.160 cu inches (2.928 hl). Sutherlandshire: peas, rye, and beans, 10,340.4 cu inches (1.695 hl); oats, barley, and malt, 14,186.876 cu inches (2.325 hl); potatoes, 24 stone (152.406 kg). Northwest—Inverness: wheat, peas, beans, rye, ryegrass-seed, and meal, 10,059.868 cu inches (1.249 hl); barley and malt, 14,076.900 cu inches (2.307 hl); oats, 17,596.125 cu inches (2.884 hl). Ross and Cromarty: wheat, rye, peas, beans, and lime, 9926.784 cu inches (1.627 hl); oats, barley, and malt, 13,235.712 cu inches (2.169 hl). Northeast--Aberdeenshire: wheat, rye, peas, beans, meal, and seeds, 10,754.016 cu inches (1.762 hl); oats, barley, and malt, 14,062.944 cu inches (2.305 hl); coal, 36 stone or 630 English lb (285.762 kg); lime, 128 Aberdeen pt (c2.20 hl); potatoes, 6 1/4 Cwt (317.512 kg). Banffshire: wheat, beans, peas, rye, and white salt, 9264.992 cu inches (1.478 hl); oats, barley, and malt, 13,476.352 cu inches (2.209 hl); potatoes, 36 stone (228.600 kg). Caithness: oats and barley, 13,623.48 cu inches (2.233 hl); oatmeal, 8 1/2 stone (53.977 kg); potatoes, 16 pk of 1 1/2 stone each (152.406 kg). Moray: wheat, rye, peas, and beans, 9384.024 cu inches (1.538 hl); barley, 13,496.128 cu inches (2.212 hl); oats, 16,870.150 cu inches (2.765 hl); barleymeal, 9 to 12 stone (57.152 to 76.203 kg); oatmeal, 8 to 9 stone (50.802 to 57.152 kg). Central—Perthshire: wheat, peas, rye, and beans, 9051.84 cu inches (1.484 hl); oats, barley, and malt, 13,356.0 cu inches (2.189 hl); barleymeal, 18 stone (114.305 kg). Stirlingshire: wheat, peas, beans, and rye, 9513.168 cu inches (1.559 hl); oats, barley, and malt,

13,752.732 cu inches (2.254 hl); oak bark, 10 stone (63.500 kg). West central--Dumbartonshire: wheat, peas, beans, and meal, 10,251.0 cu inches (1.680 hl); oats, barley, and malt, 13,668.0 cu inches (2.240 hl). West--Argyllshire: wheat, rye, beans, and peas, 10,217.608 cu inches (1.675 hl); oats, barley, and malt, 13,752.732 cu inches (2.254 hl). East--Angus: wheat, peas, and beans, 8899.552 cu inches (1.491 hl); oats, barley, and malt, 13,287.412 cu inches (2.178 hl)--both bolls average of Montrose, Forfar, Brechin, Dundee, and Arbroath bolls; meal, 8 stone (50.802 kg); potatoes, 32 stone (203.208 kg); coal, at Dundee, 56 stone (355.614 kg). Fifeshire: wheat, peas, and beans, 9099.552 cu inches (1.491 hl); oats, barley, and malt, 13,235.712 cu inches (2.169 hl). Kincardineshire: wheat, rye, and peas, 9926.784 cu inches (1.626 hl); oats and barley, 13,649.328 cu inches (2.237 hl); coal, 72 stone (457.219 kg); lime, 128 to 132 pt (c2.18 to c2.25 hl); lime shells, 85 pt (c1.45 hl); potatoes, 5 Cwt (254.010 kg). Kinrossshire: wheat, peas, and beans, 9022.0 cu inches (1.479 hl); oats, barley, and malt, 13,209.860 cu inches (2.165 hl). South--Lanarkshire, Glasgow and Lower Ward: wheat, 9256.76 cu inches (1.517 hl); peas and beans, 13,084.8 cu inches (2.144 hl); oats and barley, 13,357.6 cu inches (2.189 hl). Peeblesshire: wheat, peas, beans, and rye, 9417.6 cu inches (1.544 hl); oats, barley, and malt, 13,393.92 cu inches (2.195 hl). Southwest--Ayrshire: wheat, rye, peas, and beans, 9830.4 cu inches (1.611 hl) in Kyle and Carrick and 8601.68 cu inches (1.409 hl) and 10,178.781 cu inches (1.668 hl) in Cunningham; oats, barley, and malt,

14,487.04 cu inches (2.374 hl) and 16,128.144 cu inches (2.643 hl) in Kyle and Carrick and 16,286.048 cu inches (2.670 hl) and 17,203.36 cu inches (2.820 hl) in Cunningham. Buteshire and Arran: wheat, peas, and beans, 11,512.312 cu inches (1.888 hl); oats, barley, and malt, 17,268.468 cu inches (2.830 hl). Kirkcudbrightshire: all grain, 23,654.0 cu inches (3.877 hl). Renfrewshire: wheat, the Linlithgow standard; beans, peas, and vetches, 9616.572 cu inches (1.576 hl); oats and barley, 13,623.476 cu inches (2.233 hl). Wigtownshire: wheat, peas, and beans, 17,203.36 cu inches (2.820 hl); oats, barley, and malt, 24,605.04 cu inches (4.033 hl); meal, 16 stone of 17 1/2 lb each (127.008 kg); potatoes, 8 Cwt (406.416 kg). Southeast—Berwickshire: all grain, 12,902.52 cu inches (2.114 hl) or 13,442.52 cu inches (2.203 hl); lime, approximately 4 Winchester bu (cl.41 hl); potatoes, 476 English lb (215.909 kg), and in Berwick township, 560 English lb (254.010 kg). East Lothian: wheat, peas, and beans, 9047.848 cu inches (1.483 hl); oats, barley, and malt, 13,209.856 cu inches (2.165 hl). Midlothian: wheat, peas, and beans, 8944.444 cu inches (1.466 hl); oats, barley, and malt, 13,028.904 cu inches (2.136 hl). Roxburghshire: wheat, peas, and beans, 9200.0 cu inches (1.508 hl) and 11,374.440 cu inches (1.864 hl); oats, barley, and malt, 13,650.0 cu inches (2.247 hl) and 17,061.660 cu inches (2.797 hl); meal, 16 stone (101.604 kg). Selkirkshire: wheat, rye, beans, and peas, 9225.0 cu inches (1.512 hl) and 11,406.750 cu inches (1.869 hl); oats, barley, and malt, 12,925.0 cu inches (2.118 hl) and 16,156,850 cu inches (2.648 hl); meal, 16 stone

(101.604 kg).

bolla, bolle. BOLL

boll of bear's sowing [*]. A m-a in Caithness (c1800) equal to approximately a Scots ACRE (6150 4/10 sq yd or c0.51 ha) and used as a measure for the payment of rent (Second Rep. 10).

bolltte. BOLT

bolt—4-9 bolt; 6 bolltte, bolte, bowlte; 7-9 boult [ME bolt fr OE bolt; akin to MLG bolte, bolt, piece of linen rolled up]. A m-q (bundled or rolled) for thread, canvas, wood, and various other goods. Its dimensions generally depended on the quality and weight of the goods being shipped. In Berkshire (c1800-1900) a bolt of osiers was 42 inches (1.067 m) around and 14 inches (35.56 cm) from the butts, while in Essex, 80 bolts of osiers made a load.—**1399** trans in Cal. Close 16.371: One whole cloth and 8 'boltes' of 'worstede.' **1507** Gras 1.701: Powlld davys for saylles the bolltte; ibid 705: Vlyons for sayles the bowlte. **1509** Ibid 577: viii bolts olrons. **1545** Rates 1.8: Canuas the bolte; ibid 27: Olrons the bolte. **c1550** Welsh 82: 1 bolt canvas...1 bolt Poldavi; ibid 107: 2 bolts wood. **1612** Halyburton 331: Lyons or Pereis threid the ball contening ane hundreth boultis. **1701** Hatton 3.220: Bolt...(of Canvas) 28 Ells. **1880** Britten 168-69: Bolt, or Boult, of oziers. (Berks.), a bundle, measuring 42 inches round, 14 inches from the butts. (Ess.), a bundle, of which 80 make a load. **1956** Economist 8: Bolt of canvas...42 yards.

bolte. BOLT

boltin. BOLTING

bolting—8 boltin (OED); bolton (OED); 8-9 bolting [bolt, bundle, + -ing]. A wt of 24 lb (10.886 kg) for straw in Gloucestershire (c1800-1900) (Second Rep. 10 and Britten 134, 169).

bolton. BOLTING

bomkyn [*]. A small BARREL.

bonch, bonche. BUNCH

bondel, bondell. BUNDLE

boot. BOUT

boschell, bosel, bosshell. BUSHEL

botel. BOTTLE[1]; BOTTLE[2]

botele. BOTTLE[2]

botell. BOTTLE[1]; BOTTLE[2]

botella. BOTTLE[1]

botelle, botle, bottel, bottell, bottelle. BOTTLE[1]; BOTTLE[2]

bottle[1]—4 botel (OED); 5 bottelle (OED); 5-6 botell (Coopers), L botella (Dur. House), botelle (OED), bottell (OED); 6-7 botle (OED), bottel (OED); 6-9 bottle [ME botel, botelle fr MF boteille fr MedL buticula, butticula, dim of LL buttis, BUTT]. A m-c for liquids (c1800-1900): aqua fortis, 4 gal (c1.51 dkl), and wine, approximately 1/5 gal (c0.76 l) (Second Rep. 10 and Donisthorpe 206). It sometimes was abbreviated bot.

bottle[2]—4 botele; 4-6 botel; 5 bottelle (OED); 5-6 botell (OED), botelle; 5-7 bottell; 6 bottel (OED); 6-9 bottle; 7 botle (OED) [ME botel fr MF botel, dim of bote, bundle]. A m-q for hay or straw weighing 7 lb (3.175 kg).—**1365** trans in Memorials 324: And if they sell their hay by boteles, they are to make their boteles in proportion to the same price. **1439** Southampton 2.82: 10 botels. c**1440** Promp. Parv. 45: Botelle of hey. Fenifascis. **1474** Cov. Leet 399: And his bottell of haye of an ob. shall way vij lb. and his liter free. **1595** Powell F 2: The assise for Inholders, or any others retailing their Hey by the Bottle, Trusse, or hundreth. **1851** Sternberg 12: Bottle...A bundle of hay or straw.

boule, boull. BOLL

boult. BOLT

bounch, bounche. BUNCH

boundell. BUNDLE

boussel, bousshell. BUSHEL

bout—8-9 bout; 9 boot (OED) [perh a special sense of bought, bend or bending]. A m-c for lead ore in Derbyshire (c1800) containing 240 DISHES (c26.43 hl) (Second Rep. 10).

bout. BUTT

bovat, bovata. BOVATE

bovate—2-7 L bovata; 3-? bovate; 6, 8 bovat; 7 bovatt (OED), L bovatus; 8 boviat (OED) [MedL bovata fr L bos, bovis, ox]. A m-a which originally was believed to be the amount of land that an ox and team could plow in a year, but which, in actual practice, varied between 4 and 32 acres (1.770 to 14.160 ha), depending on the quality of the soil in any particular region. Occasionally it was used synonymously with the

VIRGATE, generally equaling 1/4 or 1/2 HIDE, but more often it was reckoned at 1/2 virgate or 1/8 hide. It was frequently abbreviated bov. in medieval MSS.—c**1130** Slade 14: In Balbegraue vj car´ iij bov´ minus de soch´ Regis. c**1153** Malcolm 192: In escambio duarum bouatarum terre in Berewyc. **1201** Cur. Reg. 9.53: Scilicet de tercia parte vij. bovatarum terre cum pertinenciis in Waberge. **1202** Feet 1.37: In una bouata terre de predictis duabus bouatis que continet xviij acras terre cum pertinenciis.... Alteram bouatam terre cum pertinenciis in Filingham continentem xxviij acras. **1204** Cur. Reg. 10.238: v. bovatas et dimidiam; ibid 239: Et Robertum de xv. bovatis terre cum pertinentiis; ibid 240: j bovata terre. **1207** Feet 1.110: Ad undecim bouatas terre unde quadraginta bouate faciunt seruicium unius militis pro omni seruicio. **1219** Eyre 258: Terciam partem ii bouatarum terre. c**1260** Clark 106: Willielmus Knotte quartam partem j. bovate cum tofto. **1327** Gray 508: Quelibet bovata continet xiii acras. c**1400** Melsa I.161: xii. bovatas terræ, videlicet unam carucatam et dimidiam terræ. c**1500** Hall 8: viii [X] xx pertice faciunt acram; duodecim acre faciunt bovatam.... ii bovate faciunt virgatam. **1599** Richmond Appendix 2.11: BOVATA. As some xv Acres, as before is declared, in some x Acres, and in some xxiv Acres, and in some xii Acres.... An Oxgang, which is called Bovat, about xv Acres. **1664** Spelman 87: Bovata, seu Bovatus terræ.... In vet. autem Statutorum M.S. ad Compositionem mensurarum, sic notatur. Octo bovatæ terræ faciunt carucatam terræ...xviii acræ faciunt bovatam terræ. Ex Skenæi autem Sententia, bovata terræ xiii acris semper constaret. **1755** Willis 361: From the Terms Hide, Carucate, Bovate, &c. so often accurring in Doomsday-Book, it appears that the primitive Husbandry here consisted chiefly in Tillage or Arable Culture; ibid 362: In Spelman's Gloss. an Account is cited from an ancient MS. that viii Bovats made a Carucate, and viii Acres a Bovate; if so, a Carucate must contain lxiv Acres of Arable; but I think otherwise...and that a carucate had no fixed Measure. **1777** Nicol. and Burn 610: Bovate...of land: as much as one yoke of oxen can reasonably cultivate in a year. **1874** Hazlitt 418: A bovate or oxgang of land contains, in general, only about fifteen acres in the county of York. **1888** Round 3.196: The carucate, divided into eight bovates. **1888** Taylor 159: Since eight Domesday bovates make one carucate, we should also expect to meet with bovates of one-eighth of these areas, viz. of 7 1/2, 15, 22 1/2, 9, 18, 27, 10, 20, 12, and 24 acres; ibid 167: A bovate of 30 acres implies a carucate of 240 acres; ibid 173: The carucate being the quantity of land tilled by one plough, and the normal plough being drawn by eight oxen, a bovate, which was originally the share of the tilled land appropriated to the owner of each of the associated oxen contributed to the cooperative plough, was normally one-eighth of a carucate; ibid 174: In the reign of Edward I. and afterwards, we occasionally find that 12 bovates went to the carucate; ibid 176: No less than 1216 bovates [in the manors of the See of Durham] of 15 acres are enumerated.... The Boldon Book enumerates 196 bovates of 12 acres; ibid 177: The Boldon Book enumerates 80 bovates of

16 acres and 70 of 8 acres...22 bovates of 20 acres and 213 bovates of 12 acres.... At Warden there were 18 bovates of 13 1/2 acres.... There are also 14 bovates of 9 acres and 2 of 18 acres. **1897** Maitland 397: The numbers of the acres in a bovate given by a series of Yorkshire inquests is 7, 7, 8, 15, 12, 6, 12, 15, 15, 6, 5, 9, 10, 10, 12, 24, 4, 16, 12, 18, 8, 6, 10, 24, 32. **1909** Curtler 16: The basis of the whole scheme of measurement in Domesday was the hide.... A quarter of this was the virgate, an eighth the bovate, which would...supply one ox to the common team. See OXGANG

bovatt, bovatus, boviat. BOVATE

bow, bowle. BOLL

bowlte. BOLT

bowsshell, bowsshelle. BUSHEL

box—5-6 boxe; 6 boxse (OED); 6-9 box [ME boxe fr OE box fr LL buxis fr Gr pyxis fr pyxos, boxtree]. A m-c originally referring to any small receptacle used for drugs and other valuable materials. However, since the eighteenth century it has included containers of any size made for the purpose of holding merchandise and personal property. It generally did not have a standard capacity except (c1700-1800) for aloes, 14 lb (6.350 kg); almonds, 25 lb (12.247 kg); camphor, 1 Cwt (50.802 kg); coals, Derbyshire, 2 1/2 striked or leveled bu (c8.81 dkl); rings for keys, 2 GROSS; quicksilver, 100 to 200 lb (45.359 to 90.718 kg); and salmon, Durham, 8 STONE (50.802 kg) (Second Rep. 10, Palethorpe sv, and Hatton 3.221). It sometimes was abbreviated bx.—**1420** Gras 1.512: iiii box[es] boras[is]; ibid 514: i boxe galbannum. **1439** Southampton 2.74: 80 boxis sitronade. **1507** Gras 1.699: Harpe strynges the boxe; ibid 703: Shomakyrs heres the boxe. **1545** Rates 1.5: Brystels the boxe. **1840** Waterston 147: Almonds...box...lbs. 25.... Camphor, box, about...cwt. 1.

boxe, boxse. BOX

boyllum, boylun. BALE

boyschel. BUSHEL

brawler [*]. A m-q, a bundle or sheaf, for straw in Somersetshire (c1800-1900) weighing 7 lb (3.175 kg) (Britten 135, 169).

bucket [ME bucket fr AF buket fr OE būc, pitcher, belly]. A m-c for chalk in Buckinghamshire and Hertfordshire (c1800-1900) containing 1 1/2 bu (c5.29 dkl) (Second Rep. 10 and Donisthorpe 206). Today, in the United Kingdom, a bucket for dry and liquid products generally contains 4 gal (18.18 l) (O´Keefe 671).

buisshel. BUSHEL

bunch—4-6 bunche; 5 bonch; 5-6 bonche, bounche (OED); 6-7 bounch (OED); 6-9 bunch; 7 bunsh [ME bunche; see WNID3]. A m-q used principally for onions or garlic, 25 heads, and glass, usually equal to 1/60 WEY or WEB or 1/40 WAW of glass.—[**1290** Fleta 120: Rasus autem alleorum continet xx flones [bunches], et quelibet flonis xxv. capita.] **1439** Southampton 2.8: M. bunchis allei. **1443** Brokage II.1: Cum ix [X] xx bonchys allei; ibid 92: Cum iiii [X] xx bonchis allei. **1478** Stonor II.73: For viij bonches of garleke. **1507** Gras 1.698: Glasse called Flemyche glasse the waw that ys to saye xl bunchys; ibid 701: Onyones the C bunches. **1545** Rates 1.19: Glasse the bonche...Garlike the C. bonches. **1590** Rates 2.17: Glasse the way...containing lx. bunches. **1612** Halyburton 308: Glasses for windows...the web contening lx bunshes. **1717** Dict. Rus. sv weigh: Of Glas 6[0] Bunches.

The bunch was also used (c1800-1900) for osiers, Cambridgeshire, a bundle 45 inches (1.143 m) in circumference at the band; reeds, Cambridgeshire, a bundle 28 inches (7.112 dm) in circumference at the band; teasels, Essex, 25 heads, Gloucestershire, 20 heads for regular and 10 heads for king´s, and Yorkshire, 10 heads (Second Rep. 10, Donisthorpe 206, Morton sv, and Britten 169).

Bunches of the following items did not have standard sizes.—**1396** Gras 1.7: Fanes the bunche; ibid 441: C bunches lini. **1402** Ibid 556: Pro xx bunche leok. c**1500** Fab. Rolls 337: Bunchys of lattes.

bunche. BUNCH

bundel, bundell, bundelle. BUNDLE

bundle—4-7 bundel; 5-6 bondel, bondell, boundell (OED), bundelle; 5-7 bundell; 6 byndle (OED); 7-9 bundle [ME bundel fr MDu bondel, bundel; akin to OE byndel, a bundle]. A m-q that varied in number or dimension according to the product, its quality, and weight (c1700-1900): barley straw, Devonshire, 35 lb (15.876 kg); bast ropes, 10; birch brooms, 1 or 2 dozen; brown paper, 40 QUIRES; glovers´ knives, 10; harness plates, 10; hogshead hoops, Gloucestershire, 36; hoops, Berkshire, 120 to 480; oat straw, Devonshire, 40 lb (18.144 kg); osiers, Gloucestershire, 1 1/4 ft (0.457 m) in circumference, Hampshire, 42 inches (1.067 m) around the lower band, and Worcestershire, 38 inches (0.965 m) in circumference; straw for thatching, Yorkshire, 1/12 THRAVE; wheat straw, Devonshire, 28 lb (12.700 kg); yarn, Hamborough, 20 SKEIN (Hatton 3.221, Second Rep. 10-11, 35, and Britten 169).—**1545** Rates 1.33-34: Paper called browne paper the hundreth bondels...Paper called browne paper the bondell.

Bundles of the following items had no consistent standard sizes throughout England.—**1443** Brokage II.27: iii bondell´ de fryyng pannys. **1507** Gras 1.696: Corke the bundelle for shyppers or ells the C. **1509** Ibid 563: iiii [X] xx bundelli papiri nigri. **1524** Ibid 195: Pro ix bundelis osyers. **1549** Ibid 627: Pro xx bundellis fannes. **1612** Halyburton 292: Bracelettis of glase the groce contening tuelf bundellis. **1721** King 282: Basket Rods...per Bundle. **1789** Topham xxvii: Steel in bars, per bundle. **1831** Pope 67: BASKET RODS, the bundle, not exceeding three feet in circumference at the band.

bunsh. BUNCH

burden—2-9 burden; 5 burdon, byrdyn; 5-6 burdyn, burdynge (Dur. House); 5-6, 8 burthen [ME burden, burthen fr OE byrthen; see OED]. A wt for steel containing either 6 or 12 sheaves of 30 GADS each and sometimes reckoned at 9 score or 180 lb (81.646 kg), and a m-q for fish (ling and mulvel or cod) numbering 20 or 22.--**1443** Brokage II.19: Cum vi brydyn piscis salsi; ibid 40: Et dimidio byrdyn piscis. **c1461** Hall 17: Also style by gadds; and euery pece of stele in hymselfe is a gadde; and xxx gaddes make a scheff, and xii scheff make a burdon. **1507** Gras 1.703:

Stelle the barelle wyche owght to be iiii [X] xx burden and vi scheffe makythe a burdyn and xxx gaddes makythe sheffe. **c1550** Welsh 63: 7 burden stile; ibid 81: 1 burden fish; ibid 275: 4 burthen steel. **1559** Remembrance 67: The xvi[th] of Octobre, John Croche of the cytye of London fyshemonger have maid pryce w[ith] Mr. Mayer...for xiis. the burdon of lyngs. **1628** Hunt B 3: Gad-steele, of which, a Burden is 9. score, 180. **1755** Postlethwayt II.191: A burthen of gad steel is 9 score, or 180 lb. **1790** Miller 18: The...FAGGOTT, GAD, BURTHEN, (the three last particularly applied to different weights of steel). **1895** Donisthorpe 206: BURDEN: of steel, 180 pounds.

burdon, burdyn, burdynge, burthen. BURDEN

buscel, buscellus, buschel, buschell, buschelle, buschellus, busellus, bushall. BUSHEL

bushel--3 L buschellus; 3-4 L bussellum, L busselum; 3-7 L bussellus; 4 boissel, bosel, boyschel (OED), buisshel (OED), L buscellus, buschel, busschel (OED), buysshel (OED); 4-5 boussel, bussell, busselle (Hall), buyschel (OED); 4-6 L busellus, busshel, busshelle; 4-7 busshell; 4-9 bushel; 5 boschell, bosshell (Southampton 2), buscel, buschelle, bussel, L busshellus, buysshell, byschelle (OED), bysshell (OED); 5-6 bousshell, bowsshel (OED), bowsshelle (OED); 5-7 bushell; 6 beyschell (OED), buschell, bushylle, buszhell (OED), buszshel (OED); 6-7 bushelle; 7 bushall (Young II) [ME busshel, boyschel fr OF boissel fr boisse, a measure of grain]. A m-c for dry products.

In England the standard or Winchester grain bu (35.238 l) contained 4 pk, or 8 gal, or 16 POTTLES, or 32 qt, or 64 pt, and was equal to 1/8 SEAM or 1/80 grain LAST. Until the sixteenth century, the bu of wheat was supposed to weigh 64 tow lb, but after the tow lb was abolished, the bu of wheat was sometimes described as weighing 56 avdp lb. Since the establishment of the Imperial system in 1824, the bu has contained 2219.360 cu inches (36.368 l) or 4 pk of 554.840 cu inches each or 8 gal of 277.420 cu inches each. Occasionally it was abbreviated bus., bush., bz., or bushl.—c**1200** Caernarvon 242: Buschellu londonia hoc est ovtavam partem quarterii. **1212** Cur. Reg. 2.322: Et duos bussellos. **1258** Wellingborough 1: Item de .xj. quarteriis et .iij. Bussellis ordei. **1290** Fleta 119: Et pondus octo librarum frumenti faciunt mensuram ialonis, et octo iaionate frumenti faciunt bussellum, de quibus octo consistit commune quarterium. **1298** Falkirk 1: iii bussellorum frumenti. c**1300** Mon. Jur. 1.80: Cest assavoir que le boissel nest mye ung potel greigneur que lestandarde de la terre. c**1300** Hall 8: Et viii galones bladi faciunt i bussellum. Et viii buselli bladi faciunt i quarterium. Et bussellus frumenti puri et bene mundati ponderabit xlviii l. sterlingorum. **1347** Rot. Parl. 2.219: De ceo q'il avoyent vendu farine & altres marchaunises par meyndre mesure que d'un bosel. **1351** Ibid 240: Et contiegne le quarter oet bussell par l'estandard. **1390** Henry Derby 6: Pro iij bussellis et l pecco auenarum. **1395** York Mem. 2.10: Unum busselum eris, dimidium bussellum et pek ligni. c**1400** Brit. Mus. 30.52v: lxiiij pyntes...the hole bousshell. c**1400** Hall 7: Et viij livres de froument font la galone de vin. Et viii galons de froument font le bussell' de Loundres, qest la oeptisme partie du quarter; ibid 12: And viij gallons of wyne make a boschell of whete; ibid 36: 8 galones faciunt 1 buschel; ibid 37: Et unus buscellus continet octo lagenas. c**1400** York Mem. 2.260: Cest assavoir des boussels, demi boussels. **1413** Rot. Parl. 4.14: C'est assavoir, viii Busselx pur la Quartre, & qe chescun Bussell contiendra oept Galons. **1443** Ibid 450: Sinoun oept busshels rasez pur le quarter. c**1440** Promp. Parv. 56: Buscel...buschelle...Modius. c**1470** Gregory 88: Ande that same yere a buschelle of whete was worthe xl d. **1474** Cov. Leet 396: & ij pyntes maketh a quart; & ij quartes maketh a Pottell; & ij Pottels makith a Gallon; & viij Gallons makith a Buysshell, and neyther hepe nor Cantell. **1495** Brit. Mus. 28.154v: & viij galones faciunt busshellum. **1496** Hall 45: That is to say: a busshell, a galown, a yerde. c**1500** Ibid 8: xii uncie faciunt libram; viij libre faciunt lagenam; viij lagene faciunt busellum. **1526** Davenport lxxxiii: Also v. buschell of wytte to be grownde. **1534** Fitzherbert 21: And if there be the .iiii. parte beanes, than wylle it haue halfe a London bushelle more: and yf it be halfe beanes, it wyll haue thre London bushels: and if it be all beanes, it wyll haue foure London busshelles fullye, and that is half a quarter.... Oone busshelle, as yf he sowed .iiii. busshelles. c**1560** Mon. Fran. 170: And this yere a bushylle of whette was at xl. d. **1586** Brit. Mus. 22.142: Of viij busshelle wheate.... And mett by the new brason busshell. **1587** Stat. 454: That the measure of the bushell conteine viii. galons of whete, and that euerie galon conteine viii. li. of wheate. **1593** Adames 13: Also if any haue and use any measures of bushelles...they are to be enquired of. **1595** Powell C: The which lvj. pound of Auoirdupois weight are and haue been accustomably vsed for the content of the bushell through this Realme. **1616** Hopton 162: All kind of graine is measured by...a gallon, whereof are made...Bushels. **1630** Cottenham 246: Tenn Bushells.... Every busshell; ibid 247: One bushelle. **1635** Dalton 145: By Statute the bushell must containe eight gallons, or 64 pounds or pintes of wheat.... And yet by the booke of Assize, imprinted An. Dom. 1597, the Bushell is to containe 56 pounds (or pintes) of Averdepois weight. **1638** Bolton 271: 56. pounds, or pints Averdepois make the Bushell of Wheate. **1665** Assize 2: But 12 ounces to the pound...and 64 pounds to the bushel. **1668** Bernardi 150: Præterea Galonem Frumentarium Angliæ, dimidium Pecci et Octantem Busselli.... Pinta denique arida, 1/8 Galonis, seu congii frumentarii, et 1/8X8 = 1/64 Brusselli aut Amphoræ Anglicanæ. **1695** Kennett Glossary sv bussellus: Bussellus was therefore first us´d for a liquid measure of wine, eight gallons.... The word was soon after transferr´d to the dry measure of Corn, of the same quantity. **1745** Fleetwood 57-58: For so quarta, quartalis, and quartalium, signifies a Peck, or the fourth Part of a Bushel. **1789** Hawney 310: And 2150.42 solid Inches is a Bushel of Corn-measure. **1819** Cyclopædia sv weights: The standard Winchester bushel is a cylinder 18 1/2 inches in diameter, and 8 inches deep. **1831** Hassler 12: The Winchester bushel being determined to 2150.42 cubic inches. **1907** Hatch 23: 4 pecks = 1 bushel (bush.) = 2219.360 cubic inches. **1956** Economist 4: Bushel: (a) United Kingdom, Imperial system = 2,219.36 cubic inches. **1969** And. & Bigg 11: 1 bu (bushel) = 0.0363687 m^3 = 36.3687 dm^3.

Local variations of the bu were numerous since it was the principal measure for dry products. Not only did local units vary in capacity, but they also varied in weight. Geographically, these variations (c1800-1900) were as follows (Second Rep. 11-13, Britten 169-70, and McConnell 10). North--Yorkshire: 1 to 3 qt (1.101 to 3.303 l) above the Winchester bu. Northeast--Durham: corn, 8 to 8 1/2 gal (c3.52 to c3.74 dkl); at Stockton, oats, 35 lb (15.876 kg), and wheat, 60 lb (27.215 kg). Northwest--Cumberland: Carlisle, triple bu of 24 gal (c10.57 dkl), and Penrith, barley, potatoes, and oats, 20 gal (c8.81 dkl); rye and wheat, double bu of 16 gal (c7.05 dkl). Lancashire: potatoes, 90 lb (40.823 kg), at Liverpool, barley, 60 lb (27.215 kg), oats, 45 lb (20.412 kg), and wheat, 70 lb (31.751 kg). Westmorland: Appleby, barley, 20 gal (c8.81 dkl); potatoes, 16 gal (c7.05 dkl). Central--Bedfordshire: 2 pt above the Winchester bu (3.64 dkl). Derbyshire: potatoes, 90 lb (40.823 kg). Leicestershire: potatoes, 80 lb (36.287 kg). Staffordshire: barley, beans, oats, and peas, 9 1/2 gal (c4.18 dkl); wheat, 72 lb (32.658 kg). West central--Worcestershire: Worcester, 8 1/2 gal (c3.74 dkl), Evesham, 9 gal (c4.11 dkl), and all other places, 9 1/2 and 9 3/4 gal (c4.18 and c4.29 dkl). West--Cheshire: barley, 60 lb (27.215 kg); oats, 45 to 50 lb (20.411 to 22.679 kg); potatoes, 90 lb (40.823 kg); and wheat, 70 to

75 lb (31.751 to 34.019 kg). Herefordshire: grain, 10 gal (c4.40 dkl); malt, 8 1/2 gal (c3.74 dkl). Shropshire: barley, peas, and wheat, 9 1/2 to 10 gal (c4.18 to c4.40 dkl). South--Bershire: corn, in some parts, 9 gal (c3.96 dkl). Oxfordshire: wheat, 9 gal and 3 pt (c4.11 dkl). Southeast--Middlesex: potatoes, 56 lb (25.401 kg). Surrey: potatoes, 60 lb (27.215 kg); turnips, 50 lb (22.679 kg). Southwest--Cornwall: eastern section, double bu of 16 gal (c7.05 dkl); and western section, triple bu of 24 gal (c10.57 dkl); potatoes, 220 lb (99.790 kg). Devonshire: barley, 50 lb (22.679 kg); oats, 36 or 40 lb (16.329 or 18.144 kg); wheat, 3 level and 1 heaped pk (c3.96 dkl). Gloucestershire: corn, usually 9 1/2 gal (c4.18 dkl), sometimes 9, 9 1/4, and 10 gal (c3.96, c4.07, and c4.40 dkl). Somersetshire: coal, 9 gal (c3.96 dkl).

There were also variations (c1800-1900) outside England: Scotland, Ayrshire bu, 2 pk (c1.81 dkl), Galloway barley bu, 46 to 53 lb (20.865 to 24.040 kg); Ireland, 1740.8 cu inches (2.853 dkl) and equal to 4 Irish pk of 435.2 cu inches each; Wales, Brecknochshire, 10 gal (c4.40 dkl), Monmouthshire, 10 to 10 1/2 gal (c4.40 to c4.62 dkl). Montgomeryshire, 20 gal (c8.80 dkl) (Second Rep. 13, Britten 169-70, and Edinburgh XII.571).

bushell, bushelle, bushylle, busschel, bussel, bussell, busselle, bussellum, bussellus, busselum, busshel, busshell, busshelle, busshellus, buszhell, buszshel. BUSHEL

but. BUTT

butress [*]. A m-c of undetermined size (c1400) for coal (Salzman 1.15).

butt--5 bout (Southampton 1); 5-6 but; 5-7 butte; 5-9 butt [ME butt fr MF botte fr OPr bota fr LL buttis, butt, cask]. Equivalent to PIPE.—**1423** Rot. Parl. 4.256: The but of Samon, xx [X] iiii [+] IIII Galons full pakked. **1443** Brokage II.1: Cum ii carectis cum iiii buttis vini; ibid 131: Cum i but romeney. **1482** Rot. Parl. 6.221: That every Butt...ordeyned for Samon, shuld conteygne...xx [X] iiii [+] IIII Galons. **1517** Hall 50: And, my lords, please hyt yow to understonde that yf the Gaskoyne ton´, pype, or hogg[eshed´] or Romney butte doo fawte of the trewe gawge & weyghte at thys tyme, God forbed that all´ the mesurys of Ynglond´ shuld´ follow thus. c**1590** Ibid 21: The pipe contenith a butt which is 1/2 of a tunne, 126 gallons; ibid 23: The but of salmone ought to be 84 gallons fully packed...the last is 6 buttes conteninge 504 gallons. **1615** Collect. Stat. 466: Nor butte of Salmon, unlesse it contain 84. gallons fully packed.... Any But of Maluesie to be sold, unlesse it doe containe in measure at the least one hundred twentie and six gallons. **1660** Bridges 26: For 63 gallons a hogshead...For 2 Hogsheads a But, & For 2 Buts a Tun. **1665** Assize 4: There is also a certain measure called a Salmon-Butt, which must hold and contain fourscore and four gallons. **1701** Hatton 3.221: Butt.... Of Sack—2 Hogsheads. Currants--15 to 22. C. **1708** Chamberlayne 210: Of these Gallons...a Pipe or Butt holds 126. **1717** Dict. Rus. sv: Butt or Pipe of Wine, contains two Hogsheads, or One hundred twenty six Gallons; and a Butt of Currans from Fifteen to Twenty-two Hundred weight. **1725** Bradley sv: Butt, or Pipe, a Liquid Measure, whereof two Hogsheads make a Butt or Pipe, as two Pipes or Butts make one Tun. **1756** Rolt sv: BUTT, is used for an English vessel, or liquid measure, either of wine or beer...otherwise called a pipe. **1773** Johnson sv: Butt...A vessel; a barrel containing one hundred and twenty-six gallons of wine...and from fifteen to twenty-two hundred weight is a butt of currans. **1829** Palethorpe sv: Butt, a large wooden cask used in England for wines, liquors, &c. holding 105 imperial gallons, or 126 gallons of the late wine measure. **1851** H. Taylor 58: The butt, from 108 (formerly 126 gallons, or 3 1/2 barrels). **1880** Courtney 161: The barrel, hogshead, tierce, pipe, butt and tun, are the names of casks. **1956** Economist 54: Butt (of ale) = 108 gallons (104 gallons in Ireland).

butta terræ. BUTT OF LAND

butte. BUTT

butt of land—5-7 L butta terre (terræ) (Du Cange) [ME but, butt, ridge of ground between two furrows, fr MedL butta, buttis; see OED and WNID3]. A m-a for land, possibly synonymous with the RIG, RIDGE, or SELION, being a strip of land or pathway between two parallel furrows of the open field.--**1409** Gray 362: Una parva butta terre vocata Pilchebutt´. **1616** Ibid 244: Item in the West Eyes xvii selions with three Butts and a geron. **1681** Acts Scotland 8.295: And that other rigg or butt of land of the samen lyand in the ffield called the Gallowbank. **1688** Holme ii(32): Smaller parcells according to that quantity of ground it containeth, both for length and breadth...3 Ridges, Butts, Flats, Stitches or small Butts, Pikes. **1695** Kennett Glossary sv buttes: BUTTES. The ends or short pieces of land in arable ridges and furrows. Gilbert Basset gave to his Priory of Burcester, viginti acras in Helle-furlong & buttes apud Ymbelowesmere. **1880** Britten 136: Butt...a ridge, or land between two furrows.

buyllon. BALE

buyschel, buysshel, buysshell. BUSHEL

bylet, byllet, byllot. BILLET

bynd, bynda, bynde. BIND

byndle. BUNDLE

byne. BINNE

byrdyn. BURDEN

byschelle. BUSHEL

bysmar. BISMAR

bysshell. BUSHEL

C

C. CENT

caas. CASE

cable length [ME cable fr ONF cable fr MedL capulum, lasso, fr L capere, to take; ME lengthe fr OE lengthu fr lang, long, long, + -thu, -th]. An Imperial maritime unit of length based on the length of a ship's cable and reckoned as 100 FATHOMS, 200 yd, or 600 ft (182.880 m) and equal to 10.133 nautical mi or 26.4 LEAGUES (Economist 8).

cabot [OF cabot (F chabot), perh fr cabo, capo, head, + ot]. A m-c for wheat in Jersey equal to approximately 3/4 English bu (c2.64 dkl). There was also (c1800-1900) a smaller cabot, 4 of which made 3 standard cabots, used for barley and corn (Second Rep. 13, Donisthorpe 208, and Britten 170).

cace. CASE

cade—4-6 L cadus; 4-9 cade; 5 cayde; 6 gag (Dur. House), gage, gagge (Dur. House) [ME cade fr L cadus, a large vessel usually of earthenware, a wine jar, a liquid measure]. A m-c, a small bbl or keg, for fish and other products: herrings, 500 to 1000; sprats, 1000.—**1392** Henry Derby 97: Et per manus eiusdem pro j cade allecium rubrorum. **1393** Ibid 208: Item in xij cadis; ibid 222: Item pro j cado pro vergws imponendo. **1439** Southampton 2.27: 3 cades allecii rubii; ibid 29: 13 cadys allecii rubii. **1443** Brokage II.89: Cum xii cades allecii relute; ibid 99: Cum xii cades shot allecii rubei. **1456** trans in Fountains xxiii: Tar, l s. a gallon, 6 s. 8 d. to 12 s. 2 d. a cade; ibid 258: Cadus corei. **1478** Stonor II.73: For iij cayde herreng, xij. s. **1502** Arnold 64]

263: XX. cadis rede hering is a last, v.C. in a cade, vi. score iiij. heringis for the C. **1532** Finchale ccccxxi: One gage of eels. c**1550** Welsh 217: 20 cados wheat...barley...oat malt. **1590** Rates 2.19: Hering red the cade conteining x [X] C...Hering red the last containing xx cades or xx thousand. **1610** Halyburton 304: Fishes called Herring-reid the Cade containing v [X] c. **1695** Kennett Glossary sv: CADE of herrings.... Memorandum that a barrel of herryng shold contene a thousand herryngs, and a Cade of herryng six hundreth, six-score to the hundreth. **1717** Dict. Rus. sv: Cade, a Cag, Cask, or Barrel.... Cade of Herrings, a Vessel or Measure containing the quantity of five Hundred red Herrings, or of Sprats a Thousand. **1829** Palethorpe sv: CADE, in commerce, a cag, cask, or barrel. A cade of herrings is a vessel containing the number 500, a cade of sprats 1000. **1883** Simmonds sv: Cade, a keg or small barrel; also a variable fish measure; 500 herrings or 1000 sprats make a cade.

cadge. CAGE

cadus. CADE

cage—5 kage (OED); 6 cadge (OED), kaig (OED); 6-9 cage [OF cage fr L cavea, cavity, cage, fr cavus, hollow]. A m-q, generally without a fixed value, for animals: quails, sometimes 28 dozen.—**1509** Gras 1.568: i cage cum xxviii dossenis quailles. **1590** Rates 2.51: Quailes the Cage.

cairata. CHARGE[2]

cais. CASE

canister--8-9 canister, cannister (OED) [L canistrum, basket for fruit or flowers, fr Gr kanastron, wicker basket, fr kanna, reed]. A m-c for tea, varying in wt from 75 to 100 lb (34.019 to 45.359 kg).—**1701** Hatton 3.221: Canister...Of Tea, 75 to 1 C. weight. **1717** Dict. Rus. sv: Canister of Tea, a quantity from Seventy-five to a Hundred Pound Weight.

cannister. CANISTER

cantel—5-7 cantel, cantell [ME cantel fr ONF cantel (MedL cantellus), dim of cant, edge, corner]. A shallow measure in which the contents did not reach the rim. Either the vessel was purposely filled this way or the merchant or seller compressed its contents. The shallow measure was limited for use in selling oats, malt, and meal by the same Edwardian statute (1325) that limited the use of the heaped measure.—**1474** Cov. Leet 396: And viij Buysshelles makith a Quarter, striken with a Rasid stryke, and neyther hepe nor Cantell. **1587** Stat. 77: No maner of graine shall be sold by the heape or Cantell, except it be otes malt and meale. **1603** Henllys 137: In all these bushells, oates and oaten mault is pressed and wrunge downe in the pecke. **1615** Collect. Stat. 464: And that the Toll bee taken by stricke, and not by heape or Cantell. **1695** Kennett Glossary sv cantredum: To sell by Cantell was an old custom of selling by the lump without tale or measure.

cantell. CANTEL

canter [*]. A m-c for ale in Bedfordshire (c1800) equivalent to a qt (c1.15 l) (Second Rep. 13).

cantrev [W cantrev, hundred TREVS]. A m-a for land in Wales (c1300) containing 2 CYMWDS or 100 TREVS or 25,600 ERWS (c9241.60 ha) (Laws Wales 998).

cap—5 cop (OED); 8-9 cap, caup (OED) [prob a later Sc form of cop; cf OE copp, cup, vessel, ON koppr, cup; see OED]. A m-c for grain in Scotland (c1600-1800): Angus, wheat, peas, and beans, 142.180 cu inches (2.330 l), oats, barley, and malt, 207.616 cu inches (3.402 l); Berwickshire, all grain, 210.04 cu inches (3.442 l); Lanarkshire, Glasgow and Lower Ward, wheat, 144.637 cu inches (2.371 l), peas and beans, 204.45 cu inches (3.351 l), oats and barley, 208.712 cu inches (3.421 l). In each case the cap was reckoned as 1/4 pk, 1/16 FIRLOT, and 1/64 BOLL (Swinton 65, 82-83, 96-98).

caracca, carack, caract, caracte. CARAT

carat--6 caracte (OED), carette (OED), carret, carrotte (OED), charect (OED), caratt (OED), carect, carrack (OED), carrot, charact (OED), charat, charract (OED), corrat (OED), karet, karrat; 7-8 carract (OED), carrat, karat; 7-9 carat [prob fr MedL carratus fr Ar qīrāt, bean or pea shell, a weight of 4 grains, a carat]. A wt of 4 gr for diamonds and other precious stones; the diamond carat equalled 3 1/5 t gr (0.207 g). Occasionally it was abbreviated c.—**1587** Acts Scotland 3.437: Of gold of xxij carret fyne. c**1590** Hall 21: Ther is 21,504 karectes of goulde waight. **1603** Ruding II.463: Gold of the standard of xxij karects. **1612** Ibid I.368: For an ounce of French crownes, being xxij carrots fine. **1617** Harpur 143: Of Gold, one ounce of the finest, without any aloy, is taken to be 24 karrats; one karrat is diuided into 4 graines.

1640 Rider Appendix sv tabula mensurarum: 8 ounces, and 1/4 of an ounce, 1 charat, 16 grains. **1646** H. Baker 269: 1 Ounce of fine Gold without any alloy, is imagined to be 24 karets. 1 Karet is divided into 4 graines. **1661** Hodder 199: An ounce of gold is divided into 24 parts called Carects. **1675** Vaughan 9: Gold, it is divided into 24 parts, which are called Carrats. **1678** Du Cange sv caracca: Parvum pondus quatuor granorum, quibus utuntur in ponderandis lapidibus pretiosis, Gallis et Anglis, Carat. **1695** Kennett Glossary sv carecta: A Carrat or Carect, us´d formerly for any weight or burden, tho´ now appropriated to the weight of four grains in Diamonds. **1707** Justice 59: Four Grains make 1 Carrat. **1728** Chambers 1.360: The Moneyers, Jewellers, &c. have a particular Class of Weights for Gold and Precious Stones, viz. Caract, Penny-weight and Grain. **1756** Rolt sv caract: The weight used in weighing diamonds, pearls, and precious stones, where it consists of 4 grains. **1778** Diderot XXVI.422: Pour peser l´or et pour les pierreries, ils se servent du karat et du grain. **1784** Ency. meth. 404: Le carat se divise en 4 grains. **1816** Kelly 84: Diamonds and other precious stones are weighed by Carats, the Carat being divided into 4 Grains, and the Grain into 16 Parts. The Diamond Carat weighs 3 1/5 Grains Troy. **1829** Palethorpe sv: CARAT, or CARACT, a certain weight used by jewellers and goldsmiths in weighing gold, silver, precious stones, and pearls. It weighs 4 grains troy weight. **1860** Britannica 808: Diamonds and pearls are also weighed by carats of 4 grains, but 5 diamond grains are only equal to 4 troy grains.... There are 150 diamond carats in the troy ounce.

carat. CHARGE[2]

caratt. CARAT

carcha. CARK

carect. CARAT

carecta, carectata, carectatum. CHARGE[2]

caretell. CAROTEEL

carette. CARAT

caritas [L caritas, charity]. A m-c for wine (c1300): Evesham, 3/4 gal (c2.84 l); Abingdon, 1 1/2 gal (c5.67 l); and Worcester, 2 gal (c7.56 l) (Prior 155). However, the caritas or ´charity´ probably originated as an allotment of wine given by an abbot to his monks over a certain period of time rather than as a definite capacity measure.

cark--4-5 kark (OED), karke (OED); 4-5, 7, 9 carke; 4-8 cark; ? L carcha (Prior), karre (Prior) [ME cark, carke, load, burden, fr ONF carque; see CHARGE[1]]. A wt for spices of 3 or 4 Cwt (136.077 or 181.436 kg), and a m-c for wood equal to 1/30 SARPLER (11.340 kg).—**1607** Cowell sv carke: Carke, seemeth to be a quantitie of wolle, whereof thirtie make a Sarpler. **1665** Sheppard 64: A Cark of Wooll is said to be a quantity, whereof 30 make a Sarplar. **1717** Dict. Rus. sv: Cark, a certain Quantity of Wooll, the thirtieth part of a Sarplar. See HUNDRED

carke. CARK

carnock—5-8 carnok; 6 cornocke; 6-7 cornock; 7 cornooke; 7, 9 cornock; 7-9 curnock; 8 carnock, cumock (prob an error for curnock); 9 cornok [see CRANNOCK]. Equivalent etymologically to CRANNOCK but metrologically to COOMB.—**1566** Recorde Kv: And in some places halfe a quarter is called a Cornocke. c**1600** Ricart 84: So that every sak be tryed and provid to be and holde a carnok, and the ij. sakkes to holde a quarter. **1616** Hopton 12: Whereof are made...Cornookes, Coombes, or halfe Quarters. **1665** Sheppard 15: And 4 Pecks make the Bushell, 4 Bushells the Coomb or Curnock; Eight Bushells the Quarter, which is two Curnocks. **1682** Hall 30: 1 Last conteynes: 10 Quarters, 20 Cornookes...80 Bushels. **1688** Holme 260: A Cornock is 2 strikes or 4 Bushels. **1707** Justice 3: 4 Bushels a Comb, or Cumock, 2 Cumocks a Quarter. **1708** Chamberlayne 109: 4 Bushels the Comb or Curnock; ibid 212: 2 Curnocks make a Quarter, Seam or Raff. **1716** Harris 2. sv measures: Carnock or Coom. **1728** Chambers 1.519: Carnock or Coom. **1755** Postlethwayt II.190: 4 bushels a coomb, or cumock, 2 cumocks a quarter, seam, or raff. **1756** Rolt sv curnock: Or Comb, an English corn measure, containing one sack. **1778** Diderot XXI.677: Carnok ou Coom. **1829** Palethorpe sv curnock: CURNOCK, a measure for corn, which contains four bushels, or half a quarter English. **1868** Eng. Cyclo. 824-25: The coomb is also called a cornook. **1877** Leigh 49: Cornok...A corn measure containing four bushels.

carnok. CARNOCK

caroteel--5 caretell; 5-9 caroteel; 9 carroteel; 7 caroteele (OED), caroteelle (OED) [ME caroteel, perh fr Ar qirtāl, coll of qirtālat, qartillat, ass´ burden, basket]. A m-c for cloves, 4 to 5 Cwt (181.436

to 226.795 kg); currants, 5 to 9 Cwt (254.010 to 457.218 kg); mace, approximately 3 Cwt (c152.00 kg); nutmegs, 6 to 7 1/2 Cwt (293.928 to 367.410 kg); and oil, 1/8 TUN (c1.19 hl).—**1439** Southampton 2.55: Pro 4 caretell´ olei continentibus 1 pipam. **1701** Hatton 3.221: Caroteel...of Cloves 4 to 5 C. weight. Currans 5 to 9 C. Mace about 3. C. Nutmegs 6 to 7 1/2 C. **1717** Dict. Rus. sv: Caroteel of Cloves 4 to 5 C. Weight. Currans 5 to 9 C. Mace, about 3 C. Nutmegs 6 to 7 and a half, C. **1721** Bailey sv: Caroteel, a quantity of some Commodities; as of Cloves, from 4 to 5 Hundred Weight. **1840** Waterston 147: Currants, carroteel, cwt. 5 to 9. **1883** Simmonds sv: Caroteel, the commercial name for a tierce or cask, in which dried fruit and some other commodities are packed, which usually averages about 7 cwt. See HUNDRED

caroteele, caroteelle. CAROTEEL

carrack, carract. CARAT

carrat. CARAT; CHARGE^2

carrata. CHARGE^2

carrect. CARAT; CHARGE^2

carrecta. CHARGE^2

carrectata. CARUCATE; CHARGE^2

carret. CARAT

carretate, carriata. CHARGE^2

carrot, carrotte. CARAT

carroteel. CAROTEEL

carruca, carrucat, carrucata, carrucate. CARUCATE

cartload [ME cart, carte + LOAD]. A m-c which generally did not have standard dimensions, but which referred to an arbitrary amount of goods loaded on a cart. Occasionally equivalent to FOTHER.

caruca. CARUCATE

carucata. CARUCATE; CHARGE[2]

carucate—1-7 L carucata; 1-3, 7 L carrucata; 3 L carruca, L caruca; 3-7 carucate; 6 carrucat (OED); 7 L carrectata; 9 carrucate [ME carucate fr MedL carucata, carrucata, plowland, plowgate, fr caruca, carruca, plow]. Equivalent to HIDE, and frequently abbreviated car., carr., or caruc. in medieval MSS. The carucate was especially prevalent in Suffolk, Norfolk, Yorkshire, Lincolnshire, and in the counties of Derby, Nottingham, and Leicester.—c**1065** St. Edmunds 3: 1 carrucate terre et dimidia. c**1130** Slade 14: In eadem villa Comes Lerc´ xj car´ et j virg´. c**1141** Malcolm 152: Et preterea duas carucatas terre. c**1153** Ibid 183: Unam carucatam terre Petioker. **1200** Feet 2.109: De una carrucata terre cum pertinentiis in Thornham. **1201** Cur. Reg. 9.52: j. carucata terre cum pertinentiis in Laleford´.... vj carucatas terre cum pertinentiis in Pihtesle.... j. carucate terre cum pertinentiis in Serdredesee. **1209** Feet 2.114: De dimidia carruca terre cum pertinentiis. **1214** Cur. Reg. 14.283: Robertus filius Maudredi per Robertum de Munceys stornatum suum petit versus Rogerum Pantolf medielatem viij. carrucatarum terre cum pertinenciis in Laweford´. c**1220** Evesham 75: Et unam carrucatam terræ. **1220** Cur. Reg. 3.151: Ad quatuordecim bovatas unde xiiij. carucate terre faciunt feodum j.

militis. **1222** St. Paul´s 135-36: Warinus de Bassingbourne tener unam carucam terræ continentem ix [X] xx acras terræ arabilis; ibid 136: Warinus de Brantone tenet unam carucam continentem vii [X] xx acras cum prato et bosco. **1340** Scrope 150: Magister Johannes...tenet ij. carucatas terræ in Blontysdene. **1362** Gray 491: Sunt ibidem tres carucate terre continentes in se cxcvi acras.... Sunt duo carucate terre in dominico que continent cxivi acre terre. c**1400** Melsa I.186: Post hæc, cum rex apud Wodestoke de singulis hidis, id est carucatis terræ, per universam terram suam duos solidos annuos ut de jure firmæ annuæ exegisset, ipse Thomas...pro populo regi resistebat. c**1420** Evesham 275: Præterea secundo anno assartavit in Ambresley duas carucatas terræ. c**1500** Hall 8: viii [X] xx pertice faciunt acram; duodecim acre faciunt bovatam; ii bovate faciunt virgatam; ii virgate faciunt carucatam. **1599** Richmond Appendix 2.10: An ancient Writer called Henry Knighton, a Cronacler of Leicester...hath these Words...Johannes Rex solemniter denunciatus, &c. & statim cepit tributum pertotam Angliam, viz. de qualibet Hida, id est Carucata terræ. **1635** Dalton 71: But a plow-land, or Carve of land, is called in Latine, Carucata terræ, that is, quantum aratrum arare potest in æstivo tempore.... And so this definition or description of Carucata terræ, sheweth that it is not of any certaine content. **1664** Spelman 126: Carrucata terræ, est ea portio quæ ad unius aratri operam designatur, a Ploughland; Mattheo Paris hida. **1678** Du Cange sv carrectata: Carrectata terræ, Modus agri. **1695** Kennett Glossary sv carucate:

CARUCATA. A plough-land...which in the reign of Rich. I. was computed at sixty acres.... Yet another...allots one hundred acres to a carucate. And Fleta temp. Edw. I. says, if land lay in three common fields, then ninescore acres to a carucate.... But if the land lay in two fields, then eightscore acres to a carucate.... In 23. Ed. III. one carucate of land in Burcester contain´d one hundred and twelve acres; and two carucates in Middleton were three hundred acres.... Caruca was sometime us´d for Carucata. **1755** Willis 358: Some hold an Hide to contain 4 yard Land, and some an 100, or 120 Acres; some account it to be all one with a Carucate or Plough-Land; ibid 361-62: The Meaning of the Word Carucate seems...to signify so much Arable Land as...would employ a Plough a Year in tilling it. And in this Sense the Measure of such a Portion of Land must be very uncertain, and different in different Places, according to the Nature of the Soil, the Difference of Husbandry, and other Circumstances. **1777** Nicol. and Burn 610: Carucate of land, from caruca a plough, signifies as much land as can reasonably be tilled in a year by one plough. **1795** Astle 3: A capital messuage and half a carucate of land, called Trumpington´s. **1874** Hazlitt 421: Carrucate or Carucate.—A plough-land, or a hide of land. **1888** Round 3.195-96: We have the hide divided into four virgates...its equivalent, the carucate, divided into eight bovates; ibid 201: In the three Domesday measures of land, the hide, the carucate, and the solin...we recognize three names for the same unit of assessment. **1888** Taylor 177: Carucates of 60 acres; ibid 178: Carucates of 80 acres....

Carucates of 96 acres.

carue--1-7 carue; 7 carve [ONF carue, plowland, fr L caruca, plow]. Equivalent to HIDE.—**1579** Rastell 35: Note that a Carue of land is a plowland. **1607** Cowell sv: Carue of land (carucata terræ) commeth of the French (charue. i. aratrum) and with vs is a certaine quantitie of land, by the which the subiects haue some time bene taxed. **1610** Norden 59: And euery plough land or carue, is foure yard land...euery yard land thirty acres. So that euery Carue or plough land contayneth a hundreth and twenty acres. **1635** Dalton 71: That a Carve, or Hyde of land (or a plow-land) which is all one, is not of any certaine content, but so much as one plow may plow in one yeer; and so in some Countrey [= district] it is more, and some other it is lesse (according to the heavinesse of their soile).... Also a Carve of land (or a Plow-land) may containe house, meadow, pasture, and wood. **1695** Kennett Glossary sv carucata: In some Countries [= districts] the word is still preserv´d a Carve of land.

carve. CARUE

cas. CASE

cascate. CASKET

case—4 caas (OED), kase (OED); 4-5 cas; 4-6 cass (OED); 4-9 case; 5 L cassa, kace (OED); 5-6 casse; 6 cace, cais (OED) [ME case, cass fr ONF casse fr L capsa, chest, box, case]. A wt, or occasionally a superficial measure, for many products: annatto, 2 1/4 Cwt (102.058 kg); apples, 40 lb (18.144 kg); liquorice juice, 1 1/2 Cwt (76.203 kg);

Normandy glass, 120 sq ft (11.148 sq m); ordinary glass, generally 1 3/4 Cwt (88.903 kg) or 196 lb based on the 112 lb Cwt; onions, 120 lb (54.431 kg); recorders, 5; sinopia, approximately 5 Cwt (c254.00 kg); steel, approximately 1 Cwt (c50.00 kg); and vermillion, approximately 2 1/2 Cwt (c127.00 kg).—**1420** Gras 1.511: xii cases canette. **1439** Southampton 2.34: 1 case saponis albi; ibid 55: Pro 1 cassa de sinapio pond. 5 C. sotile; ibid 86: 1 case de inde; ibid 89: 3 casys de canella; ibid 96: Pro 2 casys de vermylon´ pond. 500 lb.; ibid 97: 5 casys de vermylon´ pond. 1,250 lb. **1443** Brokage II.109: Cum iiii cas´ suger; ibid 251: Cum X casys suger dymyter; ibid 269: iii casys triacle kery. **1507** Gras 1.698: Glasse called Normandy the casse. **1509** Ibid 570: ii basketts cum iiii cases specktakilles. **1530** Ibid 196: Pro sex cases de glasse. **1545** Rates 1.10: Combes the cace; ibid 13: Caruinge knyues the cace. c**1550** Welsh xlvii: Steel, the case (= 1 Cwt.). c**1590** Hall 27: The casse of glass is a hundrid and 3/4 in waight, after 112 to the 100, so that the casse must be, after this ratte, 196 poundes waght haberdepoysse. **1701** Hatton 3.221: Case...Of Normandy-Glass 120 foot. Of Recorders 5. **1717** Dict. Rus. sv: Case, of Normandy Glass, a quantity consisting of 120 Foot. **1840** Waterston 147: Annatto, case, nearly cwt. 2 1/4; ibid 148: Liquorice juice, case, nearly...cwt. 1 1/2. **1956** Economist 51: Onions: 1 case = 120 lb.... Apples: 1 bushel or case = 40 lb. See HUNDRED

cask--6-7 caske (OED); 6-9 cask [Sp casco, potsherd, skull, helmet, cask, fr cascar, to break, fr L quassare, to break]. A m-c (c1700-1900) for almonds, approximately 3 Cwt (c152.00 kg); arsenic, 4 Cwt (181.436 kg); bristles, 10 Cwt (508.020 kg); butter, Caithness, 72 to 84 lb (32.658 to 38.101 kg); cider, Gloucestershire, 110 gal (c4.16 hl); clover seed, 7 to 9 Cwt (355.614 to 457.218 kg); cloves, mace, and nutmegs, approximately 300 lb (c136.00 kg); cocoa, 1 1/4 Cwt (63.502 kg); madder, 15 to 23 Cwt (762.030 to 1168.446 kg); pilchards, 50 gal (c1.89 hl); raisins, 1 to 2 1/2 Cwt (50.802 to 127.005 kg); red herrings, generally 450; soda, 3 to 4 Cwt (152.406 to 203.208 kg); sugar, 8 to 11 Cwt (391.904 to 538.868 kg); tobacco, 224 lb (101.604 kg); tallow, 9 Cwt (457.218 kg); and wheat flour, 2 Cwt (101.604 kg) (Dict. Rus. sv, Hatton 3.221, Second Rep. 13, Waterston 147, and Britten 170).

caske. CASK

casket—5-9 casket; 6 caskytt (OED); 7 cascate (OED); 9 casquet (OED) [ME casket, modif of MF cassette; see CASSET]. A m-c smaller than a CHEST, often used for precious gems and other valuable items.—**1467** Eng. Gilds 379: The same quayer to be put in a boxe called a Casket. **1664** Gouldman sv: A casket...or little coffer. **1883** Simmonds sv: Casket, a small jewel case or box.

caskytt, casquet. CASKET

cass, cassa, casse. CASE

casset [MF cassette, dim of ONF casse, case]. Equivalent to CASKET.—**1435** Southampton 1.84: Casset...suchre candy.

cast [ME vb casten fr ON kasta, prob akin to L gestare, gerere, to bear, carry]. A m-c (c1800-1900) containing 8 gal (c3.03 dkl) (Second Rep. 13

and Donisthorpe 208), and a m-q (c1600) for earthen pots, consisting of 3 in number (Rates 2.14).

castrel. COSTREL

caup. CAP

cayde. CADE

cek. SACK

celder, celdra, celdre. CHALDER

ceme. SEAM

cent—3 cente, centeine, centeyne; 3, 8-9 cental; 3-7 L centena; 3-9 cent; 4 It centinaio; 4-6 L centum; 5-9 L C.; 7 L centanarium [L centum, hundred]. Equivalent to HUNDRED. In the early modern period, it was occasionally abbreviated ct.—c**750** Brit. Mus. 11.105: C...Centum. c**1225** Coggeshall 150: xviii. centena millia marcarum argenti. **1228** Gras 1.163: De cent de canevas. c**1253** Hall 11: La centeine de bord, caneuaz et de lengeteile´ est de cent aunes, et checun cent de vi [X] xx.... Le last de arang´ est de xM., et checun mil est de X Cent, et chescun cent de vi [X] xx.... La centeine de cire, sucre, peyuer, cumin, almand, et de alume, si est de xiii peris et di., et checune pere de viii li. La sume de lib. en la centeyne, cent viii li. c**1272** Ibid 10: Item centena zucari, cere, piperis, cimini, amigdalorum, et allume continet tresdecim petras et dimidiam; et quelibet petra continet octo libras. c**1275** Gras 1.227: i cente de peaus lanes. **1290** Fleta 119: Centena vero canabi, tele...consistit ex sexies viginti; ibid 119-20: Centena vero ferri ex quinquies viginti peciis; ibid 120: Centena autem muluellorum et durorum piscium consistit ex octies viginti piscibus. c**1300** Brit. Mus. 1.148: Et consistit quelibet centena...c. et .xx. c**1300** Brit. Mus. 13.29: Centena cere. sukari. pipis. ferri [sic]. cimini. amigdalorum & alum continet .xiij: petras & dimidiam & quelibet petra continet .viij. libras. c**1303** Gras 1.159: De centum bordis...centum ulne de canobo...centum minute tele. **1323** Ibid 209: De centum pellibus leprorum...de una centena cuniculorum. **1324** Elton 274: Item in dimidia centena grossorum clauorum. c**1340** Pegolotti 255: Canovacci vi si vendono a centinaio, d'alle 120 per 1 centinaio. c**1435** Amundesham II.317: Item, centum smelte, vel centum sparlynges, vel centum welkes. **1443** Brokage II.7: iiii c canvas. c**1461** Hall 17: Also stocke fyssche ys sold by vi [X] xx.... But the rule of Doggermen ys to sell vi [X] xx and iiij fysschys for a C. **1474** Cov. Leet 396: And to this day the C. ys trewe after xx [X] v for the C.... xx [X] v for the C, the wich kepes weyght & mesure l li. the halfe C, xxv li. the quartern. c**1475** Hall 16: Also fysshis, fowles, and bestes be sold by numbyr and by dyuers C. **1496** Keith 1.24: The c is trew at this day, v score for the c. c**1510** Gross II.44: Et de centena linie albe.... Et de centena linie late mensure. Et de centena linie stricte mensure. **1524** Gras 1.194: Pro uno centum cole fysch. **1545** Rates 1.1: Almondes the .c. pounde. **1549** Gras 1.630: Pro centum horse shoes. **1678** Du Cange sv centanarium: Pondus centum librarum.... Centena ceræ, zuccari, piperis, cumini...apud Anglos, continet 13. petras et dimidiam: et quælibet petra continet 8. libras. Summa ergo librarum in centena

108; ibid sv centena: Ferri, ex 100 petris. **1708** Chamberlayne 205: Cod-fish, Haberdine, Ling, etc. have 124 to the C.... Herrings 120 to the C. **1721** King 282: Books unbound...per Ct; ibid 283: Glass-Pipes great...per Ct. **1732** J. Owen 113: C. in Number 100. **1756** Rolt sv: CENT, is an abridgment of the Latin word centum, signifying a hundred: thus cent is said of a certain fixed weight, called...quintal, or hundred, composed of 100 lb. **1880** Courtney 152: 100 lb...marked c. **1951** Trade 28: Cental = 100 pounds. **1966** O´Keefe 673: 1 cental = 45.3592 kg.

cental, centanarium, cente, centeine, centena, centeyne, centinaio, centum. CENT

ceroon. SERON

cest. CHEST

cester, cestre, cestron. SESTER

chaarge. CHARGE[1]

chaftmonde. SHAFTMENT

chain—7 chaine; 7-9 chain [ME chayne, cheyne fr OF chaeine fr L catena, chain, brace; akin to L cassis, net]. A m-l for land surveying: the Rathborn Chain—396 inches (10.058 m) or 2 PERCHES of 16 1/2 ft each, the perch divided into 10 equal parts called "primes" containing 19 4/5 inches each, and the prime sub-divided into 10 equal parts called "seconds" or links containing 1 49/50 inches each; the Engineer´s Chain—100 ft (30.48 m) or 100 links of 1 ft each; the Surveyor´s or Gunter´s Chain—792 inches (20.116 m) or 100 links of 7.92 inches each or 66 ft or 22 yd or 4 perches of 16 1/2 ft each, and equal to 1/10 FURLONG or 1/80 MILE. In Scotland Gunter's chain contained 892.8 inches (22.677 m) or 100 links of 8.928 inches each, while in Ireland it contained 1008.0 inches (25.60 m) or 100 links of 10.08 inches each. Occasionally it is abbreviated ch. or chn.—**1610** Folkingham 52-53: And to accomodate these for exact and expedite operation disme or deuide each foote of the Rule and Perch of the Chaine into decimals or Tenths, and each Tenth or Prime of the Rule into Seconds, but it shall suffice to diuide the Prime of the Chaine into two lincks, with three rings betweene euery lincke to keepe it from crossing. **1650** Leybourn 2.12: I would advise my Surveyour to have his Chain made of a good round wyer, not to contain above two Statute Poles, or Perches, or three at the most. **1653** Leybourn 1.46: The Chain which Master Rathborne ordinarily used...contained in length two Statute Poles or Perches, each Pole containing in length 16 1/2 feet, which is 198 inches, then each Pole was divided into 10 equall parts called Primes, every of which contained in length 19 4/5 Inches; again, every of those Primes was sub-divided into 10 other equall parts called Seconds, so that every of these Seconds contained in length 1 49/50 Inch, so that the whole Pole, Perch, Unite, or Commencement...was divided into 130 [sic] equall parts or Links, called Seconds; ibid 47: As every Pole of Master Rathborns Chain was divided into 100 Links, so Master Gunters whole Chain (which is alwayes made to contain four Poles) is divided into 100 Links, one of these Links being four times the length of the other. Now if this Chain be made according to the Statute, each Perch to contain 16 1/2 Feet, then each Link of this Chain will contain 7 Inches, and 92/100 of an Inch, and the whole Chain 792 Inches, or 66 Foot. **1654** J. Eyre 10: I would advise the Surveyour to have his Chain made of a good round wyre, not to containe above two statute Poles. **1677** H. Coggeshall 1.31: But to take off 10 Chaines and 56 Links. **1779** Swinton 23: Gunter Link ...7.92 [inches].... Gunter Chain...792 [inches]. **1816** Kelly 86: Land is usually measured by a Chain of 4 Poles, or 22 Yards, which is divided into 100 Links. **1829** Palethorpe 19: CHAIN, an instrument formed of 100 pieces of wire joined to each other by small rings of wire, each piece or link measuring 7 92/100 inches. It is in length 22 yards, and is used in measuring land. Ten chains long, and one chain wide make a statute acre, of 4840 square yards. **1834** Pasley 3: 100 Links...1 Chain of 22 Yards. 10 Chains...1 Furlong. 80 Chains...1 Mile. **1849** Strachan sv land measuring: 10 1/4 Imperial chains, equal to 225 1/2 yards. **1850** Alexander 20: Chain; for surveying...England...22.--yards. **1883** McConnell 12: The chain used for measuring land is 4 poles or 22 yds. long, and consists of 100 links, each link being 22/100 yd., or 7.92 in. long. **1889** Francis 11: The Irish chain is 84 ft. **1895** Donisthorpe 208: CHAIN...Scotland, 74.4 feet. **1899** Browne 118: 4 Poles or 100 Links, or 22 Yards = 1 Chain = 20.12 [m]; ibid 123: Chain or 4 Poles or 22 Yards...20.1168 [m]. **1907** Hatch 22: 1 link = 7.92 inches = 0.66 foot; 100 links = 1 Gunter's chain = 66 feet; 80 chains = 1 statute mile; ibid 35: 1 chain

= 20.116782 metres. **1951** Trade 27: Chain...22 yards. **1964** Breed 16: Professor Edward Gunter conceived the idea of taking the acre´s breadth of four perches, which he called a chain, and dividing it into 100 links. **1969** And. & Bigg 11: 1 chain (66 ft) = 20.1168 m; ibid 18: 1 engineer´s chain (100 ft) = 30.48 m.

chaine. CHAIN

chairge. CHARGE[1]

chalder—2-7 L celdra; 5 L celdre (Gras 1), chaldre, schaldre (Fab. Rolls), sheldra (Hatfield); 5-9 chalder; 6 chalderne, chaudder (OED), chauldren, chawlder (OED); 6-9 chaldron; 7 chauder (OED), chauldron; 8 chaldern, chaudron; ? celder (Prior), cheldra (Prior), childyr (Prior), seldra (Prior) [ME chalder, chaldre fr MF chaldere, kettle, pot, fr LL caldaria, caldron]. A m-c for coal, coke, and grain in England, Wales, and Scotland. It sometimes was abbreviated ch. or chd.

In England the standard chalder of coal, first regulated in 1421 under Henry V, contained 32 bu, totaling 1 ton or 2000 lb (907.180 kg), and was equal to 1/20 KEEL of 20 tons. In 1676-77 it was officially increased to 36 heaped bu, totaling 1 ton of 2240 lb (1016.040 kg) or 20 Cwt of 112 lb each. However, the chalder of sea coal varied from these standards, generally containing 48 bu (c21.62 hl) or 12 sacks of 4 heaped bu each.—**c1400** Gras 1.214: De qualibet chaldre carbonum maris. **1406** trans in Cal. Close 19.159: He shall bring to London and nowhere else 70 ´chaldres´ of coal. **1421** Rot. Parl. 4.148: & votre Custume ent est prise solonc le portage de xx Chaldres. **1439** Southampton 2.8:

2 chaldr´ carbonum. **1443** Brokage II.5: Cum i chalder carbonis. **1503** Gras 1.649: Pro vi chaldriis see colys. **1555** York Mer. 155: Item, a chalder of coles to the marchaunts owne use. **1562** Ibid 168: A chalder of coles for the merchauntes own house, meanyng so many coles as ye will spend yearlye, iii s. c**1590** Hall 24: The chalderne of sea cooles is 12 sackes of sea coole, every sacke ought to conteyne 4 bushells watter measure, the bushell hepid as much as yt will stand, so that the chalderne is 48 bushells in grosse. **1590** Rates 2.10: Coles the chauldren containing xxxvj. bushels common measure. **1603** Henllys 139: Coles are sold by the barrell [in Wales]...and not by the Chaldron as ys used in other partes of this Realme. **1664** Gouldman sv chauldron: A chauldron of coals, i. 36 bushels. **1708** Chamberlayne 213: 36 Bushels are a Chaldron of Coals. **1717** Dict. Rus. sv chaldern: Or Chaldron, a dry English Measure consisting of four Quarters or thirty-six Bushels heap´d up according to the seal´d Bushel, kept at Guild Hall, London. **1773** Johnson sv: Chalder, Chaldron, Chaudron...A dry English measure of coals, consisting of thirty six bushels heaped up.... The chaldron should weigh two thousand pounds. **1784** Ency. meth. 139: Le score de charbon donne à bord du navire, 21 chaldrons; mais il en désigne seulement 20.... Le chaldron de charbon est de 36 bushels. **1829** Brockett 64-65: Chalder...a chaldron...a measure of coals containing 36 bushels. **1850** Alexander 20: Chaldron: for coal...London...36.—bushels. **1872** Robertson 69: The old standard measure being represented by the chaldron of four quarters, 8 combs and

32 bushels. **1956** Economist 8: Chaldron...4 1/2 quarters.

The Newcastle coal chalder contained 72 heaped bu totaling 53 Cwt or 5936 lb (2692.510 kg) and was equal to 1/8 keel. This chalder, standardized at the above specifications in 1695, equaled 2 standard English chalders of 36 heaped bu each. Prior to 1695 it had weighed 42 Cwt (2133.684 kg).—c**1580** Hostmen 5: To w[hich] is answered that for the space of these seven yeres last past, a chalder of Coles Newcastell measure hath not ben raysed in price above two shillings, w[hich] is 16 d. in a London Chalder. **1600** Ibid 18: And so manye Chaldron and Chaldrons of Sea Cole, Stone Cole, or pytt Cole. **1650** Ibid 91: Mr. Ralph Davison, of this Towne, and Free Brother of this Companie, hath sould Eight Chalder of Coles to A man of warr. **1695** Ibid 39: And three such wains or six such carts shall be reckoned for one chaldron and no more. **1703** Ibid 164: For every chaldron of Coles. **1784** Ency. meth. 138: Le keel, de 8 chaldrons. **1829** Brockett 65: 8 Newcastle chaldrons make a keel. **1831** Pope 461: A Newcastle chaldron of 24 coal-bolls ought to contain 232.243 1/5 cubic inches, whereas in reality the standard weight of 53 cwt. requires only 217.989 cubic inches. **1834** Pasley 115: 1 Newcastle Chaldron of Coals (53 cwt.)...5,936 [lb]. **1850** Alexander 20: Chaldron; for coal...Newcastle...5936.—pounds.

The chalder was also used in England (c1800-1900) for products other than coal: lime, Cambridgeshire, 40 bu (c14.09 hl), Derbyshire, in some parts, 32 heaped bu (c14.41 hl), Surrey and Yorkshire, 32 bu (c11.28 hl); grindstones, Durham, 1 to 36 in number depending on their size

(Second Rep. 14 and Britten 170).

In Scotland the corn chalder contained 16 BOLLS of 4 FIRLOTS each, or 140,629.44 cu inches (23.049 hl) for wheat, peas, beans, rye, and white salt, and 205,153.53 cu inches (33.625 hl) for barley, oats, and malt (Swinton 32). Both chalders were equal to 256 pk or 1024 LIPPIES or FORPITS.--c**1150** Acts Scotland 1.311: Quando Celdra brasei ordiacei venditur pro dimidia marca lagena seruisie uendatur pro .ob.... Et de qualibet Celdra boni brasei ordiacei nisi octies viginti lagenas seruisie. **1159** Malcolm 193: Et in molendinis .xx. celdras inter farinam & frumentum. **1298** Falkirk 2: Videlicet pro celdra frumenti ij s. et celdra avene xij d. **1624** Huntar 5: 4. Firlets makes the Bow. 16. Bowes is a Chalder. **1678** Du Cange sv celdra: Mensuræ species apud Scotos. **1695** Kennett Glossary sv cellarium: Hence the old Latin Celdra a certain measure, which the Scotch call Chalder, and we retain in the measure of Coals a Chaldron. **1761** Thomson iii: As 16 bolls makes a Scots chalder. **1816** Kelly 93: 4 Firlots...1 Boll. 16 Bolls...1 Chalder. **1820** J. Sheppard 91: WHEAT, BEANS, PEAS, RYE, SALT, AND GRASS SEEDS...16 boles...1 chaldron or 140629.440 [English cu inches].... OATS, BARLEY, AND MALT...16 boles...1 chaldron or 205153.640 [English cu inches]. **1872** Robertson 67: The Scottish chalder contained 64 firlots. **1880** Britten 170: Chalder (Scotch for chaldron), nearly 12 quarters Winchester measure; of corn, 16 bolls.

chaldern, chalderne, chaldra, chaldre, chaldron. CHALDER

chappin. CHOPPIN

char. CHARGE[2]

charact, charat. CARAT

chardge. CHARGE[1]

charect. CARAT

charge[1]—4 chaarge (OED); 5-9 charge; 6 chairge (OED), chardge (OED); 7 L chargia [ME charge fr OF charge fr chargier, to load, fr LL carricare fr L carrus, wagon]. A m-c for salt generally weighing 1 1/4 Cwt (63.502 kg), but occasionally equaling 2 1/4 Cwt or 9 quarter Cwt (114.304 kg).--**1439** Southampton 2.53: 15 charg´ salis. **1678** Du Cange sv chargia: Pondus definitum, statuta mensura. **1955** Bridbury 158: 4 quarters = 1 hundredweight of 112 lb. 5 quarters = 1 charge. See HUNDRED

charge[2]—3 L carrata, L carrectata (Gras 1), L carriata (Battle), charge; 3-4 L carectatum (Gras 1), charre, L charrus; 4 L carrecta (Swinthun), L carucata (Fab. Rolls), It ciarrea; 4-7 L carecta, L carectata; 7 L cairata; ? carat (Hewitt), carrat (Prior), carretate (Salzman 1), char (trans in Fleta), charret [see CHARGE[1]]. Equivalent to and eventually (c1350) supplanted by FOTHER.—**1249** Close 5.202: Quatuor carratas plumbi. c**1253** Hall 11: Fet asauer ke la charge de plum est de xxx fotmaux, et checun fotmal est de vi pers, ii lib. meyns; checun pere est de xii lib. c**1270** Report 1.420: Duodecim pondera faciunt unam carratam majorem. c**1272** Hall 9: Charrus plumbi debet ponderare et constat ex triginta fotmellis, et quodlibet fotmellum continet [vi] pertas, exceptis duabus libris...[et] petra constat ex duodecim libris;

ibid 10: Et tunc est summa petrarum in le Charre octies viginti et octo petre. **1290** Fleta 119: Item charrus plumbi consistit ex xxx. fotmellis et quodlibet fotmellum continet vj. petras minus duabus libris, et quelibet petra ponderat xij. libras.... Et magnas charrus ex octies viginti et xv. petris. c**1300** Brit. Mus. 13.29: Notandum quod la charre plumbi constat ex .xxx. fotmals. Et quodlibet fotmal continet vj. petras duabus libris minus. Et quelibet petra xij libras. c**1300** Hall 8: Summa librarum carecte Londonie: duo milia et centum libras. c**1300** Brit. Mus. 21.60v: Sexies viginti petre faciunt charrum plumbi. **1323** Gras 1. 209: De i carectata plumbi. c**1340** Pegolotti 255: Piombo vi si vende a ciarrea, e ogni ciarrea si è di peso la montanza del peso in somma di 6 sacca peso di lana, di chiovi 52 per 1 sacco e di libbre 7 per 1 chiovo. c**1420** Evesham 309: Ac etiam tres carectatas straminis pro lectis monachorum et minutorum ibidem annuatim, videlicet duas carectatas ad festum Annunciationis et unam ad festum Michaelis per manus cellerarii. **1540** St. Mary's 61: Et quod predicti quatuor tenentes facient annuatim iiii. carectas Eurbarum in Turbaria predicta. c**1600** Brit. Mus. 32.182: Cairata 38 plumbi constat ex triginta ffotmell. **1678** Du Cange sv charrus: Mensuræ [vel ponderis] species apud Anglos. **1695** Kennett Glossary sv carecta: CARECTA. A Cart or Carriage or Waggon; ibid sv carectata: CARECTATA. A Cart-load or Waggon-load. The Prior and Canons of St. Frideswide, gave the Vicar of Oakle—duas carectatas feoni, & duas carectatas straminis.

chargia. CHARGE[1]

charract. CARAT

charre, charret, charrus. CHARGE[2]

chast, chaste. CHEST

chaudder, chauder, chaudron, chauldren, chauldron, chawlder, cheldra. CHALDER

cheopinet [dim of chopine, var of CHOPPIN]. A m-c for liquids (c1500-1600) containing 1/2 pt (c0.25 l) (Horwood 640). It was considered the English equivalent of the Scots CHOPPIN.

chest--1 cest (OED), cyst (OED); 3-5 chiste; 3-6 cheste; 3-9 chest; 4-7 chist; 5 chast (OED), chaste (OED), ciste (Southampton 2); 5-6 chyst (OED) [ME chest, chist fr OE cest, cist, cyst fr L cista fr Gr kistē, box, chest]. A m-c which did not have a standard size (c1700-1850) for products other than castle-soap, 2 1/2 to 3 Cwt (127.005 to 152.406 kg); cochineal, 1 1/2 Cwt (76.203 kg); gum arabic, 4 to 6 Cwt (181.436 to 272.154 kg); indigo, 1 1/2 to 2 Cwt (68.038 to 90.718 kg); and isinglass, 3 1/2 Cwt (158.756 kg) (Waterston 148, Hatten 3.221, and Second Rep. 14), and (c1600) for sugar, 3 Cwt (152.406 kg) (Rates 2.36). Occasionally it was abbreviated cht.--**1443** Brokage II.50: Et iii chestes candell´ de cera; ibid 111: 1 chiste cum diversis haberdasshe. **1524** Gras 1.196: Pro quinque chests de glasse. **1545** Rates 1.18: Glasses of borgoyn collored the chest; ibid 37: Suger the cheste. c**1550** Welsh 50: 4 chests sugar; ibid 83: 1 chest dry wares. c**1610** Lingelbach 112: Buckrams of Bridges by the half Chest. **1628** Hunt C: For the Bagge, Barrell, Chest, Fraile, Vessell or Caske. **1661** Acts

Scotland 7.252: Glasse called window glasse ilk three chists. **1831** Pope 43: The particular weight of tobacco or snuff in each hogshead, cask, chest, or case. See HUNDRED

cheste. CHEST

chide. SHIDE

childyr. CHALDER

chiphus. SIEVE

chist, chiste. CHEST

chopin, chopina, chopine, chopinum, choppen. CHOPPIN

choppin—3 schopin (OED); 4-5 L chopinum, choppyn, chopyn; 4, 6-9 chopin; 6 L chopina, choppen (OED), choppyne (OED); 7 choppine; 7-9 chopine, choppin; 8 chappin (OED) [ME chopyn fr MF chopine, a liquid measure, fr MLG scōpe, scōpen, scoop]. A m-c in Scotland containing 2 MUTCHKINS or 1/2 Scots pt or 51.702 cu inches (c0.85 l).—**1310** trans in Memorials 78: And they will make no false measures, such as the measures called ´chopyns´ and ´gylles´. **1322** Elton 255: Agnes vxor Iohannis clerici communiter vendidit ad denarium et tres obolos cum Choppyn et fregit assisam.... De vxore Radulfi Barkere pro eodem communiter et vendente vt supra xviij d...et quia non tulit chopin. **1331** Ibid 297: De Emma Scauelok pro eodem sexies xviij d. plegius vir eius tulit galonam potellum quartam et chopinum et pro chopino iij d. **1425** Acts Scotland 2.12: A chopyn of the auld mete ande of the new mete. **1528** Jacobus 174: ij quarte j chopin mellis. **1624** Huntar 1: Our liquid Metts, as the Choppin, the Pinte, Quart, and Gallon, for metting of Wine, Ale,

Beere, Vineger, Oyle, Aqua-vitie; ibid 4: Everie pinte is devided in 2 choppins and 4 muchkins.... The pinte doth weigh 55 ounces,..the choppine, l. pound 11 vnces and a half. **1773** Johnson sv chopin: A term used in Scotland for a quart, of wine measure. **1779** Swinton 17: 8 English wine-quarts make about [sic] 9 1/16 Scotch choppins; ibid 29: Chopin...51.702 [cu inches]. **1816** Kelly 93: 2 Mutchkins...1 Chopin. 2 Chopins...1 Standard Pint. **1820** Second Rep. 14: Chopin or Choppin...Scotland: 1/2 a pint, 2 mutchkins = 52 1/2 cubic inches, about 2 English pints. **1860** Britannica 805: Choppin, 51.7 [cu inches]. **1883** Simmonds sv chopine: The chopine was also half of a Scotch pint. **1956** Economist 8: Chopin (or choppin)...1.5 pints.

choppine, choppyn, choppyne, chopyn. CHOPPIN

chudreme--2 chudreme, L cudrinus [*]. A m-c for cheese in Scotland.—**1164** Malcolm 262: Et viginti cudrinis de caseis redditus mei de Sterling.... Triginta caseos quorum quilibet facit Chudreme.

chyde. SHIDE

chyfe. SHEAF

chyst. CHEST

ciarrea. CHARGE[2]

cipha. SIEVE

ciste. CHEST

cistern, cistra. SESTER

civerus. KIVER

claue, clava, clave, clavus, claw, clawe, cleaue, cleave, clou,

cloue. CLOVE

clove—3-4 L clava, L clavus; 3-9 clove; 5 claw, clawe, clou (Southampton 1); 5-7 cloue; 6-7 claue, clave; 7 cleaue, cleave [AF clou (MedL clavus) fr L clavus, a nail]. A wt of 6 1/2, 7, or 8 lb (2.948, 3.175, or 3.629 kg) for cheese, wool, metals, and other agricultural and nonagricultural goods. It was commonly called a half-stone.—**1228** Gras 1.157: l clove de tasels. **1290** Rot. Parl. 1.47: Unde clavus ponderat vi li. et demid, et si deficiat dim´ li. perdunt clavum. c**1300** Hall 40: Et clavi tot sunt quot sunt septimane in anno. **1304** Gras 1.303: Pro xxxii clavis lane. **1430** Rot. Parl. 4.381: Que le pois d´une weye de formage, puisse tener xxx & ii cloues; c´est assavoir, chescun cloue vii li. c**1461** Hall 13: Also woll is weyd by this weyght, butt itt is nott rekynnyd soo, for ytt is bowght odyr by the nayle.... vij lb. make a nayle; ibid 19: For thai use to by or sell most comynly odyr by the Clawe, the Nayle.... The clawe amountythe in poundes vii.... That ys to say...Claw content´ vij. **1540** Recorde 203: In Cheese...the verye weightes of it are Cloues and Weyes: so that a Cloue shoulde contayne 7 pounde. c**1590** Hall 23: 7 poundes waight haberdepoyse is the halfe stonne or clave of woole, or nayle.... 7 pounds daberdepoyse [sic] is the claue or nayle of woole.... 7 poundes waight haberdepoise is the clove of cheesse. **1606** Ibid 38: A cloue is 7 pounde. **1613** Tap 1.63: Wooll. One Last contayneth...Stones. 312 Claues. 624 Pounds. 4368. **1616** Hopton 164: And a cleaue is halfe a Stone. **1635** Dalton 149: A weigh of cheese must containe 32 cloves, and every clove 8. l. of averdepois weight. **1665** Assize 5: And every Clove to be seven pounds of Avoirdupois-weight. **1665** Sheppard 61: The Wey of Cheese must contain 32 cloves, and every Clove 8 pound of Averdepoys weight. **1678** Du Cange sv clava: Pondus quoddam apud Anglos, nimus petra. Affine videtur voci Clove, de qua sic Skinner in Etymol. Anglic.: Clove pondus quoddam casei octo libris constans. **1682** Hall 31: A Cleave half a stone. **1696** Jeake 80: Beef, in 1 Nail, 8 Pounds of common use. **1708** Chamberlayne 207: Wooll is Weigh´d by the clove, which is seven Pounds.... In Essex, they Weigh Cheese and Butter by eight Pounds to the Clove. **1710** Harris 1. sv weigh: And each Clove 8 Pound, tho´ some say but 7. **1717** Dict. Rus. sv: But in Essex, 8 pounds of Cheese and Butter go to the Clove. **1829** Palethorpe sv: CLOVE, a term used in weights of wool. It is 7 lb. or half a stone. In Essex, 8 lbs. of butter and cheese make a clove. **1850** Alexander 21: Clove; for wool...7.—pounds. **1883** McConnell 15: 7 lbs. avoirdupois = 1 clove. **1895** Donisthorpe 86: The clove or cloven stone. **1956** Economist 58: 1 clove = 7 lb. See NAIL

clue [ME clewe, clue fr OE cliewen, a ball, a globular body; akin to OHG kliuwa, a ball]. A m-l (c1800-1900) of 4800 yd (43.872 hm) for yarn or hemp (Second Rep. 14 and Donisthorpe 208).

coard, coarde. CORD

cobyte. CUBIT

coddus [perh L form of OE codd, small bag]. A m-c of uncertain size for grain.—**1678** Du Cange sv: Mensura annonaria, Anglis. Vetus Charta tom.

1. Monast. Anglic. pag. 175: De Ecclesia Hamptona 11. denarios, et 2. Coddos bladi.

coeme. COOMB

cofer. COFFER

coffer—4-6 coffre; 5-6 cofer; 5-9 coffer [ME coffre fr OF cofre, coffre fr L cophinus, basket, fr Gr kophinos]. A m-c similar in size and application to a CHEST or trunk.—**1440** Gross II.249: That there be iiii. keys of the tresor coffer; ibid 250: The said iiii keyes shall not open the same Cofer. **1509** Gras 1.571: i coffre cum xx peciis brussel. c**1550** Welsh 65: 6 coffers drywares; ibid 79: 1 coffer containing 35 tin; ibid 136: 7 coffers and fforsets of dry wares. **1590** Rates 2.48: A Cofer iiij d. **1664** Gouldman sv chest: A Chest or Coffer. Arca, cista, capsa.

coffin—3 L cofinus; 4 L cophinus; 4-6 coffyn; 4-9 coffin (OED); 5 cofyn [ME coffin, a basket, receptacle, fr MF cofin fr L cophinus, basket; see COFFER]. A m-c similar in size and application to a CHEST or trunk.—**1208** Bish. Winch. 79: In cofino empto, ij d. **1303** Gras 1.161: Cophinus racemorum...Cophinus sucre. **1439** Southampton 2.74: 1 cofyn de scamanye. **1534** Fitzherbert 115: .xii. coffyns or skyppes of fragmentes.

coffre. COFFER

coffyn, cofyn. COFFIN

colbrond [*]. A m-c of uncertain size (c1400) for coal in Cheshire (Hewitt 189).

comb, combe. COOMB

comble--2-7 L cumulata (mensura); 4-9 comble; 5 coumble [MF comble fr L cumulus, heap, summit, crown]. A heaped measure which contained an amount of grain extending above the rim. The actual amount in excess of a level measure depended on the proportions of the vessel, and it was restricted by statute in 1325 for use only in selling oats, malt, and meal.—c**1200** Rameseia III.159: Quæ mensura, sexies impleta et rasata, facit unam mensuram cumulatam. **1299** Liber xxv: Mensura cumulata. **1351** Rot. Parl. 2.240: Et soit chescune mesure de blee rasee, sanz comble. **1362** Ibid 269: & brees pur les ditz Hostelx soient mesurez par mesure acordant a l´Estandard, rasee & nient comble. **1390** Rot. Parl. 3.281: & oept Busselx pur le Quarter rasez & nient comblez. **1413** Rot. Parl. 4.14: Par force de queux Ordinaunces tiel Mesure ad este use, ove un Bussell du dit Quarter comble. **1415** Ibid 81: Q´il ne preigne pur le quarter si noun viii busselx tant soulement, racez & nient comblez. **1444** Rot. Parl. 5.103: Mesurez par mesure accordant a l´Estandard, rase & nient coumble. **1587** Stat. 77: No maner of graine shall be sold by the heape [comble] or cantell, except it be otes malt and meale. c**1590** Hall 24: Every sacke ought to conteyne 4 bushells watter measurs, the bushell hepid as much as it will stand. c**1634** Ibid 53: Item, hee further sheweth that the Sackes or Baggs for Lyme ought to conteine a Bushell heapt. **1883** Simmonds sv: Comble...a heaped measure.

come, coom. COOMB

coomb—3-4 L cumba; 5 cumb, cumbe; 5-7 combe; 5-9 comb; 6 coeme, come (OED), koome (OED); 6-7 coome; 6-9 coomb; 7 coombe, coumb (OED); 8-9 coom; ? cowme (Prior) [ME combe fr OE cumb, vessel, measure; see WNID3]. A m-c for grain containing 4 bu (cl.41 hl) and equal to 1/2 SEAM. It was commonly called a half-quarter. After the establishment of the Imperial system, the coomb increased slightly in official size (1.455 hl) because of the larger bu.—**1202** Feet 2.178: Ad quamlibet firmam sexdecim cumbas brasio auene et iiij cumbas et dimidiam de grudo...et viginti quatuor cumbas frumenti; ibid 180: Sexdecim cumbas de braseo auene. **c1320** Thorpe 3: Debent etiam parare VIII cumbas de Grudo et XVI cumbas de Braisio. **c1450** Common 168: For vij cumbe whete and ij bushell, the price of a cumbe iij s. iiij d; ibid 170: v cumb barly & a busshell. **1573** Tusser 36: Ten sacks whereof euerie one holdeth a coome...A Coeme is halfe a quarter. **c1600** Ricart 84: So that every sak be tryed and provid to be and holde a carnok, and the ij. sakkes to holde a quarter. **1613** Tap 1.61: One Last of Corne or Grain containeth Quarters 10...Coombs. 20. **1616** Hopton 12: Whereof are made...Cornookes, Coombes, or halfe Quarters. **1628** Hunt B2: Bushell, Combe, Last; ibid C: Two Bushells in a Strike: [2] Strikes in a Combe: [2] Combs in a Quarter. **1635** Dalton 144: 4 Bushels maketh the Coombe. **1665** Sheppard 7: Four Bushells make a Coomb; Two Coombs make a Quarter. **1682** Hall 30: 1 Last conteynes: 10 Quarters, 20 Cornookes...80 Bushels. **1688** Holme 260: A Cornock is 2 strikes or 4 Bushels. **1708** Chamberlayne 109: 4 Bushels the Comb or Curnock; ibid

212: 2 Curnocks make a Quarter, Seam or Raff. **1716** Harris 2. sv measures: Carnock or Coom. **1717** Dict. Rus. sv: Coomb or Comb...four Bushels, or half a Quarter. **1728** Chambers 1.519: Carnock or Coom. **1778** Diderot XXI.677: Carnok ou Coom. **1816** Kelly 88: 4 Bushels...1 Coom...140.93721 [1]. **1850** Alexander 22: Coom...dry capacity...4.—bushels. **1888** Fr. Clarke 37: 2 coombs make 1 quarter. **1931** Naft 22: 1 British Coomb...4 British bushels...1.4547 hectoliters. **1956** Economist 8: Coomb...4 bushels. **1966** O´Keefe 671: 1 coomb = 4 bu. = 1.455 hl. See CARNOCK

coombe, coome. COOMB

coorde. CORD

cop. CAP

copa. COPE

cope—4-6 L copula; 5 cupe; 5-6 L cupa; 6 L copa (Finchale), cope, copul, copule, copynett [ME cupe, cope, perh fr (assumed) OE cȳpa, basket]. A m-c of uncertain size for grain, fish, and other products.—**1303** Gras 1.351: Pro cciiii [X] xx [+] ii copulis i quarteron et di. ficuum et racimorum. **1304** Ibid 172: Pro xxi copulis ficorum. **1395** trans in Cal. Close 15.324: 850...(copulas) of figs and raisins. **1402** trans in ibid 17.545: 42...(cupas) of hides. **1404** trans in ibid 18.264: 309...(copulas) of fruit. **c1475** Gras 1.193: Of a cupe waad. **1530** Finchale ccccxxviii: One cope ficuum...One copynett ficuum. **1540** St. Mary´s 58-59: Videlicet, villata de Stantroff ix. copulos, appreciatos ad xiii. s. iiii. d. le copul; ibid 75: Et quod decime granorum de

Clonsillagh, quod est membrum de Castleknocke, numerantur xxxii. copule. c1550 Welsh xlvii: Irish cod fish, the copule...ling, the copule; ibid 58: 30 copules linges...20 copules codfish.

cophinus. COFFIN

copul, copula, copule, copynett. COPE

cord—4-5 coorde (OED); 4-7 corde (OED); 4-9 cord; 6 coarde (OED); 7-8 coard [ME cord fr OF corde fr L chorda, catgut, chord, cord, fr Gr chordē]. A m-q for wood, originally determined as the amount encompassed by a length of cord or string, equal to a double cube of 4 ft or 128 cu ft (3.624 cu m), but there were local variations (c1800): Derbyshire, 128, 155, and 162 1/2 cu ft (3.624, 4.389, and 4.601 cu m); Gloucestershire, approximately 78 cu ft (2.209 cu m); and Sussex, 126 cu ft (3.568 cu m) (Second Rep. 14). Occasionally it was abbreviated c.—**1701** Hatton 3.222: Coard...of Wood, 4 foot long, 4 foot broad, 8 foot deep. **1756** Rolt sv: Cord of wood...so called, because formerly measured with a cord, but is now measured between two stakes of wood, four feet high, and eight feet apart, being four feet broad, or deep. **1797** Winter 101: The measure of a cord of wood in Glocestershire, is eight feet four inches long, four feet four inches high, and two feet two inches broad. **1819** Cyclopædia sv weights: And 128 solid feet, that is, 8 feet long, 4 feet broad, and 4 feet deep, are a cord of wood. **1880** Britten 139: Cord, a certain (very variable) quantity of cut wood piled up. **1956** Economist 7: 128 cu. feet = 1 cord. **1969** And. & Bigg 18: 1 cord (timber) = 3.62456 m^3.

corde. CORD

corf—5 corffe (OED); 7-9 corfe (OED); 8-9 corf; 9 corve (OED) [ME corffe fr MDu corf, prob fr L corbis, basket]. A m-c for coal (c1800-1900) in Durham that contained 4 bu totaling 3 1/4 Cwt (165.106 kg) and in Derbyshire, 2 1/4 bu or 2 Cwt (101.604 kg) (Second Rep. 14 and Pasley 114-15). It was a large basket used by miners to carry coal from the underground veins to the surface. See HUNDRED

corfe, corffe. CORF

cornock, cornocke, cornok, cornook, cornooke. CARNOCK

corrat. CARAT

corve. CORF

costerell. COSTREL

costrel—3 costret (error for costrel); 4-5 costrell (OED), costrelle (OED), costril (OED), costrille (OED); 4-9 costrel; 5 costerell (OED), costrele (OED), costrylle (OED); 6 kostorell (OED); 7 castrel (OED) [ME costrel fr MF costerel fr costier, at the side, fr coste, rib, side; hence, vessel carried at a man´s side]. A m-c for wine that was made of leather, wood, or earthenware.—**1298** trans in Memorials 40: Be it remembered, that Walter of Caile, near Brestolle, came...and acknowledged that he was altogether ignorant of the usage of the City; and that he had been attached by the taking of...one costret. **1883** Simmonds sv: Costrel, a kind of bottle.

costrele, costrell, costrelle, costret, costril, costrille, costrylle. COSTREL

coture. COUTURE

coumb. COOMB

coumble. COMBLE

count [ME counte fr MF compte, conte (F compte) fr LL computus, a computation, fr L computare, to reckon, compute]. Equivalent to TALE.—**1858** Shuttleworths 792: Tale or Count. By this are counted fish, hides, paper, parchment, sables, &c.

courtceldra [court, a var of quart, for quarter fr OF quartier fr L quartarius, a fourth part, fr quartus, the fourth, + celdra CHALDER; hence, a fourth part of a chalder]. Equivalent to SEAM (c1300) (Prior 165).

couture—4-5 coture, couture [MF couture fr L culter, knife; hence, the blade or spike (coulter) on the plow which cut the roots of old vegetation or weeds during the process of plowing]. Equivalent to FURLONG.--**c1400** Henley 8: Byen sault ke vne coture deyt estre de quarante perches de long.... Ore en arrant alet xxx foys entur pur fere le reon plus estreyt e kant le acre ert pararre a donkes estes all lxxij coutures ke sunt vj lywes kar ceo fet asauoyr ke xij cotures sunt vne lywe.

cowme. COOMB

cran [Sc cran fr Gael crann, a lot, a measure]. A m-c for herrings (c1800) equal to approximately 34 wine gal (c1.29 hl) (Second Rep. 14). A standard but bottomless 30 gal herring bbl was heaped full and the bbl then lifted, leaving the herrings in a pile on the ground or floor. The present fish cran in the United Kingdom is reckoned at 37 1/2 Imperial gal (1.705 hl); the quarter cran at 9.375 Imperial gal (0.426 hl) (Economist 53 and O'Keefe 313, 671). Originating in Scotland, the cran spread to the English fishing towns during the nineteenth century.

crannacus, crannoc, crannoca. CRANNOCK

crannock--3-4 L crannocus; 4 L cranocus; 4-9 crannock; 5-7 cranoke; 6 creneoke (OED), crenneke, crennock (Shuttleworths), crenoke, crineoke, cryneoke (OED), crynoke, krenneke, krennock (Shuttleworths); 6-7 cranok (Henllys); 7 L crannoca; 8 crannoc (Ireland), crannok; 8-9 cranock, cronnog; 9 crynog; ? L crannacus (OED), crannoke (Prior), crennoc (Prior) [MedL crannocus, of Celt origin; akin to W crynog, crannock]. A m-c used principally for grain in England, Wales, Scotland, and Ireland. In Ireland the crannock of wheat varied from 8 pk to 8 bu (c0.70 to c2.82 hl) and of oats, from 7 to 14 bu (c2.47 to c4.93 hl), while in Wales the crannock generally contained 10 bu (c3.52 hl) although variations from 10 to 12 bu (c3.52 to c4.23 hl) were not uncommon.—**1228** Close 1.52: Cepit de eadem abbatia c oves et xxviij crannocos frumenti. **1297** trans in Cal. Just. I.178: And took a crannoc[us] of oats. **1315** Ireland xxxv: Quinque crannocos frumenti torelliati, boni, sicci et mundi, de mensura septem bussellorum cumulatorum pro quolibet crannoco.... Quinque crannocos avenarum, boni et mundi bladi, quindecim bussellos cumulatos, pro quolibet crannoco. **1319** Ibid xxxv: Crannoco [frumenti], videlicet, mensurato per septem bussellos rasos et octavum bussellum cumulatum.... Quolibet, videlicet, crannoco [frumenti] continente octo pecks, cumulatos.... Quilibet crannocus [avenarum] continebit quindecim pecks cumulatos boni et mundi bladi.... Quiquidem crannocus [avenarum] continebit sexdecim pecks cumulatos boni.... Crannoco [avenarum] videlicet mensurato per septem bussellos cumulatos. **1322** Ibid xxxv: Pro quolibet crannoco avenarum per quatuordecim bussellos cumulatos. c**1370** St. Mary´s 354: Cranocus frumenti vendebatur pro viginti tribus solidis, avene pro sexdecim solidis. **1452** Gross II.69: Othir salt a cranoke or within for his howssolde. **1586** Shuttleworths 558: 2 1/2 krennekes of salt at the North Wyche. **1587** Ibid: Two krennekes of salt 22 s. **1588** Ibid: A crenneke of salt. **1589** Ibid: Four crenokes of salt. **1590** Ibid: Two crineokes of salt 31 s. **1591** Ibid: 3 1/2 crynokes of salt 54 s. **1603** Henllys 137-38: Neither ys the Cranoke or Wey measures used in the selling thereof. **1678** Du Cange sv crannoca: Mensuræ genus apud Hibernos. **1787** Liber xxv: An Irish-measure, called a Crannok, containing two quarters, is mentioned. **1789** Topham 363: The cranock, or cronnog, in Irish, was a basket or hamper for holding corn, supposed to hold the produce of 17 sheaves of corn, and to be equal to a British barrel. **1816** Kelly 114: The Cronnog was a basket or hamper for holding corn, made of twigs, and lined with skins.... The term Crannock, for a barrel, corrupted, as may be supposed, from the Irish word Cronnog. **1820** Second Rep. 14: Cranock or Crynog...of lime: Cardiff, 4 llestraid = 10 W. bushels...Cowbridge and Bridge-end, 11 W. bushels...Neath and Swansea, 12 W. bushels.

crannocus, crannok, crannoke, cranock, cranocus, cranok, cranoke, creneoke, crenneke, crennoc, crennock, crenoke, crineoke. CRANNOCK

cronn—3 L cronnus; ? cronn (Prior) [*]. A m-c for grain at Worcester containing 4 bu (c1.50 hl).—c**1290** Worcester 118b: Item ad vj. Septimanas j cronn. frumenti; ibid 126a: Homini de Poywik pro Gurgite de Timberdene j. cronnum frumenti in festo S. Michaelis.

cronnog. CRANNOCK

cronnus. CRONN

cryneoke, crynog, crynoke. CRANNOCK

cubet, cubete, cubette, cubide. CUBIT

cubit—1-7 L cubitus; 4 cupet (OED), cupyde (OED); 4-7 cubite; 4-9 cubit; 5 cobyte (OED), cubete (OED), cubital; 5-6 cubet (OED), cubette (OED), cubyt (OED), cubyte (OED); 7 cubide (OED) [ME cubite fr L cubitus, elbow, cubit]. A m-l, originally the distance from the elbow to the extremity of the middle finger, which was generally taken as 18 inches (4.572 dm), or 6 PALMS or 2 SPANS.—c**1075** Hall 2: Quarum haec sunt nomina: digitus, uncia, palmus, sextas, pes, cubitus; ibid 3: Cubitos in pedes; ibid 4: Qualis erat cubitus quo Archa Testamenti, quam Moises iussu Dei fecerat, fuerat mensurata. Alius cubitus, qui et maior dicitur, quo Archa Noe demetita esse dinoscitur, qui brachio extenso toto cubito capiti prelato se esse demonstrat. c**1100** Ibid 5: Cubitus fit ex pede et semis. c**1300** Ibid 7: Pes et dimidius faciunt cubitum usualem. **1395** York Mem. 1.142: Unus pes et dimidius, cubitum usualem.

c**1400** Brit. Mus. 20.1v: Pes & dimid faciunt cubitum. c**1400** Hall 6: Tres cubiti vel quinque pedes faciunt passum. **1440** Palladius 119: And cubital let make her longitude. **1616** Bullokar sv cubite: Cubite. Halfe a yarde: the measure from a mans elbow to the toppe of his middle finger. **1624** Huntar 8: 3. Inches is a palme.... 6. Palmes is a cubite; ibid 10: A Cubite containeth...Inches—18. **1682** Hall 28: A Cubit is a foote and half. **1688** Bernardi 221: Virga Anglica. 3 Pedes, 12 Palmi, 2 Cubiti. **1701** Hatton 3.224: Cubit...18 Inches. **1708** Chamberlayne 207: 1 Foot and an half makes a Cubit. **1873** Grote 168: The cubit and the foot, having a natural standard, cannot differ very much from each other in any two countries. **1883** Simmonds sv: Cubit, a measure of length in England equal to 18 inches. **1903** Warren xiv: 18 inches correspond to 20 11/18 (the ancient cubit). **1964** Breed 6: The cubit was originally the length of the bent forearm from the elbow point to finger tip. **1966** O´Keefe 667: 1 cubit = 18 inches = 45.7 cm.

cubital, cubite, cubitus, cubyt, cubyte. CUBIT

cudrinus. CHUDREME

cumb, cumba, cumbe. COOMB

cumock. CARNOCK

cumulatus. COMBLE

cupa, cupe. COPE

cupet, cupyde. CUBIT

curnock. CARNOCK

cut [*]. A m-l for yarn (c1700-1800): Northern England, 1/12 HANK (42.656 m worsted yarn and 63.980 m cotton yarn); Scotland, Clydesdale, 120 threads (274.20 m), each thread 2 1/2 yd in length (Acts Scotland 9. 311, Second Rep. 15, and Brockett 89).

cwaer QUIRE

cymwd [W cymwd, co-mote (Laws Wales)]. A m-a for land in Anglesey (c1300) containing 50 TREVS or 12,800 ERWS (c4620.80 ha) and equal to 1/2 CANTREV (Laws Wales 998).

cyst. CHEST

cyvar [W cyvar, co-aration (Laws Wales)]. A m-a in Wales (c1800): North Wales, Anglesey and Caernarvon, 3240 sq yd (2709.063 sq m) and Merionethshire, 2430 sq yd (2031.723 sq m); South Wales, in some parts, 2821 sq yd (2358.725 sq m) or 192 LLATH or 11 1/2 sq ft (Second. Rep. 15, Donisthorpe 214, and Laws Wales 999).

cyvelin [*]. A m-l for cloth in North Wales (c1800) equal to 9 ft (2.743 m) (Second Rep. 15).

D

dacker, dacra, dacre, dacrum, daiker, daker, dakir, dakker, dakrum, dakyr. DICKER

dale [OE dāl, portion, allotment]. A m-c of unknown size used (c1400) for salt at Worcester (Prior 166).

daugh [*]. A m-a for land in Scotland (c1800-1900) varying in size according to the quality of the soil in any region (Second Rep. 15 and Donisthorpe 209).

davach—8-9 davach, davoch [Sc davach fr ScGael dabhach, vat, tub; akin to OIr dabach, tub, land measure]. A m-a for land in northern Scotland generally considered equal to 4 PLOWLANDS of 104 acres each, the plowland being divided into 8 OXGANGS of 13 acres each, or 416 acres in all (c212.16 ha) based on the Scots acre of 6150 4/10 sq yd (c0.51 ha). The actual number of acres would vary, however, depending on the quality of the land.—**1872** Robertson 135-36: The Ploughgate, or carucate of 104 acres. It was divided, as in northern England, into eight oxgates.... The equivalent of the ploughgate in northern Scotland was the Davoch, a large pastoral measure equal in actual extent to four ploughgates.... In [the] course of time the Davoch seems to have been calculated, as a measure of land, at four ploughgates. **1962** W. C. Dickinson 62: There was certainly a fiscal unit known as the davach, upon which renders of service were based and which was in turn divided into four quarters.

davoch. DAVACH

deaker, decker, dekar, deker. DICKER

desone. DOZEN

dessertspoonful [dessertspoon + -ful]. A culinary m-c containing 1/4 fluid oz or 2 fluid dr (7.103 ml) and equal to 2 TEASPOONSFUL or 1/2 TABLESPOONFUL (Stevens 3 and Economist 8).

dicar, dickar. DICKER

dicker—3 L dakrum; 3-6 diker, dyker; 3-6, 9 dacre; 3-7 L dacrum, daker, dakir; 4-5 dakyr (York Mem. 1); 4-7 L dacra; 5 dykur, dykyr; 5-6 dycer, dycker (OED); 6 daiker, deaker, dekar, deker (OED), dickar (OED), dikar (OED), dikkar (OED), dykker; 6-8 decker; 6-9 dicker; 7 dacker; 7, 9 dicar; 8 dakker (OED) [ME forms akin to MLG dēker; all fr L decuria, a division or parcel of ten, fr decem, ten]. A m-q for a variety of goods: hides, razors, etc., 10; horseshoes, 10 pairs; gloves, 10 pairs; and necklaces, 10 bundles, each bundle containing 10 necklaces.—**c1243** Select Cases 3.lxxxvi: vj. lesta correi, ij. dakeres minus. **c1253** Hall 11: Le daker de gaunz est x peyre. Le dakir de fers de chival est de xx fers. **c1272** Ibid 10: Last vero coriorum ex viginti dikeres; et quodlibet diker constat ex decem coriis.... Item diker cirothecarum constat ex decem paribus cirothecarum. **1276** Gras 1.227: ii dakers des quirs. **1290** Fleta 119: Item lastus coreorum consistit ex x. dacris, et quodlibet dakrum ex x. coreis. Dacrum vero cirotecarum ex x. paribus. Dacrum vero ferrorum equorum ex xx. ferris. **c1300** Brit. Mus. 13.29: Daker cerothecarum ex decem paribus. **c1300** Brit. Mus. 1.148: Et quodlibet dyker ex .x. coreis. **1304** Gras 1.170: Pro iiii dacris coriorum. **1305** Ibid 304: Pro i lasto i dacro coriorum. **1321** Ibid 248: l lasta v dacre. **1323** Ibid 209: De quolibet lasto coreorum siccorum unde xx dacres faciunt lastum et x corea faciunt i dacrem. **1324** Ibid 251: i lastam et i dacram et ix coria corioum. c**1350** Ibid 179: De x et di. dakeris coreoum. **1381** trans in Cal. Close 13.440: 29 dakers of oxhides and cowhides. c**1400** Gras 1.214: De qualibet dycer corei tannati. **1439** Southampton 2.24: Pro 1 dykur de hud´. **1443** Brokage II.57: 11 dyker´ coriis bovinis; ibid 170: Cum v dykyr´ et ix coriis bovinis. c**1461** Hall 17: And x hydes make a dyker. c**1475** Gras 1.192: Of a daker calf´ skynnes tanned. **1507** Ibid 702: Rede hydes containing x hydes to the decker. **1509** Ibid 565: vii dyker rasours. **1545** Rates 1.35: Redde hides the dekar conteynynge ten skynnes...Rasures the deaker. **1547** trans in Cal. Pat. 19.399: To export 300 "dykkers" of tanned leather and calf skins. **1548** trans in Cal. Pat. 19.401: Licence to Edward Vaughan...to export 200 "dyceres" of leather or tanned leather hides or backs. c**1590** Hall 23: Every dicker consisteth 10 skynns. **1590** Rates 2.30: Rasors the dicker containing tenne. **1597** Halyburton cxiv: The daker of Selche skinis contenand ten. **1597** Skene 1. sv serplaith: Ten hides makis ane daiker, and twentie daiker makis ane last. **1615** Collect. Stat. 465: And euerie dicker consisteth of ten skins. **1616** Hopton 164: There is also...Hides, Dickers, and Lasts. **1661** Acts Scotland 7.252: Hides of all sorts ilk three dacker.... Halfe lang skins ilk ten daker. **1664** Spelman 351: Last corioum...constat ex 20. dakirs, & quodlibet dakir ex 10 coriis. **1678** Du Cange sv dacre: Vel Dacrum, consistit ex 10. coriis. **1708** Chamberlayne 205: Of Hides, 10 are a Dicker; 20 Dickers a Last...of Gloves, 10 Pair a Dicker. **1717** Dict. Rus. sv: Dicker of Leather, is ten Hides or Skins...of Gloves, ten Pair; of Neck-laces ten Bundles, each Bundle containing ten Neck-laces. **1784** Ency. meth. 139: Le last de petites peaux, est de 10 deckers, ou 200 dites.... Le decker de gands, 10 paires, ou 20 dites. **1805** Macpherson I.471: 10 hides...1 dacre, 20 dacres...1 last.... 10 pairs of gloves...1 dacre.... 20 horse-shoes...1 dacre. **1868** Eng. Cyclo. 826: The dicar, or dicker, always 10. **1874** Hazlitt 424: Dicker.—A quantity of iron amounting to ten bars. **1883** Simmonds sv: Dicker, a commercial term for ten of some things...as ten skins make a dicker of hides. **1956** Economist 61: 1 dicker = 10 skins. 1 last = 20 dickers.

digit—1-8 L digitus; 6-9 digit [L *digitus*, finger]. A m-l, originally a unit of body measurement, a finger's breadth, which was equal to 1/4 PALM, 1/12 SPAN, 1/16 ft, 1/24 CUBIT, 1/40 STEP, and 1/80 PACE. Based on the ft of 12 inches, the digit was made equal to 3/4 inch (1.905 cm)—*c***1075** Hall 2: Quarum haec sunt nomina: digitus, uncia, palmus; *ibid* 3: Palmus autem iiij digitos habet. *c***1400** *Ibid* 5: Digitus rotundus est xvi pars pedis. Uncia est digitus et eius tercia pars; *ibid* 6: Quattuor digiti faciunt palmum. **1711** Beverini 117-18: Digitus, Mensurarum apud omnes Nationes est quantitas latitudine ordinarii Digiti per latum æqualis, & apud omnes fuit quarta pars Palmi. **1756** Rolt sv: DIGIT. A measure, containing 3/4ths of an inch. **1868** Eng. Cyclo. 817: The complete table of the 16th century is as follows...a digit, or finger-breadth; four digits make a palm. **1883**

Simmonds sv: As a measure the digit is three-fourths of an inch.

digitus. DIGIT

dikar, diker, dikkar. DICKER

dish [ME dish fr OE disc, plate, fr L discus, dish, disc, quoit, fr Gr diskos, quoit, platter, fr dikein, to throw]. A m-c for lead ore in Derbyshire (c1750-1900) that varied between 14 and 16 pt (1.101 to 1.258 dkl), each pint containing 48 cu inches (Rolt sv load, Second Rep. 15, and Donisthorpe 209). The dish was also a wt of 24 oz (0.680 kg) for butter (c1800) in Cheshire; the wt of 12 oz (0.340 kg) was called a half-dish (Cyclopædia sv weights).

disson, dizzen. DOZEN

dodd—3 dodd (St. Paul's), L dodda [perh fr ME vb dod, to beat, to thresh]. A m-c for grain on the St. Paul's Estate containing 1.125 Colchester SEAMS (c3.96 dkl).—**1222** St. Paul's 47: Doddas avenæ; ibid trans lxxvi: 24 doddæ equal 27 Colchester quarters.

dodda. DODD

doit. DROIT

doleum. DOLIUM

dolium—2-6 L dolium; 3-4 L doleum [L dolium, large jar]. Equivalent to TUN.—c**1150** Acts Scotland 1.312: Assisa vini secundum constitutionem regis David. Cum dolium vini fuerit ad .xx. s. lagena vini erit pro duobus .ʒ. **1228** Gras 1.157: 1 doleum vini. **1249** Close 5.200: XXX dolia vini; ibid 202: unum dolium vini. c**1270** Report 1.420: Dolium cervisiæ continet ccxl gallones. **1290** Fleta 120: Doleum vini lii sextaria vini puri continere et quodlibet sextarium quatuor ialones. **1306** Rot. Parl. 1.207: Quod omnia ligna doleorum vini. c**1330** Hall 31: Dolium vini de Moysun continet communiter cxlii justas ceruisie. Justa ceruisie continet i lagenam et dimidiam secundum standardum Regis. **1439** Southampton 2.15: Pro 1 dolio vini; ibid 16: Pro 8 doliis et 1 hoggyshed vini. c**1500** Hall 8: In dolio vini sunt ccl lagene. **1526** Jacobus 88: iij dolio ij pounsiones vinj claretj et albj. **1728** Cinque Ports 34: The Latin, Dolium, is sometime used for an Hogshead, or other large Cask, less than a Ton, but in several Translations, some of which are very ancient, it is rendered a Ton.

doosen, doozen, dosain, dosan, dosand, dosane, dosayn, dosayne, dosein, dosen, dosene, doseyn, doseyne, dosin, dossand, dossein, dossen, dosseyn, dossin, dosson, dossone, dossyn, dossyne, dossynne, dosyn, dousaine, dousayne, dousen, dousin, doussin, douzaine, douzein, douzeine, douzen. DOZEN

dozen—2 duzeinne (OED); 3 douzeine, duzeynne; 3-7 L duodena; 4 dosain (OED), dosene (OED), dozyne (OED); 4-5 doseyn, doseyne (OED), dozeyn (OED), dozeyne (OED); 4-6 dosayn (OED), dosayne (OED), dosein (OED), L duodenum; 4-7 dosen; 5 disson (OED), dossyn, dossyne, dossynne, dozene (Southampton 1), dusan (OED), dussen (OED), duzan (OED), duzsein; 5-6 dosan, dossen, dosyn, dusane; 6 desone (OED), dosand, dosane, dosin (OED), dossand, dosseyn (OED), dossin (OED), dosson (OED), dousaine (OED), dousayne (OED), dousin, douzaine (OED), douzein (OED); 6-7 doosen, dousen (OED); 6-9 dozen; 7 doozen (OED), dossein (OED), dossone, doussin (Halyburton), L dozena, dozin, dozzen (OED), L duodecim, duzen; 7-8 douzen (OED), L dudena; 8 dizzen (OED); ? duone (Durham), dusein (Prior), dussein (Prior), dusseine (Prior), duze (Langtoft) [ME forms fr OF dozaine fr doze, twelve, fr L duodecim fr duo, two, + decim fr decem, ten]. A m-q consisting of 12 of any item. In the early modern period two abbreviations occasionally are used: doz. and dz.--c**1253** Hall 11: La douzeine de gaunz et de parchemine et de suterie si est de xii peus.... La duzeynne de fer est de vi [sic] pecis. c**1272** Ibid 10: Item duodena pargamenti, in suo genere, continet duodecim pelles; et duodena cirothecarum continet duodecim paria cirothecarum. **1308** Gras 1.362: Adduxit xxi duodenas arcarum. **1393** Ibid 528: Pro xl duodenis capparum. **1396** Ibid 440: X duodenis redelassh´. c**1400** Ibid 213: De qualibet duodena pellium de jenetts; ibid 214: De qualibet duodena nigri vel albi panni monachalis. **1414** Rot. Parl. 4.52: & de chescun sis Duzseines. **1418** Wills 4: Item...ii doseyn. **1420** Gras 1.456: Et v dossenis pellium vitulinarum tannatarum. **1439** Southampton 2.2: 5 dosyn caligarum; ibid 76: 2 dosyn´ candelebrorum. **1443** Brokage II.15: iiii dosyn ropys oynyns. c**1461** Hall 12: Unces or pounds or dossynnes; ibid 13: That is ix dossyn. **1468** Stonor I.101: For ij dossyne of candelle. **1507** Gras 1.695: Bottells the dossen; ibid 700: Lether for cousschenes the dossen; ibid 701: Olld shettes called packyng shetts the dossen. **1509** Ibid 570: ii dossenas napkins; ibid 572: Di. dossenam cours cusshens. **1524** Ibid 196: Pro un´ dosan rolls bokerams. **1532** Beck 2.83: A dousin and a halfe of Spanysshe gloves. c**1549** York

Mer. 144: Item, paid for vj dossand bread. **1554** Mer. Adven. 96: And for everi two hoolle clothes or doble dossens so shipte or sowlde. **1567** Barfield Appendix XLVI: Itm payd for j dosyn of belles. **1581** Acts Scotland 3.216: And euerie dosand of clayth twelf elles allanerlie. **c1590** Hall 25: The parchement rowle is 5 dossen, conteninge 60 skynns. **1597** Halyburton cxiii: Flakonis of Tun the dosane thereof; ibid cxiv: Twell elne for the dusane. **c1610** Lingelbach 112: Napkin Canvas by the peece or dosen.... Buckrams of other makinges by the doosen or half doosen. **1612** Halyburton 294: Cartes the groce contening tuelf dozin paire.... Caskettis of steill the dossone. **1616** Hopton 164: The skins of Goats are numbered by the kippe...and Calues by the dozen 12. **1617** Young II.54: 12 dosen of tynn buttons. **1661** Acts Scotland 7.251: Bridle bits the grosse or tuelve duzen. **1665** Sheppard 57: The Dozen of Iron consisteth but of six [sic] Pieces. **1678** Du Cange sv dozena: Duodecim, duodeni.... Dudena, pro Duodena. **1756** Rolt sv: DOZEN, or Dosen, generally signifies the number 12; but several commodities have 13 or 14 to the dozen. **1883** Simmonds sv: Dozen, the number twelve.

dozena, dozene, dozeyn, dozeyne, dozin, dozyne, dozzen. DOZEN

drachime, drachm, drachme, dragm, dragma, dragme. DRAM

dram—4-8 dragme; 5 L dragma; 6 drachime (OED); 6-7 dragm (Hopton); 6-8 drachme; 7 dramme; 7-9 drachm, dram [ME dragme fr MF dragme fr LL dragma fr L drachma fr Gr drachmē, a handful]. A unit of wt in the ap and avdp systems.

An ap dr contained 3 s (3.888 g) of 20 gr each and was equal to 1/8 ap oz of 480 t gr. Comparatively the ap dr was 2.194 avdp dr, 2.017 Scots t DROPS, and 0.006302 Scots tron lb. In the Imperial system an ap fluid dr is a m-c containing 60 MINIMS or 0.216734 cu inches (3.55153 ml) or the volume of 54.6875 gr (3.544 g) of distilled water at 62° F and equal to 1/8 ap fluid oz of 1.733875 cu inches.—**c1450** Hall 33: Scrupulus 3 pars dragme. Dragma octava pars uncie; ibid 34: 3 scrupuli faciunt dragmam, vis. 60 grana ordei. **c1475** Ibid 35: A dragme is the eyghtethe part of an unce & is thus...Wryte ʒ. **c1600** Ibid 36: Scrupuli is 20 barley cornes.... 3 scrupules contain a drachme. **1616** Bullokar sv dramme: A smal weight, the eight part of an ounce: It conteineth in it three scruples. **1640** Rider Appendix sv tabula mensurarum: 3 drammes and a scruple; ibid sv weights: A weight being the second part of a dramme. **1651** Violet 95: Eight Ounces and four Drams Troie. **1688** Bernardi 137: Vel more Pharmacopolarum: Libra de Troy, 12 Unciæ ℥, 96 = 12 X 8 drachmæ ʒ. **1696** Cocker 108: ʒ a dram. **1708** Chamberlayne 205: The Apothecaries reckon 20 Grains Gr. make a Scruple ℈, 3 Scruples 1 Drachm ʒ, 8 Drachm 1 Ounce ℥. **1716** Harris 2. sv weight: Grains 20. Scruples 3. Drachms 8. Ounce. **1717** Dict. Rus. sv: Dram or Drachm, the just Weight of sixty Grains of Wheat...the eighth part of an ounce. **1737** Greaves 680: The French as well as we (and so do all physicians of all countries that I know) divide their ounce by eight drachmes. **1829** Palethorpe sv: DRAM.... In apothecaries´ weight it contains 3 scruples; and is the one eighth part of an ounce. **1832** Wilkinson Preface: 1/15 of a Drachm...or 4 grains; ibid 50: Drachm contains Three Scruples ℈. Scruple contains Twenty Grains gr. **1851** H. Taylor 45: Eight drachms of apothecaries´ weight may mean an ounce of 480 grains, whilst eight drachms fluid measure are but 437 1/2 grains, still called an ounce; ibid 45-46: It [the dram] may be 60...or 54.7 grains. **1880** Courtney 158: 3 scruples 1 drachm, marked dr. or ʒ. **1907** Hatch 24: 1 fluid drachm (fl. dr.) = 60 minims (min.) = 0.216734 cubic inch; ibid 35: 1 drachm = 3.88794 grammes; ibid 36: 1 fluid drachm = 3.55153 millilitres. **1920** Stevens 2: 1 Fluidounce, fl. oz. = 8 Fluidrams, fl. dr. = 480 Minims, ♏. **1934** Int. Traders´ 74: Dram (fluid or liquid)...United Kingdom...3.551 milliliters. **1951** Trade 22: 8 drachms...1 apothecaries ounce.... 1 drachm = 60 grains. **1969** And. & Bigg 12: 1 drachm = 3.88793 g.

The avdp dr contained 27.344 t gr (1.772 g) and was equal to 1/16 avdp oz of 437 1/2 t gr. However, in many medieval and early modern sources the avdp dr was confused with the ap dr: either the avdp dr to avdp oz ratio of 16 was incorrectly taken as the ap ratio of 8, or the ap conversion factor of 60 for dr was erroneously used instead of the avdp conversion factor of 27.344; sometimes both errors were made together. Comparatively the avdp dr was 0.919 Scots t drop and 0.002873 Scots tron lb.—**1682** Hall 29: Aver-du-pois conteynes: every pound, 16 ounces; every ounce, 8 drgmes [sic]; every dragme, 3 scruples; every scruple, 20 graines. **1688** Bernardi 135: Insuper uncia Avoirdupois pro mercibus caducis explicat 8 drachmas aut 3 X 8 = 24 scripulos Avoirdupois. **1699**

Hatton 1.19: 16 Drachms is 1 Ounce, 16 Ounces make 1 Pound. **1717** Dict. Rus. sv: Dram or Drachm, the just Weight of sixty Grains of Wheat; in Avoir-du-pois Weight, the sixteenth part of an ounce. **1724** Coles Appendix: 4 Quarters...1 Dram. 16 Drams...1 Ounce. 16 Ounces...1 Pound. **1790** Jefferson 1.985: The Pound is divided into 16 ounces; the ounce into 16 drachms. **1813** Cooke 95: 27 1/3 Grains...1 Drachm. **1829** Palethorpe sv: DRAM, or DRACHM, an English weight, 16 of which make 1 ounce avoirdupois; ibid sv pound: An avoirdupois dram contains 27.34 grs. troy. **1848** Skilling xxi: 8 drams, or 1/2 an ounce. **1850** Alexander 28: Dram...27.34 grains. **1868** Eng. Cyclo. 823: Averdupois weight...the dram is 27 grains and 11-32nds of a grain. **1907** Hatch 34: 1 drachm = 1.77185 grammes. **1951** Trade 22: 1 dram = 27.34375 grains.

dramme. DRAM

draught [ME draught, draht fr OE dragan, to draw]. A wt of 61 lb (27.669 kg) for wool in Sussex (c1850). It was 1/4 PACK of wool weighing 240 lb (108.862 kg), with 1 lb allowed for the turn of the scale (Cooper 42).

droit—7-9 droit; 8 dwit; 8-9 droite; 9 doit [*]. A moneyer's unit of wt equal to 1/24 MITE or 1/480 t gr (0.000135 g). It belonged to a series of imaginary wt used to compute exact coin wt by alternate subdivisions of 20 and 24.—**1665** Sheppard 15: 24 Droits make a Myte. **1707** Justice 4: One Mite into 24 Droites.... One Droite into 20 Perits. **1725** Bradley sv weights: The Moneyers subdivide the grain thus: 24 Blanks make 1 Perrot; 20 Perrots 1 Dwit; 24 Dwits 1 Mite; 20 Mites 1 grain.

1783 Beawes 893: 20 Perits a Droite. **1784** Ricard II.151: On divise le grain en 20 mites, 480 droits. **1819** Cyclopædia sv weight: The grain troy is divided into 20 mites, the mite into 24 doits, the doit into 20 periots, and the periot into 24 blanks. **1840** Ruding 1.411: Twenty-four droits a mite, twenty perits a droit. **1868** Eng. Cyclo. 822: In some old books a grain is 20 mites, a mite 24 droites, a droite 20 peroites.

droite. DROIT

drop [ME drope fr OE dropa; see WNID3]. A wt in the Scots t and tron systems: t, for gold and silver, 30 gr (1.944 g), or 1/16 oz of 480 t gr (31.103 g); t, for meal, meat, hemp, and iron, and tron, 29.75 gr (1.928 g), or 1/16 oz of 476 t gr (30.845 g).—**1661** Acts Scotland 7.254: Each turner weight and ane drop and ane halff. **1779** Swinton 36: In Scotland, Gold and Silver are weighed by the...[English t] ounce and pound; but the ounce is divided into 16 drops, and the drop into 30 grains; ibid 38: Troye. For Meal, Butcher-meat, Hemp, Iron...Drop. dr...29.75 [English t gr].... Trone. For home productions.... Drop. dr...29.75 [English t gr]. **1816** Kelly 92: The Scotch jewellers divide the troy ounce into 16 drops, each drop being 30 troy grains. **1896** Klimpert 81: Drop...bis 1826 eine Gewichtsstufe in Schottland = 1/16 Ounce.

dudena, duodecim, duodena, duodenum, duone, dusan, dusane, dusein, dussein, dusseine, dussen, duzan, duze, duzeinne, duzen, duzeynne, duzsein. DOZEN

dwit. DROIT

dycer, dycker, dyker, dykker, dykur, dykyr. DICKER

E

el, eline. ELL

ell—1-7, 9 eln; 2-7 L alna, L ulna; 2-9 elne; 3-6 ellen; 4 ellyn (OED); 5 ellne; 5-7 elle; 5-9 ell; 6 el, eline (OED) [ME ellen, eln, elne fr OE eln; akin to L ulna, elbow, ell]. A m-l for cloth generally containing 45 inches (1.143 m) or 5/4 yd of 36 inches, although ells of 54 inches (1.372 m) in Shropshire and 48 inches (1.219 m) in Jersey were also used. In Scotland, the standard ell contained 37 Scots inches equal to approximately 37 1/5 English inches (c0.95 m); the following were exceptions (all reckoned in English inches): 37 1/8, Langholm in Dumfriesshire and Fifeshire; 37 1/4, Angus for woolen cloth; 38, coarse linens and woolens in Inverness, home manufactures in Ross and Cromarty, and green linens in Wigtownshire; 38 5/12, plaiding in Aberdeenshire; 38 1/2, home manufactures and laborers'-tradesmen's work in Kincardineshire; 39 in some parts of Dumfriesshire; 39 1/2, plaiding and stuffs in Edinburgh; and 40-41, raw woolen cloth in Wigtownshire (Swinton 72, 79, 81, 86, 89, 115, 127 and Donisthorpe 209). From the twelfth through the fourteenth centuries L ulna was used ambiguously to indicate both ell and YARD.—c**1150** Acts Scotland 1.309: Vlna Regis Dauid debet continere in se. xxxvij. pollices mensuratas cum pollice trium hominum. scilicet ex magno. ex medio. et paruo. Et ex medio pollice hominis debet stare. aut ex longitudine trium granorum boni ordei sine caudis. Pollex autem debet mensurari ad radicem vnguis pollicis. **1208** Bish. Winch. 21: In xxx ulnis canevacio; ibid 49: In xiij ulnis de canevaz ad lanam. c**1260** Bracton II.244: Tam ulnis quam ponderibus. c**1272** Hall 7: Et xij pollices faciunt pedem; et tres pedes faciunt ulnam. **1308** Gras 1.365: Et xxx ulnas panni mixti coloris. c**1350** Eng. Gilds 352: And that euerich chaloun ouer thre ellen of lengthe out of a-syse be forfeted. **1351** Rot. Parl. 2.235: Sit longitudinis viginti & sex Ulnarum, & latitudinis sex quarteriorum infra Listas. c**1400** Hall 41: Nota quod tres pedes regii faciunt ulnam Regis. **1425** Acts Scotland 2.12: The Elne to contene xxxvij Inche. **1439** Southampton 2.6: 60 ulnis panni linii. **1440** Scrope 229: 2 pece of lynyn cloth...iiij [X] xx [+] viij. ellnes. **1443** Brokage II.245: iiii c dimidio ulnas de canvas. c**1461** Hall 14: And xii ynchis make a fote; and iij fote make a yard; and the Ynglysh ell go V qr. [quarters] off yard. **1474** Cov. Leet 397: The Elle to be v quarters of the yarde in lengthe, and hitt to be sysed and sealed and that hit be marked in iiij partes accordyng for an Eln. **1502** Arnold 204: Item a Fll [Flemish] ell conteyneth iii q´t´s of an Eng. yarde, and v. q´t´s of the Fll ell makith an Eng. ell. **1503** Acts Scotland 2.246: That pynt quarter ferlot pec elwand stane & pund be of ane quantite & mesor. **1507** Gras 1.696: Canvas called Vytory´ canvas the bale containing ii [X] c elles. **1534** Fitzherbert 25: An elne of lengthe. **1540** Recorde 207: 3 Foote and 9 Ynches, make an Elle. **1555** Acts Scotland 2.496: Pynt quart fyrlot peck elnwand stane and pund. **1587** Ibid 3.521: The eln...threttie sevin Insches. c**1590** Hall 27: The English ell is 5 quarters of an Englishe yard in lenght, conteninge 45 ynches in lenght. **1597** Skene 1.11: They ordaned and deliuered, that the Elne sal conteine thrittie seuen inche, as is conteined in the statute of King David the First, made thereupon. **1616** Hopton 165: Also an English mile is...1408 Elles. **1624** Huntar 5: 3. Foote and an inch, or 37 inches makes the Ell of Edinburgh; Ibid 6: 45. Inches is the English Ell. **1635** Dalton 150: Three foot and nine inches make an Ell. **1646** H. Baker 203: Forasmuch as 3 elles English are worth 5 elles Flemmish. **1663** Acts Scotland 7.488: The ell is designed to be thirty seven inches, Yet many vse inches by which the ell is divyded into fourty tuo inches. **1665** Sheppard 16: 3 foot and 9 Inches an Eln. **1678** Du Cange sv alna: Ulna, certæ longitudinis virga, qua telas pannosque metiuntur. **1682** Hall 28: An Elle is a yard and 9 inches. **1688** Bernardi 197: Pes est...12/45 Ulnæ Anglicæ; ibid 221: Virga Anglica...4/5 Ulnæ Anglicæ. **1708** Chamberlayne 207: 1 Yard and a Quarter makes an Ell. **1717** Dict. Rus. sv: Ell, a long Measure, consisting of 3 Foot and 9 Inches. **1742** Account 1.545: The Ell is...universally reputed equal to one Yard and a Quarter, or to 45 Inches. **1779** Swinton 17: When it is generally known, that an English yard is 36 inches, and a Scotch ell 37 2/10, it must appear to no use to keep any measure but the first. **1805** Macpherson I.316: He also directed that the measure of the eln or yard should be of uniform length throughout his kingdom; ibid 642: The parliament of Scotland decreed [in 1427], that the elne should contain thirty-seven inches, agreeable to the law of King David I. **1816** Kelly 94: 1 [Scots] Ell...37 1/5 English Inches...30 Scotch ells = 31 English yards. **1820** Second Rep. 15: Ell...Shropshire: of linen cloth, 6 quarters = 54 inches...Jersey: 4 feet = 48 inches...Scotland: the standard is 37 inches. **1822** G. Gregory II. sv measures: The Scots elwand was established by king David I. and divided into 37 inches. **1840** Waterston 145: The standard Scottish ell of 36 Scots or 37.0598 Imperial inches. **1882** Beck 1.112: ELL. A measure of length, originally taken from the arm, a cloth measure equal to 1 1/4 yds. **1887** Bonwick 346: 28 elns in length. **1956** Economist 8: Ell, English...45 inches.... Scottish...37.06 inches.

elle, ellen, ellne, ellyn, eln, elne. ELL

ench, enche. INCH

ende—5 L fynes (for L finis, end; Southampton 2); 6 ende [ME ende fr OE ende; see WNID3 sv end]. A wt for iron equal to 1/112, 1/120, or 1/132 ton (9.072 to 7.711 kg).—c**1500** Southampton 2.120: 112 endes makyth a tunne yryn and yn the kyngys custome they alowe 132 endes to the tun. **1545** Rates 1.21: Iron of spayne the ende. **1562** York Mer. 168: Iron sex skores endes to the tonne, iii s. iiij d.

enoforium. OENOPHORUM

erw [W erw, what has been tilled (Laws Wales)]. A m-a for arable land in Wales—the standard ACRE (c1300-1800)—which varied considerably in size from one region to another, but which appears most often to have averaged 4320 sq yd (0.361 ha) (Laws Wales 999, Donisthorpe 204, and Second Rep. 5).

escheppa, eschippa, eskippa. SKEP

estarium, esteria. STRIKE

estik, estika, estike. STICK

estrica, estricha, estricum, estrike. STRIKE

ethyndel. EYGHTYDELL

eyghtydell—5 ethyndel, eyghtydell (Prior), eyhtyndyl (Prior), eytendele [ME eyghtydell fr eyght, eighte, eight, fr OE eahta, + -dell, part, portion, fr OE dǣl, part]. A m-c for grain containing approximately 4 gal (c1.76 dkl) and equal to 1/2 bu or 1/8 COOMB.—**c1440** Promp. Parv. 222: Half a buschel, or eytendele (half of a bowndel, boshel, or ethyndel).

eythyndyl, eytendele. EYGHTYDELL

faat. FATT

factus [*]. A m-c on the St. Paul´s Estate containing 17 bu (c5.99 hl), or 2 SEAMS of 8 1/2 bu each.—**1283** St. Paul´s 164: Per factum bracini.

fad--7 fawde (Best); 9 fad, faud [*]. A m-q of straw equal to 1/12 THRAVE.—**1829** Brockett 110: Fad, faud, a bundle of straw, twelve of which make a thrave.

fadam, fadame, faddam, faddom, faddome, fademe. FATHOM

fadge [ME faige, perh fr (assumed) OF fais, bundle]. A m-q for sticks (c1800-1900), as a bundle of undesignated size (Brockett 110 and Simmonds sv).

fadim, fadme, fadmen, fadom, fadome, fadowme, fadum, fadym, fædm. FATHOM

fæt, fætt. FATT

fagate, fagatt, faget, fagett, faggat, fagget. FAGGOT

faggot--4 fagate, faget (OED), fagett (OED); 4-6, 8 faggott; 4-9 fagot; 5 fagatt (OED), fagott (Finchale), ffagott; 5-9 faggot; 6 faggat (OED) [ME fagot fr MF fagot, prob fr OPr fagot, perh fr VL facus, modif of Gr phakelos]. A measure for firewood, 3 ft (0.914 m) in length and 24 inches in circumference; and a wt for steel, 120 lb (54.431 kg).—**1350** trans in Memorials 254: Also...that the cart which brings firewood, [for] talwode, shall take for the hundred, at Crepulgate 6 d., and for the hundred of fagates 4 d. **1474** Cov. Leet 399: And his ffagott of wodde of an ob. schal-be iij schaftmond and a halfe a-bout and a yerde of lenthe. And his ffagott of j d. schal-be vij schaftmond a-bout, kepyng the same lenght. **1587** Stat. 171: And euerie fagotbed to conteine in length three foot. And the band...to be of foure and twentie inches about, besides the knot. c**1590** Hall 28: Euery faggot, bend or sticke ought to conteyne, in lenght, 3 foott; and the bond of euery such faggott ought to be 24 ynches about, besydes the knott. **1616** Hopton 163: Fagots should be three foot long, and the band beside the knot 24 inches made round. **1628** Hunt B3: A fagot of Stelle 120 [lb]. **1635** Dalton 149: Talwood, billet, and fagot. **1660** Bridges 31: A faggot of Steel is...120 lb. **1665** Assize 18: Item, every Faggot-band to contain in length three foot, and the band of every such Faggot to be 14 [sic] inches about besides the knot. **1682** Hall 30: Fagots must be 3 foote long, and the band 24 inches round, besides the knot. **1708** Chamberlayne 207: A Faggot of Steel is 120 Pounds. **1756** Rolt sv measures: Faggots are to be 3 feet long, and at the band 24 inches about, besides the knot. **1790** Miller 18: The...FAGGOTT, GAD. **1883** Simmonds sv fagot: A quantity of steel in bars, equal to 120 lbs.; a bundle of sticks of wood about 3 feet long and 2 feet round. See SHAFTMENT

faggott, fagot, fagott. FAGGOT

fal. FALL

faldom, faldome. FATHOM

fall—5-9 fall; 6-7 fal [fr vb fall; see first and second citations]. A m-l in Scotland containing 6 ELLS or 6.2 English yd (5.669 m) and a m-a containing 36 sq ells or 38.44 English sq yd (32.140 sq m). The latter superficial fall was equal to 0.007942 English acre.—c**1400** Acts Scotland 1.387: The aker sall contene four rude...the rude .xl. fallis The fall sall hald .vj. ellis. **1607** Cowell sv perche: Sixe elnes long make one fall...and sixe elnes long, and sixe broade, make a square and superficiall fall.... So much land as falleth vnder the rod or raipe at once, is called a fal of measure, or a lineal fall; because it is the measure of the line or length onely. **1624** Huntar 6-7: 6. Ells of the standard of Edinburgh, makes a lineall fall, wherewith land is measured; ibid 7: 6 Ells long, and 6 Ells breadth, makes a superfitiall or square fall, wherewith land is reckened. **1665** Sheppard 19: 6 Elns long make a Fall, which is the common lineal measure. And six Elns long, and six broad, make a square and superficial Fall of measured Land; ibid 19-20: And it is to be understood, That one Rodd, one Raip, one lineal Fall of Measure, is all one; for each one of them containeth six Elns in length. Howbeit, a Rodd is a Staff or Pole of wood; a Raip is made of Towe or Hemp; And so much Land as falleth under the Rodd or Raip at once, is called...a Fall of Measure or a linear fall, because it is that measure of the line or length only as the superficial Fall is the measure both of length and breadth. **1779** Swinton 24: Fall or Rood. ≐ 6.2 [English yd]; ibid 27: Fall... .0079421 [English acre]. **1816** Kelly 94: 6 Ells...1 Fall, or Rood.... 30 Scotch ells = 31 English yards; ibid 95: 6 Square Scotch Feet...1 Square Ell. 36 Square Ells...1 Square Fall. **1820** Second Rep. 16: Fall...Aberdeenshire and elsewhere: of land, 6 ells square. **1832** Edinburgh XII.570: 6 Ells = 1 Fall = 223 1/5

[English inches]. **1880** Britten 170: Fall, 1/160 of a Scotch acre, as the perch is of the English acre. **1962** W. C. Dickinson 63: The ´fall´ (which contained 6 ells...).

fan [ME fan fr OE fann fr L vannus, fan, van for winnowing grain]. A m-c, a wide, shallow wicker-basket, for chaff in Cambridgeshire and other areas of Eastern England (c1800-1900) containing 3 heaped bu (c1.35 hl) (Second Rep. 16 and Britten 142, 170).

fangot [prob fr It fangotto, var of fagotto, bundle]. A m-q (c1700) for raw silk, 1 to 2 3/4 Cwt (50.802 to 139.705 kg), and grogram and mohair yarn, 1 1/2 to 2 1/2 Cwt (76.203 to 127.005 kg) (Hatton 3.226).

fardal. FARDEL[1]

fardall. FARDEL[1]; FARDEL[2]

fardel[1]—4-5 L fardellus; 4-6 fardele (OED); 4-9 fardel, ferdel (OED), ferdell (OED), ferdelle (OED); 5 fardille (OED); 5-7 fardell; 6 ferdle (OED); 6-7 fardall (Halyburton), farthel (OED), farthell (OED), farthelle (OED); 6-9 fardle; 7 fardal (OED), ffardell; ? fardelle (Prior), ferdall (Prior) [ME fardel fr MF fardel fr OF fardel fr farde, bundle, prob fr Ar fardah, bundle]. A m-q for cloth and other items assembled as a bale or bundle of no standard dimensions.--**1308** Gras 1.361: Adduxit iiii fardellos canabi. **1324** Ibid 386: Pro ii fardellis gladiorum. **1392** Ibid 541: Pro xl worsted´ in i fardello. **1420** Ibid 456: Pro i fardello cum viii vergis panni largi sine grano. **1439** Southampton 2.13: 2 fardell´ de napere; ibid 42: 2 fardell´ pellium coniculorum continentibus 10 C. pelles; ibid 68: i fardell´ de peletory continente ii mantellis de lebard. **1443** Brokage II.25: Cum ii fardellis canvas; ibid 67: i fardello de cerico; ibid 69: ii fardellis fetherbeddes; ibid 134: ii fardellis flokkys; ibid 271: ii fardellis pellium. c**1550** Welsh 52: 16 fardels linen cloth; ibid 61: 4 fardels merchandise; ibid 67: 6 fardels frise; ibid 74: 1 fardel mercery wares; ibid 97: 3 fardels grocery, mercery and haberdashe wares. c**1555** Remembrance 72: For a fardell of canvas from the Watergat into any place above the Newe Corner or into Saynte Mihells paroche: ii d. **1590** Rates 2.7: Canuas called vetery canuas the Bale or fardle. c**1610** Lingelbach 61: Bee yt fardell, trusse Ballett maund, chest, ffat, butt, pype, barrell great or small; ibid 79: Packed or made vp into ffardells Trusses or Bales. **1756** Rolt sv scavage: A great pacquet or fardle, containing between 15 or 20 cloths. **1883** Simmonds sv: Fardel, a bundle or little pack.

fardel^{2}—2 ferdel (Prior); 4 L fardellum, L ferdellum; 6 fardall, fferdalle; 6-? fardel (OED); 7 fardell, L fardella, L ferdella; 9 ferdell; ? ferdelh (Skinner), L ferdellus (Prior) [ME ferdel fr ferde, fourth, fr OE fēartha, fēortha, fourth, + del, part, fr OE dǣl, part]. A m-a of land equal to 10 statute acres (4.050 ha) or 1/4 VIRGATE of 40 acres. Equivalent to and superseded by both dimensions of FARTHINGDALE (c1400-1500).—**1338** Langtoft 600: Decem acræ faciunt ferdellum. Quatuor fardella faciunt virgatam unam. c**1500** Brit. Mus. 6.7: j virgat...iiij fferdalles.... j fardall...x Acr. **1651** Noy 57: Two Fardells of Land make a Nooke of Land. **1664** Spelman 212: Fardella terræ (ut mihi constat è veteri MS.) est quarta virgatæ terræ. Decem acræ terræ (inquit MS.) faciunt...unam ferdellam, & 4 ferdells faciunt virgatam. **1695** Kennett Glossary sv furendellus: Fardella, Ferlingus. A fardingel, farundel, or ferling of land, i.e. the fourth part of an acre. **1874** Hazlitt 437: Ferdell, fardingdeal, or ferundell of land.

fardele. FARDEL[1]

fardelet. FARDLET

fardell. FARDEL[1]; FARDEL[2]

fardella. FARDEL[2]

fardelle. FARDEL[1]

fardellum. FARDEL[2]

fardellus. FARDEL[1]

fardendela. FARTHINGDALE

fardille. FARDEL[1]

fardingdeal, fardingdeale, fardingdela. FARTHINGDALE

fardingland. FARTHINGLAND

fardle. FARDEL[1]

fardlet--5 fardelet (OED); 5-7 fardlet (Shipley) [OF fardelet, dim of fardel, FARDEL[1]]. A m-q, a bale or bundle, smaller than a FARDEL[1], used for cloth and other items from the fifteenth through the seventeenth centuries (Shipley 258).

farlet. FIRLOT

farthel, farthell, farthelle. FARDEL[1]

farthendale, farthendel, farthendele, farthindale, farthindel,

farthing. FARTHINGDALE

farthingdale—3 feorthendele, feorthendell, ferchendell; 5 forthingdole (OED); 6 farthendel, ferendell (Gray), ferrundell (Gray); 6-7 farthendele; 6, 9 ferundel; 6-? farthingdale; 7 fardendela, fardingdela, farthendale, farthindale, farthindel, farundel, L furendellus; 7-8 fardingdeale, farthingdole, farundell; 7-9 fardingdeal; 8 farundale; 9 ferundell; ? farthing (Gras 2), farundele (Prior), L furchendellus (Prior), L furthendellus (Prior) [ME ferthing fr OE fēorthung fr fēortha, fourth, + dale, part, fr OE dāl, part, portion]. A m-a for land with two very different dimensions, one fortyfold the other. Because of its general meaning of "a fourth part," it came to be used interchangeably with FARDEL[2] and FERLING and ultimately supplanted them totally in indicating 10 statute acres (4.050 ha) or 1/4 VIRGATE of 40 acres. It also was used as the equivalent of a ROOD or 1/4 acre of 40 sq perches (0.101 ha). In Latin documents, "farthingdale" is rendered by such forms as Una Quartrona (Beamont 10), Quadrantaterræ, etc., all having the general meaning of "quarter (land)."—c**1290** Worcester 62b: Et j ferchendell in alio.... iij...feorthendeles.... Et dimidia virgata j. feorthendell. **1540** Recorde 208: A Rod of lande, which some call a roode, some a yarde lande, and some a farthendele, 4 Farthendels make an Acre. **1600** Hylles 67: A farthendele or roode of lande. **1600** Hill 67: 4. Farthendales, or 4. perches in breadth, & 40. in length make, 1. Acre of lande. **1607** Cowell sv farding deale: Farding deale alias Farundell of land (Quadrantaterræ) signifieth the fourth part of an acre. **1664** Spelman 212: Fardella, Ferdella, Fardendela, Fardingdela, Farding, Fardingel, Farthindel, Farundel, & Ferlingus.... Farding deale autem aliàs Farundel juxta Cowellum, quartam partem acræ significat, quam nos rodam vocamus. **1665** Sheppard 24: And that a Fardingdeale alias Farundell of Land (Quadranta terræ, in Latine) signifieth the fourth part of an Acre. **1667** Roberts 302: Perch 1 in breadth and 40 in length...a Farthingdole. Farthingdole 2—is half an Acre. Farthingdole 4—is an Acre. **1678** Du Cange sv roda: Anglis, Quarta pars acræ, quæ et Farding deale, seu Farundel dicitur, juxta Cowellum, ex Anglico Rodd, Pertica. **1682** Hall 29: A Roode or a Farthendale conteynes 10 day workes; that is, one pearch in breadth and 40 in length. **1695** Kennett Glossary sv furendellus: Fardella, Ferlingus. A fardingel, farundel, or ferling of land, i.e. the fourth part of an acre, which in Wiltshire is now call'd a fardingdale: and in other parts a farthindale. **1717** Dict. Rus. sv farding-land: Farundale of Land; is the fourth part of an Acre; ibid sv furendal: Fardingdeal, of Land, the fourth part of an Acre. **1784** Ency. meth. 139: L'acre de terre d'Angleterre est de 4 fardingdeales. **1874** Hazlitt 430: In a manuscript law-book, written by Ambrose Couper, Esq., a student in one of the Inns of Court, in the year 1579...it is laid down as a rule, that...ten acres make a ferundel or fardingdeal, four ferundels make a yard-land, and four yard-lands a hide; ibid 437: Ferdell, fardingdeal, or ferundell of land. See FARTHINGLAND

farthingdole. FARTHINGDALE

farthingland—3 ferthinglond, ferthinlond; 8 fardingland; ? farthingland (Gras 2), ferthinland (Prior), forland (Prior), L forlandus (Prior) [ME ferthing fr OE fēorthung fr fēortha, fourth, + land]. Equivalent to both dimensions of FARTHINGDALE.—c**1290** Worcester 101a: Walterus de Grava pro ferthinlond; ibid 101b: Ricardus Boernild pro iij...ferthinglond. **1717** Dict. Rus. sv farding-land: Farding-land or Farundale of Land; is the fourth part of an Acre.

farundale, farundel, farundele, farundell. FARTHINGDALE

fat, fate. FATT

fatham, fathem. FATHOM

fathom—1 fædm (OED); 2-4 fedme (OED); 3 fadim (OED), fadum (OED), fathum (OED); 4 fademe (OED); 4-5 fadme (OED); 4-6, 8 fadom; 4-7 fadome; 5 fadmen (OED), fadym (OED), vathym (OED), vetheym (OED), vethym (OED); 5-6 fadam (OED), fadame (OED), fadowme (OED), fathem (OED), fawdom (OED), fawdome (OED); 6 faddam (OED), faldom (OED), faldome (Durham), fauddme (Dur. House), faudom (OED), feddom (OED), vadome (OED); 6-7 fatham, fathome; 7 faddome, L fathomus; 7-8 faddom; 7-9 fathom; 8 ffatham [ME fadme fr OE fædm, fæthm, fathom, the arms outstretched]. A m-l generally containing 6 ft (1.829 m), but occasionally 7 ft (2.134 m). In Yorkshire a "fandam," a corruption of fathom, was a measure for hay stacks, and was reckoned as the distance between a man´s hands when his arms were stretched out around the stack (Britten 170). The fathom is currently used as a measure for roundwood, 216 cu ft (6.116 cu m) (O´Keefe 669). It sometimes was abbreviated fath., fm., fth., or fthm.—**1392** Henry Derby 158: Et pro xx fadom cordez. **1393** Ibid 242: Et eidem pro xl fadome corde. c**1536** Leland 107: It is in bredth a mile, and a ii. miles of lenght, and wher as it is depest a xiii. fadom. **1624** Huntar 1-2: Wee haue our Measures, for length, breadth and thicknes, as the Inch, the Foote, the Faddome, and the Ell; ibid 8: 6. Foote is a fathome; ibid 9: A Fathome containeth...Inches—72. **1625** Tap 2.C4: There is 14. Fatham depth. **1635** Dalton 150: Seven foot maketh a fadome. **1665** Sheppard 16: 7 foot a fathom. **1682** Hall 28: A Fadome is two yards. **1688** Bernardi 202: Pes Anglicus...1/6 Fathomi. **1704** Mer. Adven. 245: Ffor sorting and laying up every ffatham of lathwood. **1717** Dict. Rus. sv: Fathom, a Measure of six Foot, generally taken for the space comprehended by the utmost stretching of both arms. **1727** Arbuthnot sv English measures of length: Foot Cubit Yard Pace Faddom Pole Furlong. **1783** Beawes 913: A Fadom (or Fathom) six [feet]. **1832** Edinburgh XII.569: 6 Feet = 1 Fathom = 1.8288 [m]. **1850** Alexander 34: Fathom...2.—yards. **1956** Economist 8: Fathom...6 feet. **1969** And. & Bigg 11: 1 fathom = 1.8288 m.

fathome, fathomus, fathum. FATHOM

fatt—1 fæt (OED), fætt (OED); 2-4 fet (OED), vet (OED); 3 feat (OED); 4-8 fat, fatt, fatte; 5 faat; 5-7 fate [ME fat, fatt fr OE fæt, fætt]. A m-c for grain generally containing 9 bu (c3.17 hl), and a m-q for other products: bristles, 5 Cwt (254.010 kg); coal, 1/4 CHALDER (c3.17 hl); isinglass, 3 1/4 to 4 Cwt (147.417 to 181.436 kg); unbound books, 4 bales equal to 1/2 MAUND; wire, 20 to 25 Cwt (1016.040 to 1270.050 kg);

and yarn, 220 or 221 bundles.—**1413** Rot. Parl. 4.14: Et auxint les Marchauntz et Citezeins de la Citee de Loundres usent de prendre de chescun Vendour pur la Quarter de Furment noef Busselx par une Mesure use deins la dit Cite appelle la Faat. **1420** Gras 1.459: Pro 1 barello i fat. **1433** Rot. Parl. 4.450: Achatours des Blees en plusours autres Citees, Villes, Burghs, et Countees d'Engleterre, continuelment de jour en autre, achatont et preignont noef Bushels [fatt] pur le quarter. **1509** Gras 1.562: ii fatts i pipa cum xii grossis papiri. **1555** York Mer. 156: A fat of eles, foure pence. **1562** Ibid 168: A fatte of eles, vi d. **1587** Stat. 244: And the marchants and citizens of London do use to take of euerie seller for the quarter of wheate ix. bushels by the measure...called the fate. **1590** Rates 2.43: Painted bookes the fat. **1607** Cowell sv fate: Fate or Fat: is a great wooden vessell, which among brewers in London, is ordinarily vsed at this day, to measure mault by, containing a quarter, which they haue for expedition in measuring. **1615** Collect. Stat. 466: The Purveyors of Corne for the kings household haue taken nine bushels for the quarter.... And the Citizens of London also do the like by a measure called the fate. **1701** Hatton 3.226: Fatt...An uncertain quantity, as of Yarn 210 [sic] to 221 Bundles; Unbound Books 1/2 a Maund; Wire 20 C. to 25 C. weight; Isinglass 3 1/4 C. to 4. C. weight. **1717** Dict. Rus. sv: Fat of Ising-glass, a quantity from three hundred Weight and a quarter to four hundred Weight: Of unbound books half a Maund or four bales: Of Wire from 20 to 25 C. Weight: Of Yarn, from 220 to 221 Bundles. **1755**

Postlethwayt II.186: Buyers of corn in London, bought by a vessel called a fat, containing 9 bushels of corn. **1964** Breed 13-14: The persistence of the London corn buyers in forcing the country folk with whom they dealt to supply corn by a measure of nine bushels (called a Faat). See HUNDRED

fatte. FATT

faud. FAD

fauddme, faudom. FATHOM

fawde. FAD

fawdom, fawdome. FATHOM

fearlot. FIRLOT

feat. FATT

feddom, fedme. FATHOM

feirtlett. FITLOT

feodum. KNIGHT'S FEE

feorthendele, feorthendell, ferchendell. FARTHINGDALE

ferdall. FARDEL[1]

ferdekyn. FIRKIN

ferdel. FARDEL[1]; FARDEL[2]

ferdelh. FARDEL[2]

ferdell. FARDEL[1]; FARDEL[2]

ferdella. FARDEL[2]

ferdelle. FARDEL[1]

ferdellum, ferdellus. FARDEL[2]

ferdkyn. FIRKIN

ferdle. FARDEL[1]

ferekin. FIRKIN

ferendell. FARTHINGDALE

ferikin, ferken, ferkin, ferkyne. FIRKIN

ferlig. FERLING

ferling—3 L furlingus; 3, 7 L ferlingus; 3-7 ferling; ? ferlig (Prior) [ME ferling fr OE fēorthling, a fourth part]. A m-a of land equal to 10 statute acres (4.050 ha) or 1/4 VIRGATE of 40 acres. Equivalent to and superseded by both dimensions of FARTHINGDALE (c1400-1500).—**c1201** Salisbury 67: Nec in duobus ferlingis terræ de Cerdestok. **1208** Bish. Winch. 63: Et cum furlingo terræ de Bruges. **1214** Cur. Reg. 14.166: j. hide terre et dimidie et j. ferlingi terre cum pertinentiis. **1227** trans in Cal. Char. 1.17: Four ferlings of land in Kingeston held by Walter. **1262** trans in ibid 2.42: And one ferling and a half which Robert de Bosco holds. **1393** trans in Cal. Close 15.145: And a messuage and one ferling of land in Treuynek co. Cornwall. **1411** trans in ibid 20.244: One ferling of land in Denepriour. **1599** Richmond Appendix 2.11: A Ferling of Land is less than a Hide, a Caruc, a Yard-land, and is no more than an Oxgang. **1695** Kennett Glossary sv furendellus: Fardella, Ferlingus. A fardingel, farundel, or ferling of land, i.e. the fourth part of an acre.

ferlingata. FERLINGATE

ferlingate—3-7 L ferlingata; ? ferlingate (Skinner), L ferlingatum (Gras

2) [ferling + gate]. Equivalent to ferling.—**1200** Cur. Reg. 8.257: Alanus de Bocland´ petit versus priorem de Plinton iij. ferlingatas terre et dimidiam in Bocl´ et dimidiam ferlingatam terre in Hoo ut jus suum. **1220** ibid 2.226: De placito tercie partis xij. ferlingatarum terre cum pertinentiis. c**1310** Nicholson 81: Decem acræ faciunt ferlingatam; quatuor ferlingatæ faciunt virgatam. **1409** Gray 362: Unum toftum et unam ferlingatam terre. **1599** Richmond Appendix 2.12: Una Virgata ex quatuor Firlingatis, & una Firlingata ex decem Acris terræ. **1664** Spelman 8: Decem Acræ faciunt ferlingatam, quatuor ferlingatæ faciunt virgatam.

ferlingatum. FERLINGATE

ferlingus. FERLING

ferlong. FURLONG

ferlot. FIRLOT

ferrekyn. FIRKIN

ferrundell. FARTHINGDALE

ferthekyn. FIRKIN

ferthelett. FIRLOT

ferthinglond, ferthinland, ferthinlond. FARTHINGLAND

fertleitt, fertlett. FIRLOT

ferundel, ferundell. FARTHINGDALE

fesse [perh ME fesse fr MF fesse, faisse fr L fascia, band]. A m-q for hay, originally referring to a cord used to bind hay, smaller than a TRUSS of 56 lb (25.401 kg) but larger than a bottle of 7 lb (3.175

kg).—**1327** trans in Memorials 167: It was agreed that hay belonging to foreigners, coming to the said city [London] by land or by water, should in future not be sold in the same city by boteles, but only wholesale by shiploads...as also, by cartloads, and fesses for horses. See BOTTLE[2]

fet. FATT

ffagott. FAGGOT

ffardell. FARDEL[1]

ffatham. FATHOM

fferdalle. FARDEL[2]

fflaggon, fflagon. FLAGON

ffocher, ffodder. FOTHER

ffoot, ffoott. FOOT

ffother. FOTHER

ffotmal, ffotmel, ffotmellus. FORMAL

ffyrkyn. FIRKIN

fhote. FOOT

fidder. FOTHER

finger [ME finger fr OE finger; see WNID3]. A m-l for cloth which originally was a unit of body measurement reckoned as approximately the length of the middle finger and equal to 2 NAILS or 1/2 SPAN. Based on the ft of 12 inches, it was generally expressed (c1500) as 4 1/2 inches (1.143 dm) (Nicholson 58).

firdekyn, firikin, firken. FIRKIN

firkin—5 ferdekyn (OED), ferdkyn (Memorials), ferken (OED), ferthekyn, ferthkyn (Coopers), firdekyn; 5-6 ffyrkyn, firkyn; 6 ferrekyn (OED), firken (OED), firkine, fyrken, fyrkin, fyrkyn, fyrkyne, fyrkynge; 6-7 firkyne; 6-8 ferkin; 6-9 firkin; 7 ferekin, ferikin, ferkyne, firking (OED); 9 firikin (OED) [ME ferdkyn fr (assumed) MDu veerdelkijn, vierdelkijn, dim of veerdel, vierdel, fourth, fourth part]. A m-c for ale, beer, butter, fish, meat, and soap. Occasionally it was abbreviated fir.

The ale firkin contained 8 gal (c3.70 dkl) and was equal to 1/2 ale KILDERKIN or 1/4 ale bbl. The Irish ale firkin (c1800) contained 2176.0 cu inches (3.566 dkl) or 10 Irish gal of 217.6 cu inches each (Edinburgh XII.572).—**1517** Hall 49: And viii galons to the ale ffyrkyn. **1566** Recorde K iiij: Nowe aboue a Gallon the next measure is a Fyrkin: then a Tertian, a Kilderkin.... And by those measures are sold...Ale, Bere, Wine & Oyle.... Of Ale the fyrken containeth 8 gallons. **1587** Stat. 595: And euerie ferkin for ale viij. gallons of the kings standard gallon. **1600** Hill 67: 8. Gallons...1. Firkin of ale. c**1600** Brit. Mus. 16:70v: Now above a gallone, the next mesure is a ferkyne a Tercian, a kylderkyne or halfe barrelle.... Of Ale the Firkyne contayneth .8. gallons. **1635** Dalton 148: Ale, the measure thereof, is...Firkin, 8. **1665** Assize 9: It was also ordained in Anno 23 Hen. 8 that the Ale-Firkin should hold and contain eight gallons. **1724** Coles Appendix: 8 Gallons make 1 Firkin of Ale. **1816** Kelly 87: 8 Gallons...1 Firkin of Ale...36,9669 [1].

The beer firkin contained 9 gal (c4.16 dkl) and was equal to 1/2 beer kilderkin or 1/4 beer bbl. Since the establishment of the Imperial system the firkin of beer has been reckoned at 9 gal (4.091 dkl) everywhere in the United Kingdom except in Ireland, 8 gal (3.637 dkl).—c**1517** Hall 50: That there shuld´ be no lesse assyse for bere than...ix galons to the fyrkyn´ of the kynges standard´. **1539** Dur. House 338: Of Master Hylton: 1 barell syngyll beyr and 1 fyrkynge doubyll beayre, 4 s. 4 d. **1587** Stat. 595: And euerie ferkin for beere nine gallons of the kings standard gallon. c**1590** Hall 22: Beare measures: The firkyn conteynyth 9 galons: the kilderkyn...18 gallons: the barill contenith 36 gallons. **1595** Powell D: The beere Firkin shall holde and containe ix. gallons. **1635** Dalton 148: Beere, the measure thereof...Firkin, 9. **1665** Assize 9: The Beer-Firkin shall hold and contain nine gallons. **1732** Coles Appendix: 9 Gallons make 1 Firkin of Beer. **1816** Kelly 87: 9 Gallons...1 Firkin of Beer...41,5872 [l]. **1883** Simmonds sv: Firkin...a cask containing nominally 9 gallons of beer or 8 of ale. **1956** Economist 54: Firkin = 9 gallons (8 gallons in Ireland). **1966** O´Keefe 671: 1 firkin = 9 gal. = 40.914 l.

The firkin of butter or soap conformed to the 8 gal capacity (c3.70 dkl) of the ale firkin, but equally important was the weight of the cask: generally 6 1/2 lb (2.948 kg) before 1662 and generally 8 lb (3.629 kg) afterwards.—c**1500** Mer. Adven. 56: Anie firkine or firkins of sope. **1566** Recorde K iiij: A Fyrken...weighe emptye 6 1/2 poundes. c**1590** Hall 24: The fyrkyne wayeth of butter, caske and all, 64 poundes haberdepoise, whereof the caske wayeth 6 pounde 1/2. **1635** Dalton 149:

The empty firkin not to weigh above 6 pound and an halfe; and to containe 8 gallons. **1665** Assize 4: And every Sope-Barrel...32 gallons...and every Firkin empty shall weigh vi pound and a half...and shall hold and contain eight gallons. **1673** Stat. Charles 159: The Firkin...ought to weigh Sixty and four pounds, viz. Fifty and Six pounds of good and Merchantable Butter Neat, and the Cask Eight pounds. **1696** Cocker 112: 8 gallons is a firkin of Ale, Soap or Herrings. **1756** Rolt sv: The firkins of soap and butter are on the footing of the firkin of ale; that is, a gallon per firkin less than that of beer. **1850** Alexander 35: Firkin; for butter...weight...56.—pounds. **1880** Courtney 154: A firkin of butter was formerly 56 lbs.

The firkin of salmon, and occasionally of eels, contained 10 1/2 gal (c3.97 dkl); of herrings and eels, 7 1/2 or 8 gal (c2.84 or c3.03 dkl); and of most other fish, 8 gal ale-firkin capacity (c3.70 dkl).—**1423** Rot. Parl. 4.256: Kynderkyns, Tercianes, and firdekyns of Heryng. **1443** Brokage II.98: viii ferthekyns allecii. **1478** Stonor II.46: ij herynges barell and a ffyrkyn with salt. **1482** Rot. Parl. 6.221: That every...half Barell, ordeyned for Samon, shuld conteygne...XXI Galons.... Also it hath ben used, that every Barell for Elys, shuld hold and conteigne XLII Galons, the half Barell, and Firkyn, after the same rate. **c1590** Hall 23: The barill [of salmon] is 42 gallons; the kilderkin...21 gallons; the firkyne 10 galons 1/2. **1600** Hill 67: 8. Gallons...1. Firkin of...herring.... 10 1/2. Gallons...1. Firkin of salmon or Eeles. **1612** Halyburton 305: Sturgeoun the ferekin; ibid

330: Sturgeoun the ferikin. **1717** Dict. Rus. sv: Firkin, a sort of Liquid measure, the fourth part of a Barrel, containing eight Gallons of Ale, Soap, or Herrings...and 10 1/2 Gallons of Salmon or Eels.

The firkin of beef or pork (c1950) was 100 lb (45.359 kg); in earlier periods meat firkins did not have standard weights (Economist 53).

firkine, firking, firkyn, firkyne. FIRKIN

firlet. FIRLOT

firlot—5 ferlot, firlote; 5-9 firlot; 6 feirtlett (OED), ferthelett (OED), fertleitt (OED), fertlett (OED), L firlota, firlotte, furlet (OED), fyrlot; 7 firlet, firlott, furlat; 7-8 furlot (OED); 8 farlet (OED); 9 fearlot (Hunter) [ME ferlot fr ON fjōrthi, fourth, + hlutr, hlotr, lot, share, part]. A m-c for grain in Scotland.

The Edinburgh firlot was the standard (c1600-1800) for wheat, peas, beans, rye, and white salt, 21 1/4 Scots pt (3.612 dkl) of 103.404 cu inches each or 2197.335 cu inches in all and equal to 1.021817 Winchester bu, and the Linlithgow firlot, 31 Scots pt (5.270 dkl) or 3205.524 cu inches and equal to 1.490650 Winchester bu was the standard (c1600-1800) for barley, oats, and malt (Swinton 32 and Second Rep. 16). These firlots were computed as 1/4 BOLL, and they equaled 4 pk, or 16 LIPPIES or FORPITS.—**1425** Acts Scotland 2.12: Ande that firlote not to be maid eftir the first mesoure na eftir the mesoure now vsit.... It sal contene in breid evin ourethort xvj Inchys vndir & abone within the burdes & in deipness vj Inche the thikness of bath the burdes sal contene ane Inche and a halfe.... The firlote xlj punde. **1457** Ibid

50: And of thir saide mesures that is to say pynt and ferlot; ibid 51: viij oxin sall sawe at the lest ilk yer ane firlot of quheit. **1525** Jacobus 50: j firlota salis; ibid 102: Summa ix celdre iij bolle j firlota auenarum. **1555** Acts Scotland 2.496: That all mesouris baith pynt quart fyrlot peck elnwand stane and pund to be of ane quantitie to by with. **1597** Skene 1.11: Foure firlottes to conteine a boll. **1609** Acts Scotland 4.409: Peck or firlott. **1618** Ibid 586: And the same measure and firlot being fund agreable with the said Jedge...the saids Commissioners caused præsentlie fill the same with water which being full...they fand that the same conteined Twentie ane pincts and ane mutchkin of just Sterline Jug.... WHICH Firlot...the saids Commissioners Haue...Ordeined...For metting of Wheat/Rye/Beines/Peas/Meal/Whyt Salt; ibid 587: They haue found it expedient to cause make ane particular Measure or Firlot...for metting of Mault/Beare/and Aites.... They find the same to conteine Thrittie one Pincts...of the just Sterline Jugge. **1624** Huntar 1: Wee haue our drie Metts, as the Peck, the Firlet, and the Bow; ibid 5: The Firlet (for metting of Beere, Malt, or Oates, which were called heaped cornes,) conteines 31 pintes of water. **1696** Acts Scotland 10.77: All Malt that is sold and bought within this Kingdom shall be delivered with the Heap furlat according to the bear furlat of the place where it is delivered. **1761** Thomson vii: The wheat-firlot shall contain 21 1/4 of the Stirling jug; and...the bear-firlot shall contain 31 of the same; ibid viii: The wheat-firlot contains 2197 34/100 [cu inches]. The bear-firlot contains 3205 54/100 [cu inches]. **1813** Cooke 103: Linlithgow Bear Measure.... 1 Firlot...3205.524 [cu inches].... Linlithgow Wheat Measure. 1 Firlot...2197.335 [cu inches]. **1816** Kelly 93: 4 Pecks...1 Firlot. 4 Firlots...1 Boll.... The standard firlot for measuring wheat, pease, beans, rye, and white salt, contains 2197.333 English cubic inches; ibid 94: The standard firlot for measuring barley, oats, and malt, contains 3205.524 English cubic inches. **1820** J. Sheppard 91: STANDARD SCOTCH DRY MEASURES FOR WHEAT, BEANS, PEAS, RYE, SALT, AND GRASS SEEDS...4 pecks...1 firlot or 2197.335 [English cu inches].

There were many local variations (c1600-1800), however, on these standard Scots firlots (Swinton 53-130, Kelly 96-112, Edinburgh XII.571, and Second Rep. 16-17). North—Nairnshire: wheat, peas, beans, rye, ryegrass-seed, oatmeal, and barleymeal, 2680.221 cu inches (4.393 dkl); barley and oats, 3573.632 cu inches (5.857 dkl). Sutherlandshire: peas, rye, and beans, 2585.1 cu inches (4.237 dkl); oats, barley, and malt, 3546.719 cu inches (5.813 dkl). Northwest—Inverness: wheat, peas, beans, rye, ryegrass-seed, and meal, 2514.967 cu inches (4.122 dkl); oats, barley, and malt, 3519.225 cu inches (5.768 dkl). Ross and Cromarty: wheat, rye, peas, beans, and lime, 2481.696 cu inches (4.067 dkl); oats, barley, and malt, 3308.928 cu inches (5.423 dkl). Northeast—Aberdeenshire: wheat, rye, peas, beans, meal, and seeds, 2688.504 cu inches (4.406 dkl); oats, barley, and malt, 3515.736 cu inches (5.762 dkl). Banffshire: wheat, beans, peas, rye, and white salt, 2316.248 cu inches (3.696 dkl); oats, barley, and malt, 3369.088 cu inches (5.522 dkl). Caithness: oats and barley, 3405.87 cu inches (5.582 dkl). Moray (Elgin): wheat, rye, peas, and beans, 2346.006 cu inches (3.845 dkl); barley and oats, 3374.032 cu inches (5.530 dkl). Central—Perthshire: wheat, peas, rye, and beans, 2262.96 cu inches (3.709 dkl); oats, barley, and malt, 3339.0 cu inches (5.473 dkl). Stirlingshire: wheat, peas, beans, and rye, 2378.292 cu inches (3.898 dkl); oats, barley, and malt, 3438.183 cu inches (5.635 dkl). West central—Dumbartonshire: wheat, peas, beans, and meal, 2562.75 cu inches (4.200 dkl); oats, barley, and malt, 3417.0 cu inches (5.600 dkl). West—Argyllshire: wheat, rye, beans, and peas, 2554.402 cu inches (4.187 dkl); oats, barley, and malt, 3438.183 cu inches (5.635 dkl). East—Angus: wheat, peas, and beans, 2274.888 cu inches (3.728 dkl); oats, barley, and malt, 3321.[illegible]3 cu inches (5.444 dkl)—both firlots average of Montrose, Forfar, Brechin, Dundee, and Arbroath firlots. Fifeshire: wheat, peas, and beans, 2274.888 cu inches (3.728 dkl); oats, barley, and malt, 3308.928 cu inches (5.423 dkl). Kincardineshire: wheat, rye, and peas, 2481.696 cu inches (4.066 dkl); oats and barley, 3412.332 cu inches (5.593 dkl). Kinrossshire: wheat, peas, and beans, 2255.5 cu inches (3.697 dkl); oats, barley, and malt, 3302.465 cu inches (5.413 dkl). South—Lanarkshire, Glasgow and Lower Ward: wheat, 2314.19 cu inches (3.793 dkl); peas and beans, 3271.2 cu inches (5.361 dkl); oats and barley, 3339.4 cu inches (5.473 dkl). Peeblesshire: wheat, peas, beans, and rye, 2354.4 cu inches (3.859

dkl); oats, barley, and malt, 3348.48 cu inches (5.488 dkl). Southwest—Ayrshire: wheat, rye, and beans, 2457.6 cu inches (4.028 dkl) in Kyle and Carrick, 2035.756 cu inches (3.337 dkl) in Cunningham; oats, barley, and malt, 3621.76 cu inches (5.936 dkl) and 4032.036 cu inches (6.608 dkl) in Kyle and Carrick, 2035.756 cu inches (3.337 dkl) in Cunningham. Buteshire and Arran: wheat, peas, and beans, 2878.078 cu inches (4.719 dkl); oats, barley, and malt, 4317.117 cu inches (7.076 dkl). Renfrewshire: wheat, the Linlithgow standard; beans, peas, and vetches, 2404.143 cu inches (3.940 dkl); oats and barley, 3405.869 cu inches (5.582 dkl). Southeast—Berwickshire: all grain, 3360.63 cu inches (5.508 dkl). East Lothian: wheat, peas, and beans, 2261.962 cu inches (3.707 dkl); oats, barley, and malt, 3302.464 cu inches (5.413 dkl). Midlothian: wheat, peas, and beans, 2236.111 cu inches (3.665 dkl); oats, barley, and malt, 3257.226 cu inches (5.339 dkl). Roxburghshire: wheat, peas, and beans, 2274.888 cu inches (3.728 dkl); oats, barley, and malt, 3412.332 cu inches (5.593 dkl). Selkirkshire: wheat, rye, beans, and peas, 2281.350 cu inches (3.739 dkl); oats, barley, and malt, 3231.370 cu inches (5.296 dkl).

firlota, firlote, firlott, firlotte. FIRLOT

flaccon, flaccone, flaccoun, flackoun, flagan, flaggon. FLAGON

flagon—5 flagan (OED), flakon (OED); 5-9 flaggon; 6 flaccon (OED), flaccone (OED), flaccoun (OED), flackoun (OED), flagone (OED); 6-9 flagon; ? fflaggon, fflagon [ME flakon fr MF flacon, drinking vessel, small keg, fr LL flasco, flasconis, bottle]. A m-c for liquids, generally containing 1 gal (c3.78 l). Since the establishment of the Imperial system the flagon of beer has been reckoned at 1 qt (1.1365 l).—**1500** Relation 95: Spent, 259 flaggons...of which, to the Lord, half a flaggon. **1604** Cawdrey 58: Flagon, great wine cup, or bottell. **c1634** Hall 52: Item hee hath found out diuerse kindes of falce Measures made by Turnors and by Crooked Lane men, by Porters and by diuerse others, that is to saie 1. The Winchester Quart measure. 2. The Wine quart measure. 3. The fflaggon, Crooked Lane measure. 4. The Juggs. 5. Black pottes. 6. Woodden Canns. 7. Bottles wherein beere and alle are mixed togeather, which is not only wastfull, but alsoe very unholsome for mens´ bodies that drink it.... 9. Siluer fflagons, and many other unlawfull measures...which...are neither marked nor sealled. **1664** Gouldman sv: A flagon. Oenophorum. **1745** Fleetwood 81: I have observ´d before, that Lagena (a Flaggon) holds 4 Quarts.... Now an 132 Flaggons must, at that rate make 528 Quarts. **1756** Rolt sv flaggon: Or Flagon. A large vessel, to contain wine. **1789** Topham xxvi: Flaggons of four quarts. **1895** Donisthorpe 209: FLASK: of Pyrmont water, 3 pints, wine measure. **1956** Economist 54: Flagon = 1 quart.

flagone, flakon. FLAGON

flasce. FLASK

flask—1 flasce (OED), flaxe (OED); 6-7 flaske (OED); 6-9 flask; 7 flasque (OED) [MF flasque, powder flask, prob modif of OSp frasco, powder flask, flask for liquids; akin to OE flasce, flaxe, bottle; see WNID3]. A m-c for liquids, generally (c1800) containing 3 pt (c1.42 l) (Second Rep.

17).

flaske, flasque. FLASK

flat [prob fr adj flat fr ME flat fr ON flatr; akin to OS flat, shallow]. A m-c, a wide and shallow vessel, for vegetables (c1895) in Buckinghamshire equivalent to a bu (c36.37 l) (Wagstaff 37).

flaxe. FLASK

fleche, flicce, flicch, flicche, flick, flickke, flik. FLITCH

flitch—1 flicce (OED); 5 flickke (OED), flykke (OED); 5-6 flicche, flik (OED), flyk (OED), flyke (OED); 6 fleche, flicch (OED), flycke (OED); 6-7 flick (OED), flytche (OED); 6-9 flitch [ME flicche fr OE flicce; akin to MLG vlicke, flitch]. A m-q for cured hog meat, namely, a side.—**1509** Gras 1.581: Pro xl flicches bakonis. **1545** Rates 1.49: Bacon the fleche. c**1550** Welsh 31: 2 flitches bacon. **1590** Rates 2.41: Bacon the flitch. **1721** King 304: 6 Flitches of Bacon. **1883** Simmonds sv: Flitch, a side of bacon.

flock—6 flocke, floke; 6-? flock [ME flock fr OE flocc; akin to MLG vlocke, crowd, herd of sheep, ON flokkr, crowd, band, troop]. A m-q for any item or sets of items, generally consisting of 40 in number.—**1545** Rates 1.6: Baste rope the floke conteynynge .xl peces.... Boxes the floke conteyning .xl; ibid 12: Cannes of wode the flocke; ibid 42: Trayes the flocke conteynyng xl.... Tables the flocke; ibid 43: Tankerdes the flocke.

flocke, floke. FLOCK

flycke, flyk, flyke, flykke, flytche. FLITCH

focher, fodar, fodder, foder. FOTHER

fodmell. FOTMAL

fodra, fodre, fodyr, folder. FOTHER

fontinell. FOTMAL

fooder. FOTHER

foot—1-7 L pes; 3 fhote (OED), fott (OED); 3-4 fot (OED); 3-6 fote, fut (OED); 3-9 foot; 4-7 fute; 5 fowte (OED), foyte (OED); 5-6 fotte; 5-7 foote; 6 ffoott, fuit (OED); 6-7 foott; 7 ffoot [ME fot, foot fr OE fōt; see WNID3]. A m-l of 12 inches (0.305 m) or 1/3 yd which originated as a unit of body measurement equal to 4 PALMS. The Scots ft was equal to 12.0649 English inches, and the early medieval Welsh foot was reckoned at 9 English inches (Donisthorpe 209).—c**1075** Hall 2: Quarum haec sunt nomina: digitus, uncia, palmus...pes, cubitus. c**1100** Ibid 6: Quattuor palmi faciunt pedem. c**1300** Ibid 7: 3 palme et 3 grana ordei faciunt pedem.... Pes et dimidius faciunt cubitum usualem. **1395** York Mem. 1.142: Notandum quod tria grana ordei sicca et rotunda faciunt pollicem; et xij pollices faciunt pedem. c**1400** Hall 6: Pes cum duabus terciis faciunt cubitum; ibid 7: Et xij pollices faciunt pedem. **1425** Acts Scotland 2.12: The quantite of the burgh of xx fute the leddir. **1440** Palladius 86: iii foote or iiii in heght. c**1461** Hall 14: And xii ynchis make a fote. **1474** Cov. Leet 396: And hitt was ordeyned at the same tyme that iij barley-Cornes take out of the middes of the Ere makith a Inche; and xij Inches makith a foote. c**1475** Nicholson 77: And xij enchis makyth a foote. **1534** Fitzherbert 11: The share is a pece of yren...a fote longe. **1537** Benese 2: A foote conteyneth .xii. ynches in lengthe. c**1560** Mon. Fran. 183: xii. foote brode and iiii. fotte of hyghte. c**1590** Hall 27: Dymension longitudes of the ynche, ffoott, yard.... The foott in length is 12 inches. c**1600** Brit. Mus. 31.212: 84 ffoot of Bord. **1603** Henllys 137: Yet doeth yt agree in the ynche, foote, and yard. **1615** Collect. Stat. 464: xij. ynches make a foot. **1616** Hopton 165: Three barley cornes make an Inch, 12 Inches a foote, 3 foote a yard. **1635** Dalton 150: Twelve inches make a foot. **1647** Digges 1: Twelve inches, a Foote. **1663** Acts Scotland 7.488: No workman nor other person shall make vse of any other foot measure, then such as consists of tuelve of these inches whairof the ell containes thirty seven. **1678** Du Cange sv alna: Pes Regius est 12. pollicum. **1682** Hall 28: A Foote is 12 inches. **1688** Bernardi 202: Pes Anglicus...12 Unciæ aut pollices. **1708** Chamberlayne 207: 3 Hand a Foot. **1717** Dict. Rus. sv: Foot, a long Measure of 12 Inches. **1761** Thomson viii: Scots foot is 12 1/15 [English inches]. **1779** Swinton 24: Scotch Foot. = 12.064864 [English inches]. **1832** Edinburgh XII.569: 12 Inches = 1 Foot = .3048 [m]. **1850** Alexander 35: Foot...12.—inches. **1916** Stratton 24: 1 foot = .3048 meter. **1951** Trade 27: Foot = 1/3 yard.

foote. FOOT

foother. FOTHER

foott. FOOT

forelang, forelange, forelong, forelonge. FURLONG

forland, forlandus. FARTHINGLAND

forlang, forlange, forlong, forlonge. FURLONG

formel, formell, formella. FOTMAL

forpat, forpet. FORPIT

forpit—8-9 forpet, forpit; 9 forpat [for four part or fourth part]. The fourth part of a pk. Equivalent to LIPPY.—**1779** Swinton 32: Lippie or Forpet. **1819** Cyclopædia sv weights: At Hexham, with Rye and Peas. 4 Quarts make 1 Forpit...4 Forpits...1 Peck.... For Oats and Barley. 4 Quarts make 1 Forpit...5 Forpits...1 Peck.... At Alnwick. 3 Quarts make 1 Forpit...4 Forpits...1 Peck.... At Wooler. 4 Quarts make 1 Forpit...3 Forpits...1 Peck. **1820** Second Rep. 17: Forpet or Forpit...Scotland: the fourth part of a peck, otherwise called a lippie. **1883** Simmonds sv forpet: The fourth part of a Scotch peck; 64 lippies or forpets make one boll. **1883** McConnell 16: 1 lippie or forpat = 1/2 gal. 4 lippies or forpats = 1 peck.

forthingdole. FARTHINGDALE

fot, fote. FOOT

fother—4 fothir, fozer; 4-6 fothyr (OED), futher, futhir (OED); 4-9 fother; 5 fothre, fouthre (Southampton 1), fuddyr; 5-6 fodre (OED), fodyr (OED), fouther; 5-7 foder, fudder, fuddir (OED), fuder (OED), fudyr (OED); 5-9 fodder; 6 fodar (OED), folder, foulder, fowther (OED); 6-7 fidder; 7 ffodder, fooder (OED), foother; 8 ffother; 9 It fodra; ? ffocher (Prior), focher (Prior) [ME fother fr OE fōther, a cartload]. A wt for lead generally of 2100 lb (952.539 kg), used interchangeably with and eventually (c1350) superseding CHARGE[2]. It was reckoned in four different ways: 30 FOTMALS of 70 lb each (31.751 kg), or 168 STONE of 12.5 lb each (5.670 kg), or 175 stone of 12 lb each (5.443 kg), or 12 WEYS, each wey of 175 lb (79.378 kg). Occasionally fothers of 1950 lb (884.061 kg), 2000 lb (907.194 kg), 2184 lb (990.640 kg), 2240 lb (1016.05 kg), 2250 lb (1020.593 kg), 2340 lb (1061.401 kg), 2352 lb (1066.844 kg), 2408 lb (1092.245 kg), 2464 lb (1117.646 kg), 2520 lb (1143.047 kg), and 2820 lb (1279.124 kg) were also used.—**1350** trans in Memorials 265: Bought one fozer of lead. **1391** Henry Derby 59: Pro j fothir. c**1425** Account 3.518: Et in lucratione dictarum iiii petrae minerae ferri, duodena continente iii fotheres, videlicet de puteis de Rayley xx duodenae i fother. c**1432** Finchale ccccxxx: Lead £ 5 6 s. 8 d. per "fothre." **1435** Southampton 1.2: Fouther de plumb. **1443** Brokage II.257: Cum vi fother plumbi in v waynes. c**1461** Hall 13: Also lede ys sold by the fudder, xix c[wt] and dim make a fuddyr, after v [X] xx [+] xii to the C. c**1475** Gras 1.192: Of a foder lead. **1545** Rates 1.46: Leade the folder. **1555** York Mer. 156: A fother of leade, taken in at the crayne, twelve pence. **1562** Ibid 168: Lead, the futher, taken in at the crayne, xiiij d.; Lead, the futher, beyng lightened, xviij d. c**1580** Hostmen 5: To pay 12 d. for every Fother. c**1590** Hall 23: The fodder at the King´s Beame 19 hundred [weight] 1/2, and every 100 is 120 poundes waight, haberdepoyse, contenith 2280 [sic]. **1590** Rates 2.42: Lead uncast the foulder containing xix [X] c. [+] di. euery c. waying v [X] xx [+] xij pound. **1603** Hostmen 39: Shall for every foother so ledd and teamed att his Stayth beinge justlie proued paye xii d. c**1610** Lingelbach 67: Four skore ffodder of lead. **1612** Halyburton 338: Lead the fidder contening ij [X] m wegth. **1613** Tap 1.62: A Fodder containes...Pounds. 2184. **1615** Collect. Stat. 464-65: The load of lead doth consist of 30. formels, and euerie formell containeth 6. stone, except two pound: and euerie stone doth consist of 12 pound. **1616** Hopton 163: Lead by the pound, hundred, and fodder. **1628** Hunt C: 2184 Lb. in a Fodder of Leade. **1635** Dalton 149: Lead, the content of the pound, the stone, and the load. **1665** Assize 5: There is also a load of Lead, which consisteth of thirty Formels, and every Formel containeth six Stone wanting two pounds. **1677** Roberts 296: The Foder maketh accordingly 19 1/2 hundred of 112 l. per cent. **1682** Hall 30: A Fodder conteynes 19 hundred [weight] and an halfe; and 2184 pounds. **1704** Mer. Adven. 243: Ffor every ffother of lead. **1708** Chamberlayne 207: The tun is Twenty Hundred Weight of every thing but Lead, of which there is but Nineteen Hundred [weight] and an half to the Tun or Fodder. **1717** Dict. Rus. sv fodder: Or Fother of Lead, a Weight containing 8 Pigs, and every Pig 23 1/2 Stone, which is about a Tun or a common Wain or Cart-load: In the Book of Rates, a Fodder of Lead is said to be 2000 pound Weight; at the Mines ´tis 2200 and an half; and among the Plummers at London, 1900 and an half.' **1794** Martin 24: The weight of a fodder is different at different places.... Derby...22 1/2 [Cwt] of 112 [lb].... Gainsbro´ and Stockwith 21 1/2 [Cwt] of 112 [lb].... Hull...19 1/2 [Cwt] of 120 [lb].... London...19 1/2 [Cwt] of

112 [lb]. **1803** Triulzi 91: Il Piombo si vende ad un peso dello Fodra, ch'è in detto loco Cant. 19., e mezzo di libre 112. per Cantaro. **1819** Cyclopædia sv weight: The fother of lead is generally 19 1/2 cwt. at Newcastle, 21 cwt. at Stockton, 22 cwt. **1820** Second Rep. 17: Fodder or Fother...of lead, a ton = 20 cwt.... With miners, 22 1/2 cwt; with plumbers, 19 1/2.... Derbyshire: Mill fodder, at the smelting houses, 2820 lbs; when shipped at Stockwith-on-Trent, 2408.... Hull, 2340 lbs.... London, 2184 = 19 1/2 cwt. **1834** Pasley 115: 1 Fodder of Lead with Plumbers...2,184 [lb].... 1 Fodder or Ton of Lead...2,240 [lb].... 1 Fodder of Pig Lead, in Northumberland (21 cwt.)...2,352 [lb].... 1 Fodder of Derbyshire Lead, shipped at Stockton on Trent (21 1/2 cwt.)...2,408 [lb]...at Newcastle, sometimes 22 cwt...2,464 [lb]...with Miners (22 1/2 cwt.)...2,520 [lb]...1 Mill Fodder of Lead at the smelting houses in Derbyshire...2,820 [lb]. **1878** Wedgwood 274: Fother...properly a carriage load, but now only used for a certain weight of lead. See CARTLOAD; HUNDRED; LOAD; TON

fothir, fothre, fothyr. FOTHER

fotinel, fotinell, fotinellum, fotmæl. FOTMAL

fotmal—3 L fotinellum (error in manuscript often made), L fotmella, L fotmellum, L fotmellus (Prior); 3-8 fodmell; 3-9 fotmal; 4 ffotmall; 4-7 fotmel; 6-7 formel, formell; 7 L ffotmellus, L formella; 8 fotinell; ? ffotmel (Prior), fontinell (Thor. Rogers 1), fotinel (Salzman 1), fotmæl (Thor. Rogers 1), L fotmelus (Prior) [OE fotmæl, foot measure; see OED]. A wt for lead of 70 lb (31.751 kg) equal to 1/30 FOTHER of 2100 lb

(952.539 kg).—**1230** Close 1.348: Godricus de Novo Castello lator presentium, regi liberavit apud Portesmue vij [X] xx et j fotmella plumbi. c**1253** Hall 11: Fet asauer ke la charge de plum est de xxx fotmaux, et checun fotmal est de vi pers, ii lib. meyns; checun pere est de xii lib...la sume de lib. en le fotmal, lxx lib. c**1269** Report 1.420: Carrata minor continet xxiv fodmelles, unum fodmell continet LXX libras. c**1272** Hall 9: Charrus plumbi debet ponderare ex constat ex triginta fotmellis, et quodlibet fotmellum continet [vi] petras, exceptis duabus libris...petra constat ex duodecim libris. **1290** Report 1.419: Item charrus plumbi consistit ex triginta fotinellis, et quodlibet fotinellum continet sex petras minus duabus libris. c**1300** Brit. Mus. 1.148: La charre plumbi constat ex. xxx. ffotmals. c**1300** Brit. Mus. 13.29: Et quodlibet fotmal continet vj petras duabus libris minus. c**1300** Brit. Mus. 5.150v: Et quodlibet ffotmall continet sex petras duas libris minus. c**1375** Prior 91: Sex Waxpunde makiet. j. ledpound. xij.. ledpunde. j fotmel. **1495** Brit. Mus. 28.156: Libræ in le fotmal...lxx lb. **1595** Powell C2: There is also a lode of Lead, which consisteth of thirtie formels, and euery formell conteineth sixe stone wanting two pounds. c**1600** Brit. Mus. 32.182: Et ffotmellus. **1615** Collect. Stat. 465: The load of lead doth consist of 30 formels, and euerie formell containeth 6 stone, except two pound. **1664** Spelman 245: Formella...Ponderis genus apud Anglos, cujus rationem habes in Stat. de Ponderibus...Le Charre (hoc est, carrionus) de plumbo constat ex 30 formellis, & quælibet formella continet 6 petras exceptis duabus libris, & quælibet petra constat ex 12 libris. **1665** Assize 5: There is also a load of Lead, which consisteth of thirty Formels, and every Formel containeth six Stone wanting two pounds. **1678** Du Cange sv formella: Ponderis genus apud Anglos. **1758** Report 1.420: Seventy Pounds make a Fotinell or Fodmell. **1805** Macpherson 1.471: 5 stones 10 pounds...1 fotmal. **1883** Simminds sv: Fotmal, a commercial term for 70 lbs. of lead.

fotmel, fotmella, fotmellum, fotmellus, fotmelus. FOTMAL

fott, fotte. FOOT

foulder. FOTHER

fourlonge. FURLONG

foust. FUST

fouther, fouthre. FOTHER

fow [perh a Sc dial var of FULL]. A m-c for grain (c1600-1800) in the Scottish shire of Ayrshire at Cunningham, synonymous with the FIRLOT: wheat, rye, peas, and beans, 2035.756 cu inches (3.337 dkl) and equal to 1/5 BOLL of 10,178.781 cu inches (1.668 hl); oats, barley, and malt, 2035.756 cu inches (3.337 dkl) and equal to 1/8 boll of 16,286.048 cu inches (2.670 hl) (Swinton 59-60).

fowte. FOOT

fowther. FOTHER

foyte. FOOT

fozer. FOTHER

fraell, fraelle, fraellus, fraiel. FRAIL

frail—4 L fraellus; 4-5 fraell (OED), fraelle (OED), fraiel (OED), frayel (OED); 4-7 frayle; 4-9 frail; 5-6 frale (OED); 6 frayl; 6-7 fraile; 7 freal (OED) [ME forms fr MF fraiel, freel, frael; see WNID3]. A m-c for fruit and small manufactured items. It was a basket, made of rushes, that could generally hold between 30 and 75 lb (13.608 to 34.019 kg) of merchandise.—**1304** Gras 1.169: Pro vi fraellis ficorum. **1394** trans in Cal. Close 15.324: They laded in a ship called ´la Petre´ of Caen, Peter Robert of Caen master, 850 barrels...of figs and raisins, two frails thereof making always a barrel. **1507** Gras 1.697: Fygges the sortte that ys to saye iii frayles for the sortte. c**1550** Welsh 98: 1 frayle spurs...1 frayl horseshoes. **1628** Hunt C: For the Bagge Barrell, Chest, Fraile, Vessell or Caske. **1695** Kennett Glossary sv frayle: FRAYLE of figs. A basket in which figs are brought from Spain and other parts. **1701** Hatton 3.226: Frail...of Raisins about 75 pounds. **1756** Rolt sv: FRAIL...denotes a certain quantity of raisins, being about 75 lb. **1829** Palethorpe sv: FRAIL, a basket made of rushes, or the like, in which are packed up figs, raisins, prunes, &c. It signifies also a certain quantity of raisins weighing 75 lbs. avoirdupois. **1840** Waterston 147: Figs (Faro), frail, lbs. 32. **1883** Simmonds sv: Frail, a package or basket made of rushes, in which dried fruit is occasionally imported.

fraile, frale, frayel, frayl, frayle, freal. FRAIL

frundel, frundele. FURENDAL

fudder, fuddir, fuddyr, fuder, fudyr. FOTHER

fuit. FOOT

full [prob fr adj full fr ME ful, full fr OE full; employed here in the sense of a heaped or ´full´ container or receptacle]. A m-c in Selkirkshire for oats, barley, and malt (c1600-1800) containing 1615.685 cu inches (2.648 dkl) and equal to 8 LIPPIES of 1 61/64 standard Scots pt each or 1/2 FIRLOT of 31 4/16 standard Scots pt or 1/10 BOLL of 156 4/16 standard Scots pt (Swinton 120).

furchendellus. FARTHINGDALE

furelang, furelange, furelonge. FURLONG

furendal—7 furendel; 7-8 frundel; 8 furendal; ? frundele (Prior) [prob fr OE fēortha, fourth, + dāl, part, portion]. A m-c in northern England generally containing 2 gal (0.881 dkl) and equal to 1/4 bu.—**1695** Kennett Glossary sv furendellus: And in the North a Furendel or Frundel of corn is two gawns or gallons, i.e. the fourth part of a bushel. **1717** Dict. Rus. sv: Furendal or Frundel of Corn, contains two Gawns or Gallons, i.e. the fourth part of a Bushel.

furendel. FURENDAL

furendellus. FARTHINGDALE

furlang, furlange. FURLONG

furlat. FIRLOT

furleng FURLONG

furlet. FIRLOT

furlingus. FERLING

furlong—1 furlang (OED); 1, 7 furlung; 2 furleng (OED); 3-5 furelang

(OED), furelange (OED), furlange (OED); 4 ferlong (OED), fourlonge (OED); 4-5 forelang (OED), forelange (OED), forlang (OED), forlange (OED), furelonge (OED); 4-6 forelonge (OED), forlong, forlonge; 4-7 furlonge; 4-9 furlong; 5 forelong; 7 L furlongus [ME furlong fr OE furlang, the length of a furrow, fr furh, furrow, + lang, long]. A m-l equal to 1/8 mi and generally containing 660 ft (2.012 hm) or 40 PERCHES of 16 1/2 ft each. The Scots furlong contained 40 FALLS or 240 ELLS and was equal to 744 English ft (226.771 m) (Swinton 24). The Irish furlong contained 840 ft (256.032 m) or 280 yd or 40 perches of 21 ft each and was equal to 1/8 Irish mi of 6720 ft. Occasionally it was abbreviated fur.—**1387** Higden IV.431: Over the thridde forlong. c**1400** Henley 8: Byen sault ke vne coture deyt estre de quarante perches de long. c**1440** Promp. Parv. 183: Furlonge. Stadium. c**1450** Higden IV.431: Halfe of a forlonge. c**1461** Hall 14: And there go viij forelonges to a myle, in Yngland. **1561** Eden xviii: viii furlonges one myle. **1616** Hopton 165: Also an English mile is 8 Furlong. **1635** Dalton 150: Fortie pole in length maketh a furlong. **1661** Hodder 32: 1 furlung be 40 poles.... 1 mile be 8 furlungs. **1664** Spelman 474: Quarentena...Stadium, Angl. a furlonge. **1665** Assize 6: Plinie Lib. 2. Cap. 23. deriveth Stadium to be a furlong. **1682** Hall 29: A Furlong is 40 pearches in length. **1688** Bernardi 202: Pes Anglicus...1/660 Stadii aut Furlongi, et 1/5280 Milliaris Anglici. **1708** Chamberlayne 207: 40 Perch make a Furlong.... 16 Foot and a half make a Perch. **1717** Dict. Rus. sv: Furlong, a Measure which in most Places contains 40 Poles or Pearches in length, being the eighth part of a Mile. **1784** Ency. meth. 138-39: La mille d'Angleterre, suivant un édit du roi Henri VII, est de 8 furlongs, 1760 yards, ou 5280 pieds (feet) de longueur. **1805** Macpherson II.203: Each furlong containing forty poles or perches, and every pole to contain sixteen feet and a half in length. **1832** Edinburgh XII.569: 40 Poles = 1 Furlong = 201.1632 [m]. **1878** Wedgwood 285: Furlong...A furrow-long, the length of a furrow. **1892** Andrews 103: The length of the acre was a furlong or furrow-length, which was as much as a man could plough without turning, and without becoming weary; this length soon became fixed at 220 yards for the common acre. **1907** Hatch 35: 1 furlong = 201.16782 metres. **1951** Trade 27: Furlong = 220 yards. See COUTURE; STADIUM

furlonge, furlongus. FURLONG

furlot. FIRLOT

furlung. FURLONG

furthendellus. FARTHINGDALE

fust—5-? fust; 6 foust (OED) [OF fust, cask, log, tree trunk, fr L fustis, stick, staff]. A m-c for wine (c1450), a cask of unknown size (Shipley 287).

fut, fute. FOOT

futher, futhir. FOTHER

fynes. ENDE

fyrken, fyrkin, fyrkyn, fyrkyne, fyrkynge. FIRKIN

fyrlot. FIRLOT

G

gad—3 L gaddus; 4-7 gadd; 4-9 gad; 5-6 gadde; 8-9 gaud (OED); gawd (OED) [ME gad, gadd fr ON gaddr, a string, nail, spike]. A wt for steel of uncertain poundage equal to 1/30 SHEAF, and a m-l for land varying from 9 to 25 ft (2.743 to 7.620 m) and synonymous with the PERCH.—**1297** Elton 64: Et in vna pecia ferri et iij gaddis aceri emptis in quadragesima pro eadem Caruca iiij d. **c1440** Promp. Parv. 184: Gad, to mete wythe londe (gadde, or rodde). **c1461** Hall 17: Also style by gadds; and euery pece of stele in hymselfe is a gadde; and xxx gaddes make a scheff. **1502** Arnold 173: In dyvers odur placis in this lande they mete ground by pollis gaddis and roddis some be of xviij. foote some of xx fote and som xvi fote in lengith. **1507** Gras 1.703: And xxx gaddes makythe sheffe. **1696** Phillips sv: Gad, or Geometrical Pearch, a Measure of Ten Foot, and in some places but Nine Foot. **1790** Miller 18: The FAGGOTT, GAD BURTHEN. **1868** Eng. Cyclo. 824: 5 1/2 yards a pole, perch, or gad. **1883** Simmonds sv: Gad...a bar of metal.

gadd, gadde. GAD

gag, gage, gagge. CADE

gait [perh a special use of gate, a way, fr ME gate fr ON gata, road, path]. A m-c for water in Northamptonshire (c1850) containing 2 buckets (c2 bu or c70 l) (Sternberg 39).

gallandde, gallande, gallante. GALLON

gallon—3 L ialon, L jalo (Swinfield); 3-4 galun (OED); 3-7 L galo, L galona; 4-5 galoun; 4-7 galon; 5 galloun, galloune, galoune, galown; 5-6 gallone, galone; 5-9 gallon; 6 gallandde, gallande, gallonde (OED), gallunde (OED), galne (OED), galond (OED), galonde (OED); 6-7 gallond; 7 gallante (OED); 8-9 gaun, gawn, goan; ? gullyn (Prior), jalon (Prior) [ME galon, galun, a liquid measure, fr ONF galon fr MedL galeta, jug, pail, a liquid measure, of obscure origin]. A m-c for many liquid and dry products. It sometimes was abbreviated gall.

The ale gal, of varying dimensions prior to its standardization at 282 cu inches (4.621 l) under Elizabeth I, contained 4 qt or 8 pt and was equal to 1/8 ale FIRKIN, 1/16 ale KILDERKIN, and 1/32 ale bbl. It, together with the beer gal of 282 cu inches, was equal to 0.340896 Scots gal; the latter being equal to 2.933447 English ale or beer gal.—**1379** Rot. Parl. 3.64: & de cervoise certeines Mesures; C´est assavoir, Galon, Potel, & Quart. **1390** Henry Derby 6: Clerico Buterie super servisia, per manus diuersorum pro v galonibus servisie, x d. **1392** Ibid 157: xxiiij galones, galo ad j d. ob., xij s. c**1517** Hall 49: That nevyr shalbe no lesse than viii pyntes to the galon´. c**1549** York Mer. 144: Item, paid for xxx gallanddes ayell, x s. **1557** Scrope 327: Videlicet, there beste ale under the herseve for iij. d. a galone; there stalle ale for iiij. d. a gallone. **1682** Hall 29: But Ale hath no more than 32 gallons to the barrell. **1694** Beilby 2: There is allowed 282 cubical inches to an Ale-Gallon. **1707** Forbes 54-55: But now by the 7 Article of the Union the thirty four Gallons English Barrel of Beer or Ale...amounting to 12 Gallons of present Scots Measure. **1716** Harris 2. sv measures: The Beer or Ale Gallon holds 282 solid Inches. **1789** Hawney 310: That 282 solid Inches is a Gallon of Ale. **1816** Kelly 87:

ALE AND BEER MEASURE...4 Quarts...1 Gallon...282 [cu inches]...4,6208 [1].

The beer gal, also of varying dimensions until it was standardized at 282 cu inches (4.621 l) under Elizabeth I, contained 4 qt or 8 pt and was equal to 1/9 beer FIRKIN, 1/18 beer KILDERKIN, and 1/36 beer bbl.—**1613** Tap 1.61: MEASVRES CONCAVE OF Beere.... One barrell containeth...Gallonds. 36. **1682** Hall 29: I Barrell conteynes: 2 Kilderkins, 4 Firkins, 36 Gallons, 72 Pottles, 144 Quarts, 288 Pints. **1708** Chamberlayne 210: 2 Pottles make a Gallon, a Gallon of Beer, or the Measure containing 282 Solid Inches, and holds of Rain Water 10 Pounds, 3 Ounces 240/1000 Avoirdupois. **1789** Hawney 310: That 282 solid Inches is a Gallon of Ale or Beer...measure.

The corn or grain gal was standardized at 268.8 cu inches (4.404 l) under Elizabeth I. Although it usually contained 4 qt or 8 pt and equaled 1/2 pk, 1/8 bu, or 1/64 SEAM, its actual capacity varied from approximately 272 1/4 to 282 cu inches (c4.46 to c4.62 l) before, and sometimes after, its standardization.—c**1272** Report 1.414: Et octo libre frumenti faciunt galonem. **1290** Fleta 119: Et pondus octo librarum frumenti faciunt mensuram ialonis. **1351** Rot. Parl. 2.240: Soient les Mesures, c´est assaver bussell...galon. c**1400** Hall 36: 8 libre faciunt I galon´, 61,440 grana. **1413** Rot. Parl. 4.14: & qe chescun Bussell contiendra oept Galons. **1474** Cov. Leet 396: & ij Pottels makith a Gallon; & viij Gallons makith a Buysshell. **1496** Hall 45: That is to say: a busshell, a galown. **1540** Recorde 204: 8

pounde (or 8 pyntes) doe make a Gallon. **1587** Stat. 454: And that euerie galon conteine viii. li. of wheate of troie weight. c**1590** Hall 20: 2 galons makith a pecke.... So that 8 gallons...makith the bushell. **1607** B. J. 20: The Corne measure of Bristow is 8. gallonds Winchester. **1615** Collect. Stat. 468: And euery gallon contain eight pounds of wheate, of Troy weight. **1635** Dalton 144: 8 pintes/4 quarts/2 pottles maketh the gallon. **1665** Assize 3: The full and just weight of xii. ounces Troy in Wheat do make a concave or hollow measure, named a pint; and viii. of the same pints do make the gallon for...Corn. **1688** Bernardi 150: Præterea Galonem Frumentarium Angliæ, dimidium Pecci et Octantem Busselli.... Aediles autem et moderatores fori eundem Galonem siccum unciis pedis Anglici 272 1/4 in. solidum construunt.... Pinta denique arida, 1/8 Galonis, seu congii frumentarii et 1/8 X 8 = 1/64 Brusselli. **1710** Harris 1. sv measures: Now a Vessel thus made will contain 2150.42 Cubick Inches; and consequently the Corn Gallon can be but 268 4/5 Cub. Inches. **1717** Dict. Rus. sv dry measure: To measure dry things, as Corn, or Grain, we have first the Gallon, which is bigger than the Wine-Gallon, and less than the Ale or Beer-Gallon; containing 272 and a quarter cubick Inches, and 9 Pounds, 13 Ounces, 12 Drams and a half of Avoirdupois-Weight. **1780** Paucton 811: Le gallon de bled est, suivant M. Arbuthnot, de 272 1/4 pouces solides Anglois; selon Chambers, de 272; selon Edouard Bernard, cité par Eisenschmid, de 272 55/100: je l'ai trouvé de 272 9/10, en déduisant sa cubature de son poids en eau pure. **1789** Hawney 310: 268.8 solid Inches is a

Gallon...of Corn-measure. **1805** Macpherson I.471: 8 pounds of corn 1 gallon, 8 gallons 1 bushel of London. **1816** Kelly 88: DRY MEASURE...1 Gallon...268.8 [cu inches]...4,40428 [1].

The wine gal was standardized in 1707 at 231 cu inches (3.785 1). Although usually containing 4 qt or 8 pt and usually equaling 1/18 RUNDLET, 1/42 TIERCE, 1/63 HOGSHEAD, 1/84 PUNCHEON, 1/126 PIPE, or 1/252 TUN, its actual capacity varied prior to 1707, with 282 cu inches (4.621 1) and 224 cu inches (3.671 1) being the most common. The oil and honey gal conformed to the specifications of the wine gal, as did those for beef, fish, and pork. The Scots wine gal, also used for all other liquid and dry products, contained 827.232 cu inches (c13.60 1) or 4 qt, or 8 pt, or 16 CHOPPINS, or 32 MUTCHKINS, or 128 GILLS (Swinton 29); it was equal to 3.581091 English wine gal, while the latter was equal to 0.279244 Scots gal. The Irish wine gal, also used for all other liquid and dry products, contained 217.6 cu inches (c3.57 1) and equaled 2 Irish POTTLES or 4 Irish qt or 8 Irish pt (Edinburgh XII.572).—**c1200** Caernarvon 242: Et octo libræ frumenti faciunt galonem Vini. **c1300** Hall 8: Et viii libre ponderant unam galonem vini. **c1330** Gross II.215: Il auera un galoun de vyn. **c1435** Amundesham II.312-13: Item, pro vino Domini Abbatis...tres galones. **c1461** Hall 7: Et viii livres de froument font la galone de vin; ibid 12: And xii unces make a lb. of Troy: and of all thys weyghts viij lb. make a galon of wyne; ibid 15: Off the mesure of Lycoure.... The tonne...ii [X] c [+] l galouns. The pipe...i [X] c [+] xxv galounes. The hogg[eshed]...lxii gallounes and dim. The almer...l galouns. The barell...xxxi gallones 1 quart, there sesteryn...iiii gallouns. c**1549** York Mer. 144: Item, for x gallandes wyne, xiij s. iiij d. c**1590** Hall 22: 8 pound troy is a gallone in waighte, or 4 quartes. c**1600** Brit. Mus. 16.70: And .8. gallonds of wyne. **1665** Sheppard 7: Of liquor, 12 ounces make a pound; 8. pound make a Gallon of Wine. **1678** Du Cange sv galo: Galona, Mensura liquidorum apud Anglos, quarum unaquæque octo continet pintas Anglicanas. **1688** Bernardi 149: Quinetiam Galonem vinarium in pedis Anglici corporeas uncias 231. **1701** Hatton 3.9: Of Wine Measure. 4 Quarts, or 231 solid Inches, is 1 Gallon. **1707** Seventh Rep. 36: That any round vessel, commonly called a cylinder, having an even bottom and being 7 inches diameter throughout, and 6 inches deep from the top of the inside to the bottom, or any vessel containing 231 cubical inches and no more, shall be deemed and taken to be a lawful wine gallon. **1708** Chamberlayne 210: So that 4 Gallons of Beer Measure are almost 5 Gallons of Wine...and each Gallon of Wine is 231 Cubical Inches, 8 Pound, 1 Ounce, and 11 Drachms Avoirdupois of Rain-Water. **1710** Harris 1. sv measures: The Beer and Ale Gallon is larger than the Wine Gallon, in proportion to the excess of the common Pound Averdupois above the true Pound Troy; that is, as 12 [to] 231...so 14 12/20 to 281 1/2, which is very near the Cubick Inches in the Ale Gallon. **1717** Dict. Rus. sv wine-measure: The English Wine-Measures are smaller than those of Ale and Beer, and hold proportion as about 4 to 5...and each Gallon of Wine is 231 cubical Inches. **1791** Keith 2.3: In England there are also two

Wine measures, the Guildhall gallon of 224 inches, and the statute Wine gallon of 231 inches. **1831** Hassler 5: The wine gallon was generally merely stated by its legal capacity of 231 cubic inches. **1860** Britannica 805: Scotland...gallon, 827.23 [cu inches].

Locally, the butter gal weighed 12 lb (5.443 kg) at Shropshire and in Wales and 16 lb (7.257 kg) at Bridgenorth.—**1717** Dict. Rus. sv gawn [of butter]: Or goan, a Word us'd in some Parts of the Country for a Gallon. **1819** Cyclopædia sv weights: The gawn [of butter], which signifies 12 lbs. of 16 oz. in Shrewsbury, and 16 lbs. of 16 oz. at Bridgnorth. **1820** Second Rep. 18: Gaun or Gawn, Shropshire and Wales: a corruption of gallon, applied to butter containing 12 lbs...Bridgnorth: of butter, 16 lbs.

Since the establishment of the Imperial system in 1824, the gal both for liquid and dry products has contained 277.420 cu inches (4.546 l). In the Imperial ap system the gal of 277.420 cu inches is reckoned as 8 ap pt of 34.6775 cu inches each or the volume of 70,000 gr (4536.0 g) of distilled water at 62° F.—**1851** H. Taylor 51: The imperial gallon now in general use was established by Act 5, George IV, in which it is declared to contain 10 pounds, Avoirdupois, of distilled water. **1855** Jessop 26: The imperial gallon = 277.274 cubic inches. **1889** Francis 31-32: The imperial gallon contains 10 lbs. avoirdupois of distilled water, weighed in air at 62° Fahrenheit, the barometer being at 30 inches. **1907** Hatch 23: Imperial Measures of Capacity, both Liquid and Dry.... 4 quarts = 1 gallon (gal.) = 277.420 cubic inches; ibid 24: 1

gallon (C.) = 8 pints = 277.420 cubic inches.... 1 gallon (C.) is the volume of 70,000 grains of distilled water at 62° F; ibid 35: 1 gallon = 4.5459631 litres. **1920** Stevens 2: 1 Gallon, C. = 8 Pints, = 160 Fluidounces. **1934** Int. Traders´ 75: Gallon...United Kingdom...277.418 cubic inches. **1956** Economist 4: Gallon: (a) United Kingdom, Imperial system = 277.420 cubic inches. **1966** O´Keefe 670: 1 gallon...4.54596 l.

gallond, gallonde, gallone, galloun, galloune, gallunde, galne, galo, galon, galona, galond, galonde, galone, galoun, galoune, galown, galun. GALLON

garb—3-4, 7 L garba; 3-7 garbe; 6-9 garb [ONF garbe (OF jarbe), of Gmc origin; see OED]. A wt for 30 pieces of steel. It was of uncertain poundage and perhaps was synonymous with the GAD. The garb was also a BUNDLE or SHEAF of corn and other grain products.—c**1253** Hall 11: La garbe de ascer est xxx pecis. c**1272** Report 1.414: Garba afferis constat ex triginta peciis. **1297** Neilson 5: Et in vii garbis asceris et dimidia emptis, v s. ix d. **1324** Ibid 56: In i. garba asceris, x d. **1495** Brit. Mus. 28:156v: Garba vero afferis constat ex triginta petris. **1607** Cowell garbe: Garbe (garba) commeth of the French (garbe, alias, gerbe...fascis.) It signifieth with vs, a bundle or sheafe of corne. **1695** Kennett Glossary sv garba: GARBA. A sheaf of corn, of which twenty four made a Thrave. **1820** Second Rep. 18: Garb...of steel, 30 pieces.

garba, garbe. GARB

gaud. GAD

gaun. GALLON

gavael [W gavael, a hold (Laws Wales)]. A m-a for land in Wales (c1300) containing 4 RHANDIRS or 64 ERWS (c23.10 ha) (Laws Wales 1000).

gawd. GAD

gawn. GALLON

gill—4 gille (OED), jille (OED); 4-5 gylle; 6 gyll; 7-9 gill; 9 jill [ME gille, perh fr MF gille, gelle, vat, tub; see WNID3]. A m-c for liquids generally equal to 1/4 pt (c0.12 l) or 1/32 gal, and frequently called a QUARTERN. In some of the shires, however, it equaled 1/2 pt (c0.24 l) and the measure of 1/4 pt was called a jack or jackpot. The Scots gill equaled 1/16 pt, 1/8 CHOPPIN, 1/4 MUTCHKIN, or 6.463 cu inches (0.106 l) (Swinton 29 and Donisthorpe 210). Since the establishment of the Imperial system in 1824, the gill has contained 8.669 cu inches (0.142 l) and equals 1/4 pt of 34.6775 cu inches. Occasionally it was abbreviated gi.—**1310** trans in Memorials 78: Such as the measures called "chopyns" and "gylles." c**1440** Promp. Parv. 194: Gylle, lytylle pot. c**1590** Horwood 640: A quarter of a pint, sometimes called a gyll. **1790** Jefferson 1.982: The gill, four of which make a pint. **1834** Pasley 42: 4 Gills...1 Pint; ibid 44: 4 Gills, or Quarterns. **1849** Dinsdale 70: Jill...a small measure, the fourth part of a pint. **1907** Hatch 23: 4 gills = 1 pint (pt.) = 34.6775 cubic inches; ibid 35: 1 gill = 1.42061 decilitres. **1956** Economist 4: Gill: (a) United Kingdom, Imperial system = 8.6694 cubic inches. **1969** And. & Bigg 11:

1 gill = 0.142065 dm^3 = 0.142 litre.

gille. GILL

glanet [*]. A wt for 30 pieces of steel (c1350). It was of uncertain poundage and perhaps was synonymous with the GAD (Hewitt 190).

glean—8-9 glen; 9 glean, glene [MF glene, glane; see WNID3 vb glean]. A m-q comprising 25 herrings and equal to 1/15 REES. In Essex and Gloucestershire it was a m-q for teasels, consisting of one BUNCH.—**1805** Macpherson I.471: 25 herrings 1 glen, 15 glens 1 rees. **1820** Second Rep. 18: Glean, Glen, Glene of teazles, Essex and Gloucestershire; a bunch. **1880** Britten 169: Bunch.... (Ess.), of teazles, 25 heads, otherwise a glean. (Glouc.), of teazles, 20; a glen. **1895** Donisthorpe 210: GLENE: of herrings, 25.

glen, glene. GLEAN

goad—4-6, 8 gode; 5 goode; 6-7 goade; 7 goadde; 7-9 goad [ME gode fr OE gād, goad, arrowhead, spear point]. A m-l for cloth, containing 4 1/2 ft (1.371 m). Occasionally it was a m-l for land synonymous with the PERCH.—c**1461** Hall 14: Thai mete grownd by the Polys, Goodys, and Roddys; and sum of thame be of xviij fote, sum of xx fote, and sum of xxi fote. **1590** Rates 2.41: Cottons the C. goades containing v [X] xx. c**1600** Brit. Mus. 16.70: They meete lande by poles, goaddes, and Rooddis, some be off xviij. foote some of .xx. foote, and some .xxj. foote in length. **1677** Roberts 35: The Goad for Frizes, Cottons, and the like; ibid 300: A Goade, only used in Welch Frizes. **1696** Jeake 65: In I Goad...4 1/2 Feet, a Measure in some places for Land and Cloth received by Custom. **1721** King 291: Cottons and Plains...per 100 Goads. **1784** Ricard II.155: La gode, dont on mesure les bayes, les frises, & autres étoffes. **1868** Eng. Cyclo. 824: A goad is an old name for a yard and a half. **1897** Maitland 372: The measuring rod that was used for land had so many names, such as perch, rod, pole, goad, lug.

goade. GOAD

goan. GALLON

gode. GOAD

goney. GUNNY

goode. GOAD

grain—3-6 greyn (OED), greyne (OED); 3-7 L grana, L granum; 4 grein (OED), greine (OED); 4-7 grayn, grayne; 5 grane (OED); 5-9 grain; 6 grene (OED); 6-7 graine [ME grain, grein fr MF grain, grain, kernel, seed, fr L granum, grain, seed]. The smallest unit of wt (c0.06 g), equal to 2/875 avdp oz, 1/450 merc and tow oz, 1/476 Scots tron oz, 1/476 Scots t oz for meal, meat, hemp, and iron, 1/480 Scots t oz for gold and silver, and 1/480 ap and English t oz. In the moneyer's imaginary subdivisions of the gr, there were 20 MITES, 480 DROITS, 9600 PERITS, or 230,400 BLANKS. It was also used as the basis for the standardization of the inch, which was defined as the length of 3 medium-sized barleycorns placed end to end (c2.54 cm).—c**1200** Caernarvon 242: xxx...& duo g[ra]na frumenti in medio spici. c**1272** Hall 7: Sciendum quod tria grana ordei, sicca et rotunda, faciunt pollicem. **1290** Fleta 119: Sterlingus...qui debet ponderare xxxij. grana frumenti mediocra. c**1461**

Hall 14: The lengythe of iii barly cornys make an ynche, so that barly growe in comyn soyle, not to lene, nodyr to muche compost abowte. **1474** Cov. Leet 396: That xxxij graynes of whete take out of the mydens of the Ere makith a sterling other-wyse called a peny.... And hitt was ordeyned...that iij barley-Cornes take out of the middes of the Ere makith a Inche. **1540** Recorde 202: After the statutes of Englande, the least portion of waight is commonly a Grayne, meaning a grayne of corne or wheate, drie, and gathered out of the middle of the eare. c**1590** Hall 21: And the content of a bushell of wheat is 430,080 grayns of wheat; ibid 27: 3 grayns of Barly, dry and rotund, do make an ynche. c**1600** Ibid 36: Grana...a grain is a barley corne taken in the midst of the eare. **1616** Hopton 159: And this Troy weight containes in euery pound 12 ounces, euery ounce 20 peny weight, euery peny weight 24 graines. **1624** Huntar 2: A corne or pickle of wheat, taken out of the middest of an eare of wheate, is the foundation of a graine weight. **1635** Dalton 150: Three barley cornes measured from end to end (or 4 in thicknesse) maketh one inch. **1678** Du Cange sv granum: Grana, Angl. Grain. **1688** Bernardi 135: Uncia Anglica de Troy...480 grana argenti triticive. **1708** Chamberlayne 205: In Troy-Weight, 24 Grains of Wheat make a Penny-Weight Sterling. **1717** Dict. Rus. sv troy-weight: The smallest Denomination is a Grain, which is the Weight of a Grain of Wheat, gathered out of the middle of the Ear well dryed. **1829** Palethorpe sv: GRAIN, the name of a small weight, the 1/20th part of a scruple in apothecaries' weight, and the 1/24th part of a dwt. troy.

1883 McConnell 9: IMPERIAL TROY WEIGHT. .003961 cub. in. of water = 1 grain. **1883** Simmonds sv: Grain...The smallest British weight in troy or avoirdupois weight. **1907** Hatch 34: 1 grain = 64.79891824 milligrams. See BARLEYCORN

graine, grana, grane, granum, grayn, grayne, grein, greine, grene, greyn, greyne. GRAIN

groce, groos, gros, grose. GROSS

gross—5 groos (OED); 5-9 groce, grose; 6 gros (OED); 6-7 grosse; 6-9 gross [ME groos, groce fr MF grosse fr fem of gros, thick, coarse]. A m-q of any item: a small gross generally consisted of 12 DOZEN or 144 in number, while a large or great gross was 12 small gross, or 144 dozen, or 1728 in number. The great gross was employed especially for wholesale selling of buttons, beads, cap-hooks, playing cards, various cases and combs, chess pieces, points of thread and silk, and tobacco pipes. Occasionally it was abbreviated gr. or gro.—c**1461** Hall 17: Also there ys a Numbyr that ys called a Grose, and itt cont[aineth] xij doss[en]; and thereby be sold poynyes, laces, purces, knyvys, balles, strenges and odyr dyuers thynges mo. **1507** Gras 1.696: Coper gowle the grose; ibid 698: Gyrdelles of thred the grosse; ibid 703: Sporres the grosse. **1524** Ibid 194: Pro un´ grosse knythose; ibid 196: Pro iiii grosse de cards...Pro sex grosse de combes. **1545** Rates 1.2: Abces the groce. c**1550** Welsh 64: 1 gross girth web. **1590** Rates 2.1: The groce containing xii. dosen. c**1610** Lingelbach 113: All other small wares by the Grosse. **1612** Halyburton 288: The groce contening tuelf dozen.

1628 Hunt B2: A Grosse, or 12. dozen 144. **1701** Hatton 3.11: 12 Pieces or things is 1 Dozen. 12 Dozen...1 Small Gross. 12 Sm. Gross...1 Great Gross; ibid 16: By the Great Gross...are bought and sold...Mettal, Glass, Thread, Silk, Handkercher, and Hair, Buttons...Cap-hooks...Playing Cards...Comb and Spectacle Cases...Lightwood and Box Combs...Chess-men...Thread and Silk Points...Tobacco Pipes. **1717** Dict. Rus. sv: A Gross, is the quantity of Twelve Dozen. **1721** King 282: Bracelets or Necklaces of Glass...37 Small Groce. **1756** Rolt sv scavage: Playing-cards, the small gross containing 12 dozen pair.... Lute-strings, called catlings, the great groce containing 12 small groce of knots. **1820** Second Rep. 18: Groce or Gross...Commonly 12 dozen.... Of bracelets or necklaces, 12 dickers or bundles of 10 make a small gross.... Of pill boxes, 12 dozen nests of 4 boxes each. **1840** Waterston 1: 144, the number forming a gross. **1885** People's Cyclo. 1986: 12 doz. (144) 1 gross (1,728) 1 great gross. **1956** Economist 8: Gross...144 or 12 dozen. Gross great...1,728 or 12 gross.

grosse. GROSS

gullyn. GALLON

gunny—8 goney (OED); 8-9 gunny [Hind goṇī fr Skr goṇī]. A m-c (c1800) for saltpeter, 1/4 Cwt (12.700 kg), and cinnamon, 3/4 Cwt (36.741 kg), contained in a sack (Second Rep. 18). See HUNDRED

gwaith [*]. A m-a for peat in North Wales (c1800-1900), containing 150 sq ft (13.935 sq m) (Second Rep. 18 and Donisthorpe 210).

gwyde [perh fr W gwyniad, a white-fleshed fish, fr gwyn, white]. A m-q for eels containing 10 STICKS or 250 in number.—c**1461** Hall 17: Also Elys be sold by the stike, that ys xxv elys; and x styckys make a gwyde.

gybe [perh fr E dial gib, a hooked stick]. A m-c for wool (c1430) containing 2 POKES or bundles (Southampton 1.88).

gyll, gylle. GILL

gyllot [dim of GILL]. A m-c for liquids (c1500-1600) reckoned equal to 1/2 GILL (Horwood 640).

habardepayce, habardepayse, habardepayx, habardepoix, habardipoys, habardypeyse, haberdepase, haberdepayes, haberdepoies, haberdepois, haberdepoise, haberdepoiz, haberdepoyie, haberdepoys, haberdepoysse, haberdipoys, haberdupois, habertypoie, haburdepeyse, haburdepoyse, haburdypeyse, haburdypoyse. AVOIRDUPOIS

haddock. HATTOCK

hakere. ACRE

half-barrel. KILDERKIN

half-coomb. STRIKE

half-quarter. COOMB

half-quartern. STONE

half-stone. CLOVE

hamper—4-5 hampere (OED); 6 hampier (OED); 6-9 hamper; 7 hampire (OED) [ME hampere, alter of hanaper fr MF hanapier, a case to hold hanaps, fr hanap, a drinking vessel, + -ier]. A m-c for dry goods. It was a large basket of wickerwork, usually with a cover, used as a packing case.—c**1550** Welsh 172: 1 hamper 6 bags dry wares; ibid 264: 1 trunk and 1 hamper household stuff; ibid 282: 2 trunks 4 hampers felts. **1607** Clode 311: For 2 hamper of guodlings...For a hamper of pyppyns. **1664** Gouldman sv: A hamper or basket of osiers. Calathus. **1883** Simmonds sv: Hamper, a wicker-work pannier.

hampere, hampier, hampire. HAMPER

hanc, hanck, hancke. HANK

hand—1-5 hond (OED); 1-9 hand; 4 haunde (OED), hoond (OED), hoonde (OED); 4-6 honde (OED); 4-7 hande; 7-9 handful, handfull [ME hand fr OE hand, hond; see WNID3]. A m-l, originally a unit of body measurement reckoned as the breadth of the palm including the thumb, made equal to 4 inches (10.16 cm).—**1561** Eden xviii: Foure graines of barlye make a fynger: foure fingers a hande: foure handes a foote. **1635** Dalton 150: Foure Inches maketh an handful. **1638** Bolton 274: Fower Inches make the handful. **1665** Sheppard 16: 4 Inches a handful. **1707** Justice 4: 4 Inches 1 Hand, or Hand's-Breadth. 3 Hands 1 Foot. 1 1/2 Foot, or 4 1/2 Hands, 1 Cubit. **1708** Chamberlayne 207: 4 Inches make a Hand. **1717** Dict. Rus. sv handful: A Measure of four Inches. **1728** Chambers 1.520: The Measure for Horses, is the Hand or Handful; which, by the Statute, contains four Inches. **1756** Rolt sv: HAND, or Handful, is...a measure of four inches.... The hand, among jockeys, is four fingers breadth; by the which the height of horses is measured. **1820** Second Rep. 18: Hand or Handful...4 inches. **1832** Edinburgh XII.569: 4 Inches = 1 Hand = 0.1016 [m]. **1931** Naft 13: 1 Hand...0.333 foot...4 inches. **1966** O'Keefe 667: 1 hand = 4 inches = 10.2 cm.

hande, handful, handfull. HAND

hank—4-9 hank; 6 hanc (OED); 6-7 hanke (OED); 7-9 hanck (OED), hancke (OED) [ME hank, of Scand origin; cf Dan hank, handle, Sw hank, a band or tie, ON hanki, clasp, hönk, hangr, hank, coil, skein]. A m-q for yarn (c1800-1950) containing 7 WRAPS, 12 CUTS, or 560 yd (5.121 hm) for worsted and 7 SKEINS or 840 yd (7.681 hm) for cotton or spun silk

(Brockett 89, Economist 58, and Bonwick 359).

hanke. HANK

hasp [ME hasp, haspe fr OE hæsp, hæspe; akin to MHG haspe, hasp, ON hespa, and perh to L capsa, chest, case]. A m-q for linen yarn (c1800-1950) containing 3600 yd (3291.840 m) or 6 HEERS or 1/4 SPINDLE (G. Gregory II. sv measure and Economist 58).

hattock—7-9 haddock, hattock, huttock (OED) [hat (t) + -ock; see description]. A m-q for grain in northern England consisting of 10 or 12 SHEAVES, and similar to a SHOCK or STOOK. Eight or ten of the sheaves were placed in an upright position; the two remaining sheaves then were placed on top of the others, rising to a peak in the center with their heads sloping downwards at both ends so as to carry off rain. These covering sheaves were called "head-sheaves" or "hoods;" hence the name.—**1674** Ray 24: Hattock, a Shock containing 12 Sheaves of Corn. **1756** Rolt sv: HATTOCK. A shock of corn, containing twelve sheaves. **1880** Britten 146: Haddock (Yks.)...Stook. In Cumb. ten sheaves are a hattock, and twelve a stook.

hauerdepiz. AVOIRDUPOIS

hauncere. AUNCEL

haunde. HAND

haverdepoise, haverdepous, haverdupois, haverdupoiz, haverdupoize. AVOIRDUPOIS

heap [ME heep, hepe, heap, multitude, fr OE hēap]. A m-q for limestone in some parts of Scotland (c1800) containing 4 1/4 cu yd (3.249 cu m) and weighing 5 tons (Second Rep. 20).

hear. HEER

heer—8-9 hear, heer [ME (Sc dial) heir, hair, hair]. A m-q for linen yarn containing 600 yd (548.640 m) or 2 LEAS or 1/6 HASP.—**1840** Waterston 148: Yarn (Linen), thread...inches 90.... Heer of 2 cuts or 240 threads...yds. 600. **1956** Economist 58: Linen...1 lea (or cut) = 300 yards...2 leas = 1 hear (or heer)...6 hears = 1 hasp.

hīd, hida. HIDE

hide—1 hīd (OED); 1-9 hyde; 1-? hide; 2-7 L hida, L hyda; 7 hilda [ME hide, hyde fr OE hīd, hīgid, originally land enough to support a family, fr stem of hīwan, hīgan, members of a household]. A m-a which probably originated as an amount of land needed to support a peasant family for a period of one year and, at the same time, as a unit for tax assessments. But, beginning in the eleventh century, the hide was usually expressed in terms of acres, with 60 (c24.30 ha), 64 (c25.92 ha), 72 (c29.16 ha), 80 (c32.40 ha), 96 (c38.88 ha), 100 (c40.50 ha), 120 (c48.60 ha), 140 (c56.70 ha), 160 (c64.80 ha), and 180 (c72.90 ha) acres being the most common. Seldom was it larger than 180 acres. In addition, it was occasionally expressed as a division of land containing a certain number of VIRGATES, most often as one of the following: a hide of 2 virgates, each virgate containing 2 BOVATES of 12 acres each, and thus 48 acres (c19.44 ha) in all; a hide of 3 virgates, no standard acreage established for the virgate; a hide of 4 virgates, each virgate generally containing 12 (c4.86 ha), 15 (c6.07 ha), 20 (c8.10 ha), 24 (c9.72 ha), 28 (c11.34 ha), 30 (c12.15 ha), 34 (c13.77 ha), 40 (c16.20 ha), 44 (c17.82 ha), 48 (c19.44 ha), or 64 acres (c25.92 ha); a hide of 4 virgates, each virgate containing 4 FARTHINGDALES of 10 acres each, and thus 160 acres (c64.80 ha) in all; and hides of 5, 6, 6 1/2, 6 3/4, 7, and 8 virgates, no standard acreage established for the virgate. Sometimes it was abbreviated h. or hid.—**1086** Sussex 4: De isto manerio habet Engeler ii hidas de rege; ibid 6: T. R. E. se defendebat pro xii hidis et modo pro iii hidis et iii virgis et dimidia. **1086** Barfield Appendix IV: Tunc pro v hidis, modo pro ij hidis et dimidia. c**1100** Bello 11: Octo itaque virgatæ unam hidam faciunt. c**1155** Henrici 176: Hida autem Anglice vocatur terra unius aratri culturæ sufficiens per annum; ibid 207: Et inquirere fecit per jusjurandum quot hidæ, id est, jugera uni aratro sufficientia per annum. c**1175** Clerkenwell 11: Et dono Willelmi de Sancto Georgio terram de Haselingefeld quam Robertus Ruffus tenuit cum managio scilicet vnam hidam. **1191** Salisbury 56: De dimidia hyda terræ et duabus acris in uno prato in Lavintone. **1200** Cur. Reg. 8.145: De duabus hidis et dimidia in Pepewell´ et in Waresle.... j. hide terre cum pertinenciis in Leghe. **1201** Cur. Reg. 9.53: Unde xxvij. hide faciunt feodum j. militis. **1204** Cur. Reg. 10.209: Jordanus filius Avicie petit versus Rogerum filium Berte j. hidam terre et xxviij. acras cum pertinentiis in Crikeshee sicut jus suum et hereditatem. **1220** Cur. Reg. 3.151: De viginti et una hidis terre cum pertinentiis.... Ad quatuordecim bovatas unde xiiij. curucate terre faciunt feodum j. militis. **1222** St. Paul´s 135-36: Warinus de

Bassingbourne tenet unam carucam terræ continentem ix [X] xx acras terræ arabilis; ibid 136: Warinus de Brantone tenet unam carucam continentem vii [X] xx acras cum prato et bosco. c1230 Red Book 188: Walterus de Cliford debet in Wirecestrescira servitium quintæ partis j militis pro una hida quam tenet. c1250 Rameseia III.208: In Comitatu Huntingdoniæ. Upwode, cum Ravele...Viginti acræ faciunt virgatam. Quatuor virgatæ faciunt hidam. Wistowe...Triginta acræ faciunt virgatam. Quatuor virgatæ faciunt hidam.... Haliwelle...Octodecim acræ faciunt virgatam. Et quinque virgatæ faciunt hidam. Soca de Slepe...Sexdecim acræ faciunt virgatam. Et quinque virgatæ faciunt hidam. Hougtone, cum Wittone...Octodecim acræ faciunt virgatam. Sex virgatæ [faciunt] hidam. In Wittone, Viginti acræ [faciunt] virgatam. Quinque virgatæ [faciunt] hidam; ibid 209: Hemmingforde...Sexdecim acrae faciunt virgatam. Et sex virgatæ [faciunt] hidam. Dillingtone...Triginta et tres acræ et dimidia faciunt virgatam. Sex virgatæ faciunt hidam. Westone...Viginti et octo acræ faciunt virgatam. Et quatuor virgatæ faciunt hidam. Bringtone...Triginta et quatuor acræ faciunt virgatam. Et quatuor virgatæ faciunt hidam. Bitherne...Quadraginta et quatuor acræ faciunt virgatam. Et quatuor virgatæ faciunt hidam; ibid 210: Elingtone...Viginti et quatuor acræ faciunt virgatam. Sex virgatæ faciunt hidam.... Stiveclee...Viginti et quatuor acræ faciunt virgatam. Et quatuor virgatæ faciunt hidam; ibid 211: Bernewelle...Triginta et sex acræ faciunt virgatam. Septem virgatæ faciunt hidam.... Cranfelde...Quadraginta et octo acræ faciunt virgatam. Quatuor virgatæ faciunt hidam; ibid 212: Shittlingdone cum Pekesdene...Duodecim acræ faciunt virgatam. Quatuor virgatæ faciunt hidam; ibid 213: In comitatu Hertforddiæ. Therfelde...Sexaginta et quatuor acræ faciunt virgatam. Et quatuor virgatæ faciunt hidam. In comitatu Suff[olciæ]. Laushulle...Quinquaginta acræ faciunt virgatam. Tres virgatæ [faciunt] hidam. In comitatu Norff[olciæ]. Brauncestre...Quadraginta acræ faciunt virgatam. Quatuor virgatæ [faciunt] hidam; ibid 214: In comitatu Cantebr[igiæ]. Ellesworthe...Triginta acræ faciunt virgatam. Quatuor virgatæ [faciunt] hidam. c**1283** Battle xiii: Quatuor virgatæ seu wystæ faciunt unam hydam. c**1289** Bray 9-10: Memorandum quod in campis de Herleston sunt viginti septimae virgatae terrae per hidam, quarum de ffeodo domini regis quatuor virgatae et dimidia, de ffeodo de Berkhamsted duae virgatae, de ffeodo de Doddesforde decem virgatae, de ffeodo de Neubotle decem virgatae et dimidia. **1338** Langtoft 600-01: Decem acræ faciunt ferdellum. Quatuor fardella faciunt virgatam unam. Quatuor virgatæ faciunt hidam unam; ibid 601: Fardellum Acræ X/virgata XL./hida. CLX. c**1350** Higden VIII.176: Hyda, id est carucata, terræ. c**1375** Hyda 237: And he by qwath his wyf ten hydys of lond at Manyngforde.... An hyde of londe at Upton. c**1450** Gray 487: I halfe hyde of londe in Gaihampton conteynynge xxv acres of land in on feelde and also many in an othyr feelde. c**1500** Brit. Mus. 6.7: j hide...iiij virgat.... clx acr faciunt...j hidam. **1599** Richmond Appendix 2.10: And yet for the better proof that a Hide of Land was both reputed before the Conquest and since Six Score Acres, I find mentioned in a Book entituled, Restauratio Ecclesiæ de Ely...these Words: Et non invenerunt de terra quæ mulieris jure fuisset, nisi unam Hidam per sexies xx Acras, & super Hidam xxiv Acras. **1607** Gray 432: Terram arrabilem in le Hide vocatam Hutchins Hilles. **1635** Dalton 71: An Hyde of land doth containe...480 acres; ibid 150: One hundred acres is an hide of land. **1664** Spelman 291: Hida, & Hyda: Scotis, Hilda.... Terræ portio, quæ vel ad alimonium unius familiæ, vel ad annuum pensum unius aratri designatur. **1665** Sheppard 21: A Hide of Land, (in Latine, Hida terræ) is a certain measure or quantity of Land...that may be plowed by one Plough in a year. Or (as others says) it is 100 acres: Or (as others would have) as much as will maintain a Family. Some say it consisteth of 100 Acres, every acre in length 40 Perches, every Perch 16 foot and a half. And again, some say, Eight Hides are 800 Acres. **1678** Du Cange sv hida: Hida, et Hyda, ex Saxon. hyd, Terræ portio, quantum sufficit ad arandum uni aratro per annum. **1682** Hall 29: A Hyde of land is fiue yards of land. **1695** Kennett Glossary sv hide: Hida Anglice vocatur terra unius aratri culturæ sufficiens, whence our term of Plough-land. The quantity of a hide was never expresly determin'd. Gervase of Tilbury makes it one hundred acres. The Malmsbury MSS...computes it as 96. acres, one hide four virgates, and every virgate 24. acres. And yet the History of the foundation of the Abby of Battle...makes eight virgates go to one hide.... One hide of land at Chesterton 15. Hen. II. contain'd sixty-four acres. **1708** Chamberlayne 208: An Hundred Acres are accounted an Hide of Land. **1755** Willis 358: Some hold an Hide to contain 4 yard Land, and some an 100, or 120 Acres; some account it to be all one with a Carucate or Plough-Land; ibid 359: The distribution of England by Hides of Lands is very antient, mention being made thereof in the Laws of Ina, a West Saxon King, about the Year 690, Cap. 14. **1867** C. I. Elton 126: It is frequently stated in ancient records that the hide (often called carucata) contained eight oxgangs, each of fifteen acres, so that it equalled 120 acres. **1888** Round 3.195: I hold that we have in the Domesday Survey three equivalent units of assessment—the hide, the carucate, and the solin; ibid 213: If what I have termed the objective hide...was reckoned...at 120 acres, that is four virgates of 30 acres each, then the subjective (or geld) hide would naturally be similarly reckoned as containing four virgates, or 120 acres. **1897** Maitland 393: Broughton 1 H. = 6 1/2 V. = 208 A.... Weston 1 H. = 4 V. = 112 A. Brington 1 H. = 4 V. = 136 A. Bythorn 1 H. = 4 V. = 176 A.... Cranfield 1 H. = 4 V. = 192 A; ibid 394: Therfield 1 H. = 4 V. = 256 A.... Graveley.... 1 H. = 6 3/4 V. = 135 A. **1904** Salzman 3.92: There is sufficient evidence to justify the positive assertion that the Sussex hide contained eight virgates. See CARUCATE; CARUE; and PLOWLAND

hilda. HIDE

hlæst. LAST

hobaid. HOBED

hobbet—8 hobbett; 8-9 hobbet, hobbit [E dial hobbet, hobbit, a measure of

2 or more bu, of unknown origin]. A m-c for wheat and other dry products generally totaling 168 lb (76.203 kg).—**1790** Miller 18: The COOMB, SEAM, HOBBETT. **1883** McConnell 11: Hobbet of 168 lbs., at Denbigh. **1896** Wagstaff 36: A ´hobbet´ of old potatoes in Flintshire = 200 lbs. A ´hobbet´ of new potatoes...210 lbs. **1956** Economist 50: Hobbet: Wheat = 168 lb. **1966** O´Keefe 130: Wheat was sold at a price per "hobbit," a term used in Wales to express a quantity consisting of 4 pecks, each peck weighing 42 lb., i.e., a total of 168 lb.

hobbett, hobbit. HOBBET

hobed—8-9 hobaid, hobed [perh akin to E dial hobbet, hobbit, a measure of 2 or more bu]. A m-c in South Wales for lime, 4 pedwran of 5 or 6 qt each (c2.20 to c2.64 dkl), and in North Wales for lime, 2 STOREDS or 4 bu (c1.41 hl), and for wheat, approximately 173 lb (78.471 kg).—**1820** Second Rep. 19: Hobaid or Hobed of lime: S. Wales, 4 pedwran, or quarters, of 5 or 6 quarts each...Anglesia and Caernarvonshire: 2 storeds = 4 bushels; ibid 31: N. Wales; of wheat, 1 1/2 hobaid, to weigh 260 lb.

hogeshead, hogesheade, hogeshed, hogesheved. HOGSHEAD

hoggat—9 hoggat; ? hoggett (OED) [var of HOGSHEAD]. A m-c for grain in Ireland (c1800) containing 10 bu (c3.52 hl).—**1816** Kelly 114: The Hoggat and the Bow are terms made use of for certain measures in the county of Down, and some other northern parts, and are equal to ten bushels, or two barrels and a half of the Bristol measure.

hoggeshead, hoggesheade, hoggeshed, hoggeshedde, hoggeshede,

hoggesheed, hoggesyde. HOGSHEAD

hoggett. HOGGAT

hoggishede, hoggisheed, hoggshed, hoggyshead, hoggyshed, hoggyshede, hoggyssed, hogheid, hogishead. HOGSHEAD

hogshead—4 hoogeshed; 4-6 hoggeshed; 5 hogesheved (Finchale), hoggeshede, hoggesyde (OED), hoggishede (OED), hoggyshead, hoggyshed, hoggyshede, hogyshede; 5-6 hogyshed; 6 hogesheade, hogeshed, hoggesheade, hoggesheed (OED), hoggisheed (OED), hoggshead, hoggyssed, hogheid, hogshed, hogsheed (OED); 6-7 hoggeshead (OED), hoggeshedde; 6-9 hogshead; 7 hogeshead, hoggshed, hogishead, hogsheade, hogshede (OED) [ME hoggeshed fr hogges, poss of hogge, hog, + hed, head; the reason for the name is uncertain]. A m-c for many products: ale, 48 ale gal (c2.22 hl) equal to 1 1/2 bbl, or 3 KILDERKINS, or 6 FIRKINS; beer, 54 beer gal (c2.49 hl) equal to 1 1/2 bbl, or 3 kilderkins, or 6 firkins; cider, Guernsey and Jersey, 60 gal (c2.27 hl), Herefordshire and Worcestershire, 110 gal (c4.16 hl); fish, mostly herrings and pilchards, generally reckoned at 3500 in number, but variations from 3000 to 4000 were not uncommon; honey, oil, and wine, 63 wine gal (c2.38 hl) equal to 1/2 PIPE, or 1/4 TUN; lime, Dorsetshire, 4 bu (c1.41 hl), Devonshire, sometimes 11 1/2 heaped bu (c4.58 hl); molasses, 100 gal (c3.78 hl); and oats, Cornwall, 9 bu (c3.17 hl). It was frequently abbreviated hhd. Since the establishment of the Imperial system the hogshead of beer has been reckoned at 54 gal (2.455 hl) everywhere in the United Kingdom except in Ireland, 52 gal (2.364 hl).—**1391** Henry Derby 23: Clerico panetrie per manus Fyssher pro ij barellis et j hoogeshed vacuis per ipsum pro floure imponendo xviii d. **1423** Rot. Parl. 4.256: The Hoggeshede xx [X] iii [+] III galons. **1439** Southampton 2.11: 1 hoggyshed de glassis; ibid 50: 1 pipa et 1 hoggyshed de alym...3 hoggyshedys...de haberdasshe...1 hoggyshead de naperye; ibid 62: Pro 6 hoggyshedys sulfuris. **1443** Brokage II.42: Et i hogyshede saponis; ibid 46: Cum iii hogyshedys hony; ibid 49: Cum hogyshede de horsseshoue...cum 1 hogyshede olei; ibid 171: 1 hogyshede caudorons veteris...1 hogyshede pelewys. **1444** Rot. Parl. 5.114: That every Tonne contene xx [X] xii and XII Galons, and every Pipe xx [X] vi [+] VI Galons...and every Hoggeshede LXIII Galons. c**1500** Brit. Mus. 24.16: Hoggesheades of wyne.... The hogesheade 63 gallons. **1507** Gras 1.699: Iryne wyer the hoggeshed. **1509** Ibid 590: Pro iii hog[eshedes] beere. **1517** Hall 49: And the contente of the Gascoyne hoges[hed´] shuld´ be, yf hyt kepe gawge, iii [X] xx & iii galons. **1533** Gross II.73: A hoggyssed or a bott of wine for his owne drinkine. **1545** Rates 1.44: Wyer the hoggesheade. **1547** Cal. Pat. 19.397: Buttes, pypes, hoggesheddes, pontions or barrelles. **1572** Mer. Adven. 97: By hoggeshed or hoggesheds. **1587** Stat. 267: The tunne of wine CC.lii. galons. the pipe C.xxvi. galons...the hoggeshed three score and three galons. c**1590** Hall 21: The hogshed which is 1/4 of a tunne contenith 63 gallons. c**1590** Horwood 640: Barrell, hoggshead, pipe. **1595** Powell C: Euery hogshead to holde and conteine threescore and three gallons. **1597** Halyburton cx: Beiffe ye hogheid thairof. c**1600** Brit. Mus. 16.70: Hoggeshedde [of wine], is .2.

barrelles. **1600** Hill 67: 63. Gallons...1. Hogsheade. 2. Hogsheades...1. Pipe or Butte. **1607** Clode 307: For 2 hogesheads of gasconie wine...For one hogeshead of cunnock wine. **1612** Halyburton 311: Olives the hogishead. **1615** Collect. Stat. 467: And euerie Hogshead to containe threescore and three gallons. **1619** Young II.152: 2 Hoggsheds of bere. **1635** Dalton 148: Wine, Oyle, and Honey: their measure is all one...Hogshead, 63. gallons. **1682** Hall 29: 1 Tunne conteynes...2 Pipes or Butts...4 Hogsheads. **1708** Chamberlayne 210: 1 Barrel and half, or 54 Gallons make a Hogshead. **1717** Dict. Rus. sv: Hogshead, a Measure or Vessel of Wine or Oil, containing the fourth part of a Tun or 63 Gallons; two of these Hogsheads make a Pipe or Butt. **1740** Barlow 457: 8 cubic Foot of Water make a Hogshead and 4 Hogshead a Ton. **1811** Carew 2.103: To make them really good fish, fit for a foreign market, and to bear all accidents of weather, they ought to lie in bulk three weeks...allowing at least three Winchester bushels of salt to every hogshead, generally computed at 3500 fish; ibid 104: They still mark the number of the fish on the head of each hogshead, which generally contain between 3 and 4000, according to the size of the fish. **1816** Kelly 87: WINE MEASURE...63 Gallons...1 Hogshead...238,4509 [1]. **1820** Second Rep. 19: Hogshead...Formerly of ale 48 gallons; of beer 54...of mollasses, 100 gallons...Herefordshire and Worcestershire: of cider, 110 gallons...Guernsey and Jersey: of cider, 120 pots, 60 gallons. **1850** Alexander 43: Hogshead; for wine, etc...63.—gallons. **1880** Britten 171: Hogshead.... (Cornw.), of oats, 9 Winchester bushels. (Dev.), of lime...sometimes 11 1/2 heaped bushels, Winchester. (Dors.), of lime, 4 bushels. **1956** Economist 54: Hogshead = 54 gallons (52 gallons in Ireland).

hogsheade, hogshed, hogshede, hogsheed, hogyshed, hogyshede. HOGSHEAD

hond, honde. HAND

honderd, hondered, hondert, honderte, honderyd, hondird, hondred, hondret, hondreth, hondryd, hondrythe. HUNDRED

hoogeshed. HOGSHEAD

hoond, hoonde. HAND

hoop—3 L hopa, L hoppa (St. Paul´s); 5 hop, hope; 6-9 hoop [ME hop, hoop fr OE hōp; akin to MDu hoep, ring, band, hoop]. A m-c for grain: Durham, 1/4 pk (c2.20 l); Montgomeryshire, 5 gal (c2.20 dkl), also called a PECCAID; Shropshire and Worcestershire, 1 pk (c8.81 l); and St. Paul´s Estate, 1 bu (c3.52 dkl). In Kendal a hoop of vegetables (c1895) equaled 1 Imperial gal (4.546 l) (Wagstaff 36).—**1208** Bish. Winch. 7: Et de xlv quarteriis j estrica j hopa de Alta Clera. **1467** Cov. Leet 334: Also they have ordenyd that the wardens Make ij strikis, ij halfe strykis, ij hopes, & let the salters have hem with-owt eny money.... Also they woll that no retaylers in the Cete take no hyr for the lone of strykis, half-strykis nor hopus lande [loaned] to the salters. **1819** Cyclopædia sv weights: The quarter bushel [in Shropshire] is called a hoop or peck. **1820** Second Rep. 19: Hoop—Durham, 1/4 peck, Shropshire, a peck, Montgomeryshire, 5 gallons, called also a peccaid.

hop, hopa, hope, hoppa. HOOP

houndred, houndret, hownderd, howndrythe, hunderd, hundered, hunderet, hundereth, hunderit, hunderith, hunderyd, hundird. HUNDRED

hundred—1-9 hundred; 3 hunndredd (OED); 3-4 hondret (OED), houndret (OED), hundret (OED); 3-5 hondred; 3-7 hundered (OED); 4 hondird (OED), houndred (OED), hunderet (OED), hunderit (OED), hunderyd, hundird (OED), hundryd; 4-6 hundride (OED), hyndyrd (OED); 4-8 hunderd; 4-9 hundrid; 5 honderd (OED), hondert (OED), honderte (OED), hondryd, hownderd, howndrythe, hundurd, hundyrt (OED); 5-6 hondered (OED), honderyd (OED); 6 hundereth, hunderith, hundrede, hundrethe, hundrith, hundrythe; 6-7 hundreth; 7 hondreth, L hundredus [OE hundred fr stems of hund, hundred, + -red; akin to Goth rathjo, number, reckoning]. A m-q, the CENT (C), and a wt, the Cwt, for many products.

The C generally numbered 100, but larger amounts were not uncommon: 106, lambs and sheep in Roxburghshire and Selkirkshire; 120, the long- or great-hundred for balks (called "great," "middle," or "small"), barlings, boards (barrel, clap, and pipe), bomspars or boom-spars, bowstaves, cant spars, canvas, capravens, cattle, cruises, deals, eggs, faggots, herrings, hogshead staves, lambskins, linen cloth, nails, oars, pins, poles, reeds, spars, stockfish, stones, tile, and wainscoats; 124, cod (sometimes 120), ling (sometimes 120), haberdine, and saltfish; 132, herrings in Fifeshire; 160, "hardfish"; and 225, onions and garlic.—c**1375** Hyda 68: Fowr hundyrd pund, everyche of hem an hunderyd pund. **1387** Higden I.57: Sex hondred paas. c**1450** Common 168: vij hundurd and di. c**1461** Hall 18: A man mak in hys couenawnt to haue the gret hondrythe. **1507** Gras 1.696: Clapp owlde the grett howndrythe and every c ys £iii. **1519** Mer. Adven. 57: And of every hundreth shepe skynnes, ij d., and of every hundreth lam fells, j d. **1545** Rates 1.4: Bowstaues the hundrith; ibid 16: Fysshe...the hunderith; ibid 19: Hedlak the hundereth elles conteynynge .xii. score elles. **1555** York Mer. 156: A hundreth waynescotts, six shillings and eyght-pence. **1560** Remembrance 72: For a hundrethe of Newelande fyshe so carryd...ii d. For drye lynge & stockefyshe the hundrith...ii d. **1562** York Mer. 168: Clabbord the small hundrythe, x d.; Waynskotte the small hundrythe, x. s. **1578** Mer. Adven. 100: For everie hunderd skinnes so bowght. **1581** Acts Scotland 3.216: Euerie hundreth skynis sex scoir. c**1590** Hall 27: The hundred of canvas and of lynnen clothe is and contenith 120 to the hundrid; ibid 28: The Hundred consisteth of 15 ropes and euery rope 15 heades; so that the 100 of onyons and garlike consisteth 225. **1603** Henllys 139: Hearings are sold freshe by the meise, w[hich] is five hundred, eche hundred contayninge vj [X] xx. **1613** Tap 1.67: Ling, Codde or Haberdine, 124 to the hundreth. Stockfish, 120 to the hundreth. **1615** Collect. Stat. 465: A hundred of Garlik consisteth of 15. ropes, & euery rope containeth 15. heads. **1616** Hopton 162: Herrings...at 120 to the hundred; ibid 164: Ling, Cod, or Haberdine hath 124 to the hundred. **1635** Dalton 149: Six score herrings shall goe to the hundred.... The hundred of hard fish must containe eight score; ibid 150: Also all other headed things, as nailes, pins, &c. are sold six score to the hundred. **1638** Bolton 274: Cattell and fish are sould sixscore to the hundred, and yet the hundred of hard fish must containe eightscore. **1678** Du Cange sv centena: Ferri, ex 100. petris. **1682** Hall 29: Ling, Cod or Haberdine, 124 to the Hundred; ibid 30: Coney, Kid, Lambe Bulge, Catt, etc.: 5 Skore to the hundred. **1701** Hatton 3.15: 120 in Number is the Hundred of Balks of all sorts...Barlings...Barrel Boards...Bomspars...Bowstaves...Cant Spars...Hogs-heads Staves. **1704** Mer. Adven. 245: Ffor sorting and laying up every hundred hogshead staves belonging to a ffreeman. **1708** Chamberlayne 205: Cod-fish, Haberdine, Ling, etc. have 124 to the C.... Herrings 120 to the C.... Filches, Grays, Jennets, Martins, Minks, Sables, 40 Skins is a Timber...other Skins, five Score to the Hundred. **1784** Ricard II.152: Le hundred, ou cent, ou quintal, qui est de 112 lb, avoir du poids; ibid 155: Le hundred, ou la centaine de poissons secs, est compté pour 124 pieces. **1805** Macpherson I.471: 120 herrings 1 hundred. **1819** Cyclopædia sv weights: The hundred of six score, such as hop-poles, faggots, &c. **1820** Second Rep. 19: Hundred...eggs, oars, spars and stone, 120...of mullets, 8 score = 160...of faggots, 6 score...of bunches of reeds, 6 score...Fifeshire: of herrings, 132...Roxburghshire and Selkirkshire: of sheep or lambs, sometimes 106. **1880** Britten 171: Hundred, of balks, deals, eggs, faggots, bunches, &c., generally 120.

The Cwt generally weighed 112 lb (50.802 kg) and was equal to 1/20 ton of 2240 lb (1016.040 kg), but, like the C, it had several variations: 100 lb (45.359 kg), aloes, angelica, annatto, antimony, arsenic, asafetida, benjamin, brass manufactured items, bugle, capers, cloves, copal, cotton, crossbow thread, down, galingale, gentian, ginger, ginseng, gum guaiac, gunpowder, indigo, isinglass, manna, myrrh, pepper (long), pimento, plums, raw linen, saccharum, sarsaparilla, thrums, tobacco, turmeric, and verdigris; 104 lb (47.173 kg), filberts in Kent; 108 lb (48.988 kg), almonds, alum (sometimes 112 lb), cinnamon, nutmegs, pepper, sugar, and wax; 113 lb (51.256 kg), cheese in Salisbury, and at Bridgenorth in Shropshire; 120 lb (54.431 kg), cheese in Cambridgeshire, Cheshire, Derbyshire, Hampshire, Leicestershire, Staffordshire; potatoes in Essex; hay in Cheshire and Derbyshire; coal in Derbyshire and in part of Shropshire; tin, occasionally and called the "Stannery or Stannary Hundred"; and iron at the king's scales in Cornwall; and 121 lb (54.884 kg), cheese in Shrewsbury. It was sometimes abbreviated _hund. wt._—_c_**1253** Hall 11: La centeine de cire, sucre, peyuer, cumin, almand, et de alume, si est de xiii peris et di., et checune pere de viii li.... La sume de lib. en la centeyne, cent viii li. _c_**1272** _Ibid_ 10: Item centena zucari, cere, piperis, cimini, amigdalorum, et allume continet tresdecim petras et dimidiam; et quelibet petra continet octo libras. **1290** Fleta 119: Item centena cere, xucarii, piperis, cumini, amigdolarum et aloigne continet xiij. petras et dimidiam, et quelibet petra continet octo libras. _c_**1340** Pegolotti 255: Mandorle, e riso...e stagno...e ferro, e tutte cose grosse si vendono in Londra a centinaio, di libbre 112 per 1 centinaio. c**1461** Hall 13: And by this weyght [112 lb] be all maner of merchaundyse bought and sold, as tynne, lede, iron, coper, style, wode...madder...laces, sylks, threde, flex, hempe, ropys, talowe.... Also lede ys sold...after v [X] xx [+] xii to the C.... And other warys that be sold by the lb., as peper, saffryn, clowys, mace, gynger and other suche, thes be called Sotyll Warys and they wold be rekynnyd after v [X] xx to the C. **1474** Cov. Leet 396: And to this day the C. ys trewe after xx [X] v for the C.... xx [X] v for the C, the wich kepes weyght & mesure l li. the halfe C, xxv li. the quartern. **1517** Hall 48: The juste halfe hondryd weygthe. **1569** Remembrance 109: xii hallfe hownderds, i quarter, i xiiii [stone] and vii li [nail]. **1577** D. Gray 7: The hundreth waight at the Common Beame in London containeth 112. lib. haberdepoiz. c**1590** Hall 22: The hundred waight of gunpowder is but fyve skore poundes waight, haberdepoyse, to the hundrid; ibid 23: The 100 of tynne at the marchantes of London is but 112 poundes haberdepoyse; ibid 24: But at the Kings beame at Cornwall yt is 120 poundes waight [for iron] to the 100; ibid 25: Item waxe...sugare, peper, cinamond, nuttmegs contaynith 13 stone 1/2; and euery stonne 8 to the hundrid; so that the hundrid contenith 108; ibid 27: The load of hay is but 18 hundredes...and euery hundred 112 poundes waight.... The casse of glasse is a hundrid and 3/4 in waight, after 112 to the 100. **1590** Rates 2.2: Antimonium the C. lb. containing v [X] xx...Arsenick the C. containing v [X] xx lb; ibid 4: Beniamin the C. containing v [X] xx pounde; Cloues the c. li. containing v [X] xx;

ibid 16: Galingale the c. containing v [X] xx. pound; ibid 23: Long pepper the hundreth containing v [X] xx. **1595** Powell C2: Two hundreth weight, is a leuen score and foure poundes. **1597** Halyburton cxiv: The j [X] c wecht of Casnet suker in barrellis. c**1600** Brit. Mus. 31.213: 7 lb is the 1/16 of a hondreth...56 lb is the 1/2 of a hondreth...98 lb is the 7/8 of a hondreth.... Hondreth Waight: which is 112 lb. **1600** Hill 66: 112. Poundes...maketh 1. hundred weight. **1607** B. J. 19: Note that in most parts of Spaine, their Kintall is 100. li. and containeth of our English waight but 102 li. So as our hundredwaight is 10. in the 100. greater than theirs. **1615** Collect. Stat. 465: A hundred of ware, sugar, pepper, cynamome...containeth...108 l. **1616** Hopton 163: Tinne, Copper, and Lattine haue 112 pounds to the hundred. **1635** Dalton 149: Sugar, spices, and wax...108. li. maketh the hundred.... Hops, five score and twelve pounds maketh the hundred. **1646** H. Baker 211: Our Hundreth waight here at London, which is after 112 lib. for the C. **1665** Assize 5: But the weight of the Wey of Essex-Cheese or Butter, is three hundred pounds weight, after the rate of five score and twelve pounds of Avoirdupois-weight.... The sack of Woll is three hundred twenty eight pounds, and a hundred and twelve pounds to every hundred weight. **1678** Du Cange sv centanarium: Pondus centum librarum.... Centena ceræ, zuccari, piperis, cumini...apud Anglos, continet 13. petras et dimidiam: et quælibet petra continet 8. libras. Summa ergo librarum in centena 108. **1688** Bernardi 137-38: Libra equidem Avoirdupois...1/112 Hundredi. **1701** Hatton 3.16: Things of which five Score is reckoned a Hundred Weight. Crossbow-thread. Ginger...Indigo. Thrums. Capers...Brass and Lattin Manufactures. **1717** Dict. Rus. sv: Hundred-Weight, the quantity of 112 Pounds in Aver-du-pois greater Weight. **1755** Postlethwayt II.188: For wax, sugar, spices, and allum...108 pounds, made the Hundred weight. **1783** Beawes 893: But besides this hundred Weight there is another called the Stannery Hundred, by which Tin...is weighed to the King.... Stannery Hundred of 120 Pounds. **1811** Carew 2.45: Note, the stannary weight is 120 lb. to the hundred. **1819** Cyclopædia sv weights: Cheese is sold by the cwt., which, at Shrewsbury, means 121 lbs., and 113 lbs at Bridgnorth.... Some articles are sold [in Cheshire] by...the long hundred of 120 lbs. Cheese is one of these. Hay, too, is generally sold by the cwt. of 120 lbs. **1820** Second Rep. 20: Hundred-Weight...properly 112 lbs = 4 quarters = 8 stone; but of aloes, angelica, annatto, asafaetida...capers, cotton, down, gentian, ginseng...gum guaicum, indigo, isinglass, manna, myrrh, long pepper, pimento, plums, saccharum...sarsaparilla, tobacco, turmeric, verdigris and raw linen yarn, 100 lbs are to be reckoned a hundred weight.... Kent: of filberts, 104 lbs. **1834** Pasley 113: 1 Hundredweight of Aloes, Angelica, Annotto, Assafœtida, Bugle...100 [lb]; ibid 114: 1 Hundredweight of Cheese at Bridgnorth, Shropshire...113 [lb]. 1 Hundredweight, usually called the long hundredweight, for Cheese in Cambridgeshire, Cheshire, Derbyshire and Leicestershire; for Hay, in Cheshire and Derbyshire; for Coals in Derbyshire and in part of

Shropshire; for Potatoes in Essex.... 1 Hundredweight of Cheese at Shrewsbury...121 [lb]. **1850** Alexander 43: Hundred weight; nett...England: 1300...100.—pounds.... for sugar and wax...108.—pounds. **1880** Britten 171: Hundred Weight (Camb.), of cheese, 120 lbs. (Ches.), of cheese...120 lbs.... (Derb.), of cheese, among diarymen, 120 lbs. (Ess.), of potatoes, 120 lbs. (Hunts.), of Leicester cheese, 120 lbs.... (Leic.), of cheese, 120 lbs. (Sal.), of cheese, Bridgenorth 113 lbs., Shrewsbury 121 lbs. (Staff.), of cheese, at Wolverhampton, 120 lbs. **1880** Courtney 153: Cwt. is formed from c., centum, wt., weight. **1882** Jackson 413: Hundredweight = 112 pounds. **1896** Wagstaff 41: A hundredweight may mean 100 lbs., 112 lbs., or 120 lbs. **1907** Hatch 34: 1 hundredweight = 50.802352 kilograms. **1951** Trade 28: Hundredweight = 112 pounds. See CENT; QUINTAL

hundrede, hundredus, hundret, hundreth, hundrethe, hundrid, hundride, hundrith, hundryd, hundrythe, hundurd, hundyrd, hundyrt, hunndredd. HUNDRED

hutch [ME huche fr OF huche, huge fr LL hutica; see OED]. A m-c in Renfrewshire (c1800-1900), a chest or coffer containing 2 Cwt (101.604 kg) of copperas or pyrite stone (Second Rep. 20 and Donisthorpe 211). See HUNDRED

huttock. HATTOCK

hyda, hyde. HIDE

hyle [perh a special sense of hill fr ME hill, hul fr OE hyll; hence, a large pile or stack]. A m-c for flax in Hampshire (c1800-1900)

198] hyle

containing 10 sheaves (Second Rep. 20 and Britten 172).

I

ialon. GALLON

iarre. JAR

incast [in + cast (after vb cast in)]. A quantity of some commodity "thrown in" or given in addition to the requirements of a particular measure; for example, an extra lb of wool in a STONE of wool (Britten 172 and Donisthorpe 211).

ince. INCH

inch—1 ince (OED), ynce (OED); 1-7 L pollex, L uncia; 3 unche (OED); 4-6 ench, enche (OED); 4-7 ynch, ynche; 4-9 inch; 5-7 inche; 6 insch, insche, intch, unch (OED), ynsh [ME inch, inche, ynch fr OE ince, ynce fr L uncia, the twelfth part, inch, ounce]. A m-l (2.54 cm) which originally was a unit of body measurement commonly associated with a thumb´s breadth. In the Roman duodecimal system it was equal to 1/12 ft. During the Roman occupation it was introduced into Britain, where it became part of the English system of weights and measures. Throughout the Middle Ages and the early modern period the inch was defined as the length of 3 medium-sized barleycorns placed end to end. The Scots inch equaled 1.0054 English inches.—c**1075** Hall 2: Quarum haec sunt nomina: digitus, uncia, palmus; ibid 4: Tantum enim precellit pes manualis pedem naturalem, quantum pollex in longitudinem protendi potest. c**1150** Acts Scotland 1.309: Et ex medio pollice hominis debet stare. aut ex longitudine trium granorum boni ordei sine caudis. **1220** Clerkenwell 140: Prima occidentalis cum solario continet in fronte iuxta vicum regium in latitudine tres vlnas et duos pollices.

c**1272** Hall 7: Sciendum quod tria grana ordei, sicca et rotunda, faciunt pollicem.... xij pollices faciunt pedem. c**1300** Ibid 7: Nota quod tria grana ordei de medio spice faciunt pollicem. **1395** York Mem. 1.142: Notandum quod tria grana ordei sicca et rotunda faciunt pollicem. c**1400** Hall 5: Uncia est digitus et eius tercia pars. c**1461** Ibid 14: The lengythe of iij barly cornys make an ynche. **1474** Cov. Leet 396: Also hitt was ordeyned...that iij barley-Cornes take out of the middes of the Ere makith a Inche. c**1475** Nicholson 77: It is to mete that iij Barly Cornys in the myddis of the Ere makyth one ynche, And xij enchis makyth a foote. c**1500** Hall 7: iii grana ordei, de medio spice, faciunt pollicem. **1537** Benese 3: Therefore ye shall take the lengthe of an ynche moost trulye upon an artificers rule, made of two foote in length, after the standarde of London, the which rule doth conteyne xxiiii ynches in lengthe. c**1550** Remembrance 23: Item under xx ynshis goth iii fisshis for one...Item under xx intches three goeth for one. **1587** Acts Scotland 3.521: The eln...threttie sevin Insches; ibid 522: The deipnes sevin insches and half insche. c**1590** Hall 27: Dymension longitudes of the ynche, ffoott, yard...accordinge to the statut and standart of England.... The inche, 3 grayns of Barly, dry and rotund. **1602** More 14-15: And first note that for this purpose, I call an ynch that which is an inch broad and twelue ynches long, of which ynches, twelue doe make a foote. **1603** Henllys 137: Yet doeth yt agree in the ynche, foote and yard. **1615** Collect. Stat. 464: It is ordained that three graines of barley drie and round do make an ynch.

1616 Hopton 165: Three barley cornes make an Inch, 12 Inches a foote. **1618** Acts Scotland 4.586: The which Firlot...shall contein nyneteen Inches/and sext parte inche. **1635** Dalton 150: Three barley cornes measured from end to end (or 4 in thicknesse) maketh one inch. **1647** Digges 1: Wherein is ordained three Barly cornes dry and round to make an Inch. **1664** Gouldman sv: Inch. Pollex. uncia. **1665** Assize 6: Uncia est in pede pars XII. **1678** Du Cange sv alna: Pes Regius est 12. pollicum. **1685** Acts Scotland 8.494: That three barly Corns set lenthways, shall make ane Inch. **1688** Bernardi 192: Uncia. Pollex transversus. 1/12 Pedis eujusque. **1708** Chamberlayne 207: The smallest Applicative Measure is a Barley-Corn, whereof three in length make a Fingers breadth or Inch. **1717** Dict. Rus. sv: Inch, a known Measure, the twelfth part of a Foot, containing the space of three Barley-corns in length. **1779** Swinton 24: Scotch Inch. = 1.0054054 [English inches]. **1843** Strachan 88: 8 acres = 50181120 inches. **1882** Jackson 282: Inch = 3 barleycorns. **1907** Hatch 35: 1 inch = 25.39997 millimetres. **1916** Stratton 24: 1 inch = 2.54 centimeters. **1951** Trade 27: Inch = 1/36 yard.

inche, insch, insche, intch. INCH

ioust. JUST

iug. JUG

iuste, iuyste. JUST

jag—6-9 jagg; 8-9 jag; 9 jaug (OED), jog [*]. A m-c for hay; it was smaller than a load of 20 Cwt or 2240 lb (1016.040 kg).—**1717** Dict. Rus. sv load: Load of Hay, contains about two thousand weight, being a good load; but a small load of Hay is called a Jagg. **1829** Brockett 166: Jag...a cart load—York. Moor has jag, a waggon load. **1880** Britten 149: Jog, a small load of hay or corn. **1883** Simmonds sv: Jag, a small load of hay.

jagg. JAG

jalo, jalon. GALLON

jar—4 L jarda; 5 jare (Southampton 1), jarre; 5-8 jarr; 6-7 iarre (OED); 7-9 jar [MF jarre fr OPr jarra fr Ar jarrah, earthen water vessel]. A m-c for dry and liquid products: green ginger, 100 lb (45.359 kg); oil or olives, 12 to 26 gal (c4.54 to c9.84 dkl); green vinegar, 100 lb (45.359 kg); and wheat, 52 lb (23.587 kg).—**1303** Gras 1.356: Pro ii jardis olei...pro l jarda olei. **1443** Brokage II.30: Cum ii jarrys olei; ibid 67: iiii jarrys lymons; ibid 156: i jarre olei continente xii lagenas; ibid 226: i jarr´ vini. **1509** Gras 1.563: l parv´ jarres olei. **1701** Hatton 3.227: Jarr...of Oyl, Olives...18 to 26 Gal. Green-ginger about 100 pounds weight. **1717** Dict. Rus. sv jarr: Of Oil, an earthen Vessel containing from 18 to 26 Gallons. A Jarr of green Ginger, is about 100 Pounds weight. **1756** Rolt sv: JAR...an earthen pot, or pitcher.... The jar of oil is from 18 to 26 gallons; and the jar of green ginger is about 100 lb. weight. **1820** Second Rep. 20: Jar...of Lucca oil, 25 gallons...of green vinegar, 100 lbs...of wheat, 52 lbs. **1829** Palethorpe sv: JAR, in commerce, is the name of a vessel or measure or fixed quantity of divers things. The jar of oil contains from 18 to 26 gallons, the jar of green ginger is about 100 lbs. **1840** Waterston 148: Olive oil, jar...imp. galls. 25.

jardum, jare, jarr, jarre. JAR

jaug. JAG

jill, jille. GILL

joust. JUST

jowcat—6 jowcat, jowcatt [perh Sc dial var of JUG]. A m-c for dry products in Scotland containing approximately 1/4 to 1/2 pt (c0.42 to c0.85 l).—**1587** Acts Scotland 3.521: The same extendis to nyntene pyntis and a Jowcat; ibid 522: The firlot to be augmentit and the standert thereof...to contene nyntene pyntis and tua Jowcattis and this to be the measor of all wictuall.

jowcatt. JOWCAT

jug—6-9 jug; 7 iug; 8 jugg [perh fr jug, nickname for the name Joan]. Equivalent to PINT (Scotland).—**1618** Acts Scotland 4.586: Twentie ane pincts and ane mutchkin of just Sterline Jug. **1624** Huntar 4: The Scottish pinte or standerd Iug of Sterling. **1707** Forbes 70: Our general standing Weights and Measures were the Eln, Firlot, Stone-weight, and Jugg; whereof the first was kept in Edinburgh, the second in Linlithgow, the third in Lanerk, and the fourth in Stirling. **1761** Thomson vi: The standard pint-jug in the custody of the burgh of Stirling, is made of brass, in form of a frustrum of a cone.... [It] contains 103 404/1000 cubic inches. **1779** Swinton 8: They made the Stirling pint or jug the unit of liquid measure.... This standard jug was committed to the keeping of the borough of Stirling. **1791** Keith 2.3: In Scotland the Pint by the Stirling Jugg is 103.404 [cu inches]. **1813** Cooke 47: The Stirling jug (containing one Scotch pint)...103.404 cubic inches. **1816** Kelly 93: The Stirling pint jug is the unit of both the liquid and dry measures of Scotland. **1820** J. Sheppard 91: The standard Stirling jug, in the custody of the dean of Guild of Edinburgh...contains 103 404/1000 English cubic inches. **1895** Donisthorpe 215: The standard jug, which was entrusted in 1621 to the care of the magistrates of Stirling, appears to contain only 103 3/4 cubic inches, or 103 7/10.

jugg. JUG

jugum terre. YOKE OF LAND

just—3-4 L justa; 5 iuste; ? ioust (OED), iuyste (OED), joust (Prior), just (OED) [OF juste fr MedL justa (mensura), right measure (of drink)]. A m-c, a large-bellied pot with handles, for ale, beer, and wine generally containing 1 1/2 gal (c6.90 l).—c**1220** Evesham 209: Et numeri conservationem cochlearium, ciphorum, justarum, manutergiorum, et aliorum utensilium; ibid 218: Percipiet etiam quilibet fratrum quotidie duas justas de cerevisia, et certæ mensuræ. c**1290** Worcester 124b: Justæ. c**1330** Hall 31: Dolium vini de Moysun continet communiter cxlii justas ceruisie. Justa ceruisie continet i lagenam et dimidiam secundum standardum Regis. c**1440** Promp. Parv. 268: Iuste, potte.

justa. JUST

K

kace. CASE

kage, kaig. CAGE

karat, karect, karet. CARAT

kark, karke. CARK

karrat. CARAT

karre. CARK

kase. CASE

keaver. KIVER

keel—5-7 keill (OED), kele; 5-8 keil; 6 keyle (OED), keyll; 6-7 keele, keile; 7 keell; 7-9 keel; 8 kiell (OED) [ME kele fr MDu kiel, ship, boat]. A m-c for coal, the capacity of a barge or flat-bottomed ship. It was commonly called a barge-load.

When the CHALDER was standardized in 1421 at a capacity of 32 bu totaling 1 ton of 2000 lb (907.180 kg), the keel contained 20 of these chalders or 20 tons (18,143.600 kg). After the chalder was increased in 1677 to 36 heaped bu totaling 1 ton of 2240 lb (1016.040 kg), the number of chalders in the keel was changed to 16 (35,840 lb or 16,256.640 kg).—**1421** Rot. Parl. 4.148: & sont ascuns gentz qi ont tielx Keles del portage de XXII ou XXIII Chaldres, & votre Custume ent est prise solonc le portage de XX Chaldres tant soulement, en deceite de Vous, tres soverain Sr. **1555** York Mer. 155: Peter Hudelesse, Richard Plaskett...of the saide cytye of Yorke, owners of certeine keles, bootes, and lighteners. **1562** Ibid 168: Provided alwaies that the merchaunts shall pay his frgyt within two days next after the keyll shall be delyvered, wytheowte any further delaye. **1787** Hale 230: It appears by the parliament roll 9. H. 5...commissions are directed to be issued to examine the quantity of the keels, in which such coals are laden, which should contain just twenty chaldrons.

The Newcastle chalder, however, was much larger than the standard chalder, and when it weighed 42 Cwt (2133.684 kg), 10 Newcastle chalders or 420 Cwt (21,336.840 kg) made a keel. After this chalder was fixed at a capacity of 72 heaped bu totaling 53 Cwt (2692.510 kg) in 1695, the capacity of the keel was changed to 8 chalders totaling 424 Cwt or 47,488 lb (21,540.177 kg).—**1603** Hostmen 19: No free brother of the saide ffelloshipp of Hostmen, shall from henceforth sell or lode in any shipp...any kynde of Coles, by lesser or greater measure then the true and accustomed measure of the Keeles or Lighters; ibid 36: And that from henceforth there shall no Coles att all be brought from aborde of any shipp, Hoie, or other vessell in any Keele or Lighter whatsoever, except yt be the sweepings, and that not to exceed in any one Keell or Lighter above two smale maunds. **1604** Ibid 54: To the said owners of Keles. **1650** Ibid 91: Whereas it appeareth by good and sufficient Testimonye that Gilbert Ellet...hath sould Eight Chalder of Coles to A man of warr.... It is therefore, Ordered That the said Gilbert Ellet shall not, dureing the time of one whole yeare, worke or serve...any Brother...in any keele or boat. **1656** Ibid 109: And whereas also the usual faire for each keele carryinge Eight Chalder of coles to the shipes was heretofore but seaven shillinges. **1679** Ibid 139: The said

Customehouse officers threaten to seize the keiles that are measured by stoke nales. **1704** Mer. Adven. 243: Ffor takeing forth of every tonne of wine from a keel or boat. **1706** Hostmen 169: And they conceived for avoideing all fraudes in the admeasure[ment] of Keiles, That the Com[missioners]...be applyed to...and that one, Two, or more persons be appointed to Examine and give Acc[ount] of the same and of all Screwed upp and Stoaked Keils, the same being a very great fraud. **1784** Ency. meth. 138: Le keel, de 8 chaldrons. **1820** Second Rep. 20: Keel of coals: Newcastle, 8 Newcastle Chaldrons = 21 ton 4 cwt = 424 cwt. **1829** Brockett 65: 8 Newcastle chaldrons make a keel; ibid 171: Keel-of-Coals, 8 Newcastle chaldrons, 21 tons, 4 cwt. **1850** Alexander 47: Keel...Newcastle...47488.—pounds. **1883** Simmonds sv: Keel...contains about 8 Newcastle chaldrons = 15 1/2 London chaldrons or 21 tons 4 cwt. See HUNDRED

keele, keell. KEEL

keever. KIVER

keil, keile, keill. KEEL

keippe. KIP

kele. KEEL

kemp—4-5 kemp (OED), kempe [*]. A m-c, a barrel or cask, of undetermined size generally used for fish.—**1391** Henry Derby 77: Pro ij kempes de rubiis allecibus. c**1440** Promp. Parv. 270: Kempe of herynge, or spyrlynge.

kempe. KEMP

kempkin. KILDERKIN

kemple—7 kimple (OED); 8-9 kemple [alter of earlier kimple, of Scand origin; akin to ON kimbull, bundle, Icel kimbill, small bundle, small haystack]. A m-c for straw in Midlothian (c1800-1900) containing 40 windlens of 5 to 6 lb each (90.718 to 108.862 kg) (Second Rep. 37 and Britten 149, 172).

kenning [*]. A m-c for corn in Durham and Northumberland (c1800-1900) containing 2 pk (c1.76 dkl) and equal to 1/2 bu (Second Rep. 21, Cyclopædia sv weights, and Britten 172).

kental, kentall, kentle. QUINTAL

kepe. KIP; KIPE

keuer, kevere. KIVER

keyle, keyll. KEEL

kibin [*]. A m-c for grain in Anglesey and Carnarvon (c1800-1900) containing 2 pk (c1.76 dkl) or 1/2 bu (Second Rep. 21 and Donisthorpe 211).

kiell. KEEL

kiever. KIVER

kilderkin—4 kynerrkyn, kynerkyne; 4-7 kilderkyn; 5 kynderkyn; 5-6 kylderkyn; 5-9 kilderkin; 6 kilderking, kilderkynne, kinderkind (OED), kinterkin, kinterkyn, kylderken, kynterkyn (OED); 6-7 kinderkin (OED); 7 kylderkyne; ? kempkin (Shipley), kinkin (Shipley) [ME kilderkin, kilderkyn fr MDu kindekijn, kinnekijn; see WNID3]. A m-c for ale, beer, butter, fish, and soap. It was commonly called a half-barrel.

The ale kilderkin contained 16 gal (c7.39 dkl) and was equal to 2 ale FIRKINS or 1/2 ale bbl. The Irish ale kilderkin (c1800) contained 4352.0 cu inches (7.133 dkl) or 20 Irish gal of 217.6 cu inches each and equal to 2 Irish ale firkins (Edinburgh XII.572).—**1420** Coopers 9: Qe nul braceour ne braceresse vendroit cervoise en groos a nully par barel ne kilderkyn, sinon qe tielx barelx et kilderkyns serroient primierement merchez en la Guyhalle par les deputees del chamberleyn. **1517** Hall 49: xvi galons´ to the ale kylderkyn. **1587** Stat. 595: Euerie kilderkin for ale xvi. gallons. c**1590** Hall 22: The firkin is 8 gallons...the kilderkyn, or 1/2 barill, contenith 16 gallons; ibid 23: And so the kilderkynne, firkyn, and tertione fully packed. **1595** Powell D: For euerie such barrell, Kilderking or Firkin, of ale or beare. c**1600** Brit. Mus. 16.70v: A kylderkyne or halfe barrelle.... Of Ale the kylderkyne contayneth .16. gallons. **1635** Dalton 148: Ale, the measure thereof is...Kilderkin, 16. gallons. **1665** Assize 9: The Kilderkin sixteen gallons. **1708** Chamberlayne 210: 8 Gallons a Firkin of Ale...2 such Firkins make a Kilderkin. **1756** Rolt sv: KILDERKIN. A kind of liquid measure, which contains two firkins, or eighteen gallons of beer measure, and sixteen of ale measure.

The beer kilderkin contained 18 gal (c8.32 dkl) and was equal to 2 beer firkins or 1/2 beer bbl. Since the establishment of the Imperial system the kilderkin of beer has been reckoned at 18 gal (8.183 dkl) everywhere in the United Kingdom except in Ireland, 16 gal (7.274 dkl).—**1517** Hall 50: xviii galons to the kylderkyn´. **1587** Stat. 595:

Euerie kilderkin for beere xviii. gallons. c**1590** Hall 22: The firkyn contenyth 9 galons: the kilderkyn...contenith 18 gallons: the barill...36 gallons. c**1600** Brit. Mus. 16.70v: Of Beare the kilderkyn contayneth .18. gallons. **1616** Hopton 160: Kilderkins, or halfe Barrels. **1635** Dalton 148: Beere, the measure thereof...Kilderkin, 18 gallons. **1717** Dict. Rus. sv: Kilderkin, a kind of Liquid Measure, that contains two Firkins or eighteen Gallons, and two such Kilderkins make a Barrel. **1816** Kelly 87: 2 Firkins...1 Kilderkin...83,1744 [1]. **1883** Simmonds sv: Kilderkin, a beer cask, containing 2 firkins, or 18 gallons. **1956** Economist 54: Kilderkin, half-barrel or rundlet = 18 gallons (16 gallons in Ireland). **1966** O´Keefe 671: 1 kilderkin = 18 gal. = 81.827 l.

The kilderkin of butter or soap conformed to the 16 gal capacity (c7.39 dkl) of the ale kilderkin, but equally important was the weight of the cask: generally 13 lb (5.897 kg) before 1662 and 20 lb (9.072 kg) afterwards.—**1566** Recorde Kiiij: Sope measures, both Fyrkin, Kylderken and Barrell, shoulde be all equall to Ale measures.... Half Barrell...weighe emptye 13 poundes. **1587** Stat. 595: And euerie halfe barrell emptie to be in weight xiii pounds. c**1590** Hall 24: The halfe barill of butter, or kilderkin, caske and all, is 128 poundes waight haberdepoise; in clean butter, but 115 poundes waight haberdepoise. **1635** Dalton 149: Sope, halfe barrell...shall be of the same content that ale is.... Butter also shall be of the same measure that sope is of. **1673** Stat. Charles 159: Every Kilderkin of Butter ought to weigh

One hundred thirty and two pounds gross at the least, that is to say, One hundred and twelve pounds of Neat Butter, and the Cask Twenty pounds.

The kilderkin of salmon, and occasionally of eels, contained 21 gal (c7.95 dkl) or 2 firkins of 10 1/2 gal each, while the kilderkin for most other fish, including eels, conformed to the ale kilderkin capacity of 16 gal (c7.39 dkl).—**1392** Henry Derby 96: Et per manus eiusdem Ricardi et Willelmi Harpeden pro iij kynerkynes de salmone salso per ipsos emptis ibidem, xxxvij scot; ibid 97: Et per manus eiusdem pro j kynerkyn anguillarum per ipsum empt´ ibidem, xj scot. **1393** Ibid 158: Clerico coquine per manus Johannis Bounche de Linne pro j kilderkyn di. de storgon, xvj s. viij d. **1423** Rot. Parl. 4.256: Kynderkyns, Tercianes, and firdekyns of Heryng. **1443** Brokage II.87: A kylderkyn allecii. **1482** Rot. Parl. 6.221: The kilderkin or 1/2 barill 21 galons.... Also it hath ben used, that every Barell for Elys, shuld hold and conteigne XLII Galons, the half Barell...after the same rate. **c1550** Welsh 58: 1 kinterkyn herrings. **c1590** Hall 23: That every...half Barell, ordeyned for Samon, shuld conteygne...XXI Galons. **1635** Dalton 149: Herring...the halfe barrell...shall be the same content that ale is. **1665** Sheppard 60: For...Herring the Barrel, half Barrel, and Firkin, is to be of the same content that Ale is.

The kilderkin was occasionally used for other products.—**c1550** Welsh 172: 12 kinterkins...dry wares.

kilderking, kilderkyn, kilderkynne. KILDERKIN

kimple. KEMPLE

kinderkin, kinderkind, kinkin. KILDERKIN

kintal, kintall. QUINTAL

kinterkin, kinterkyn. KILDERKIN

kip—6 keippe, kepe, kyppe; 6-9 kip; 7 kipp, kippe [cf MLG kip, bundle of hides, MDu kip, kijp, pack or bundle, ON kippi, bundle]. A m-q for skins: lamb, 30, and goat, 50.—**1507** Gras 1.698: Golde skynes the kyppe. **1525** Percy 355: ij Keippe and a half [of lamb skins] after xxx Skynnes in a Kepe. **1613** Tap 1.67: Goats the kippe, 50. **1616** Hopton 164: The skins of Goats are numbered by the kippe, which is 50. **1660** Bridges 31: Of Goat skins, 50 a Kip. **1682** Hall 30: Skins-Goates: 50 to a Kipp.

kipe—4, 9 kype; 6 kepe (OED); 8-9 kipe [ME kype fr OE cȳpe, cȳpa, basket; akin to MLG kīpe, basket]. A m-c, an osier basket, for fish containing approximately 1 bu (c3.52 dkl).—**1706** Phillips sv: Kipe, a Basket made of Osiers, broader at Bottom, and narrow'd by Degrees to the Top, but left open at both Ends; which is used for taking of Fish. **1880** Britten 150: Kype (Glouc.), a wicker measure about a bushel. **1883** Simmonds sv: Kipe, a basket for catching fish.

kipp, kippe. KIP

kishon [Manx kishan, Ir cisean, dim of cis, kish, basket, hamper]. A m-c on the Isle of Man (c1800-1900) containing 8 qt (c8.81 l) or 1 pk (Second Rep. 21 and Britten 172).

kiver—3 L civerus (Bish. Winch.); 5 kevere (OED); 7 keaver (OED), keuer

(OED), kiever (OED); 8 keever (OED); 8-9 kiver [ME kevere, alter of keve, kive, a keeve, tub, vat, fr OE cȳf; see WNID3 sv keeve]. A m-c in Derbyshire and Cheshire (c1800-1900) for corn; a shallow wooden vessel or tub that contained 12 SHEAVES (Second Rep. 21 and Britten 149, 172).

knightes ffee. KNIGHT´S FEE

knight´s fee—3-7 L feodum (militis); 5 knyghtes fee, knyghts fee; 6-? knight´s fee; 7 knightes ffee [knight fr ME knight, boy, youth, knight, fr OE cniht, cneoht, boy, youth, attendant, military follower; fee fr ME fee, fief, payment, fr OF fé, fié, fief, of Gmc origin; akin to OHG fihu, cattle]. A m-a which probably originated as an amount of land needed to support a knight and his family for a period of one year. In this sense, the knight´s fee (also called knight´s service or servicium militis) was regarded as a unit of income for a fighting man just as the HIDE was probably a unit of income for a working man or serf. But, certainly as early as the thirteenth century, the knight´s fee was expressed as a land division containing a definite number of BOVATES, VIRGATES, or hides, and, even though there was little uniformity, the following were the most common: a knight´s fee of 2, 2 1/2, and 3 hides, no standard acreage established for the hide; of 4 hides, each hide containing 120 acres, or 480 acres (c194.40 ha) in all; of 4 hides of 16 virgates, each virgate containing 4 FARTHINGDALES of 10 acres each, or 640 acres (c259.20 ha) in all; of 5 hides of 20 virgates, each virgate containing 24 acres, or 480 acres (c194.40 ha) in all; of 5, 5 1/2, 6, 6 1/2, 7, 7 1/2, 8, 8 1/2, 9, 10, and 12 hides, no standard acreage established for the hide; of 12 hides totaling 600 acres (c243.00 ha); and of 14, 15, 16, 18, 19, 20, 21, 22, 24, 26, 27, 48, and 60 hides, no standard acreage established for the hide.—**1201** Cur. Reg. 9.53: Unde xxvii. hide faciunt feodum j. militis. **1202** Ibid 177: xlviii hide faciunt feodum j militis. **1206** Cur. Reg. 11.120: Unde sex hide faciunt feodum unius militis; ibid 284: In Stodham, unde vj. hide et dimidia faciunt seruicium unius militis. **1208** Feet 2.124: Unde V carrucate terre et dimidia faciunt seruicium unius militis; ibid 148: Unde sexdecim carrucate terre faciunt seruicium unius militis in eadem uilla pro omni seruicio. **1220** Cur. Reg. 3.151: Unde xiiij. carucate terre faciunt feodum j. militis. c**1230** Red Book 431: Osbertus Archidiaconus tenet xj carucatas terræ, unde xiiij carucatæ faciunt feodum militis.... Rogerus Tempestas, iij carucatas et ij bovatas, unde xiiij faciunt feodum militis; ibid 735: Robertus de Everingham, xv bovatas in Haytone, unde ix carucatæ faciunt feodum. Adam de Blainville, j carucatam in Kane, unde xij faciunt feodum.... Thomas de Aslakeby, j carucatam in Boultone, unde xvj faciunt feodum.... Johannes de Selford, iij carucatas in Langeleythorpe, unde xviij faciunt feodum. Hugo de Arderne, iij carucatas in Hundesburtone, unde xxj faciunt feodum.... Rogerus de Stapeltone, ij carucatas in Wath, unde xx faciunt feodum. Willelmus de Besingeby, ij carucatas in Hovingham, unde x faciunt feodum; ibid 736: Willelmus Camerarius, ij carucatas in Asserlay, unde xxij faciunt feodum.... Abbas de Fontibus, iij carucatas in Swyntone et alibi, unde lx faciunt feodum.... Alanus Pistor, dimidiam carucatam in Treske, unde xv faciunt feodum. c**1250** Rameseia III.47-48: Modus qualiter relevium liberorum tenentium domini Abbatis Rameseiæ debet solvi et exigi de feodis militum, et qualiter feodum integrum componitur ex certis hidis, hidæ ex certis virgatis, et virgata ex certis acris; scilicet, quod quatuor hidæ faciunt feodum integrum, quatuor virgatæ hidam; ibid 48: Una hida, quæ est quarta pars feodi.... Una virgata terræ, quæ est quarta pars hidæ; ibid 209: Quinque hidæ...pro uno feodo; ibid 210: Sex hidæ...pro uno feodo. **1253** Greenstreet 7: Johannes de Tapintone tenet in eadem vnum feodum militis de predicto Willelmo. c**1283** Battle xiii: Nota quod virgata terræ et wysta idem sunt et unum significant: Virgata seu wysta est sextadecima pars unius feodi militis: Quatuor virgatæ seu wystæ faciunt unam hydam: Quatuor hydæ faciunt unum feodum militis. **1304** Swinfield 221: Alanus de Walynton' tenet .j. hydam et dimidium apud Walynton' et Masinton' per militiam pro quarta parte unius feodi. **1338** Langtoft 600-01: Decem acræ faciunt ferdellum. Quatuor fardella faciunt virgatam unam. Quatuor virgatæ faciunt hidam unam. Quatuor hidæ feodum unum faciunt; ibid 601: Fardellum Acræ X./virgata XL./hida. CLX./feodum unum CCCCCCXL. **1454** Scrope 221: Johannes Monpyson armiger tenet hydam terræ in villa de Wyly...per quintam partem unius feodi militis. **1494** Fabyan 246: viij. hydes make a knyghtes fee, by the whiche reason, a knyghts fee shuld welde c.lx. acres. **1603** Henllys 135: X plowlands make a knightes ffee being...640 acr. **1610** Norden 59: There is some difference of the quantity of a Knights fee, as the custome of the places doe differ in measure of land, for in the Duchy of Lancaster, a Knights fee containeth foure hides of land.... But after some computation, a Knights fee contayneth fiue hydes of land. **1610** Folkingham 60: A knights Fee.... M. Camden recordes it to be 680 acres or 800 acres. After some computations it containes 5 Hydes of land, each Hyde 4 Yard-land at 24 acres. **1651** Noy 57: Four Yard Lands made a Hide of Land, and foure, and some say eight Hides make a Knights Fee. **1664** Spelman 126: Unde octo Carucatæ faciunt feodum unius militis.... Unde 48. carucatæ faciunt feodum unius militis.... Pro duabus carrucatis terræ, unde xii faciunt feodum.... Unde xxvii. carrucatæ terræ faciunt feudum unius militis. **1665** Sheppard 22: And again, some say, Eight Hides are 800 acres, and make a Knight´s Fee. **1695** Kennett Glossary sv feodum: Militis vel militare. A Knights fee, which by vulgar computation contain´d 480. acres, as 24. acres made a virgate, four virgates one hide, and five hides one Knights fee.... In 3. King Steph. at Ottendon Com. Oxon. five virgates made the fourth part of a Knights fee. **1755** Willis 360: And as to Knight´s Fees, so frequently occurring in this History, it denotes so much Inheritance as is sufficient to maintain a Knight with convenient Revenues, which in Henry the IIId´s Time was 15 £ per Annum; but Sir Thomas Smith rateth it as 40 £ . A Knight´s Fee contained 12 Plough-Land or 600 Acres of Land. **1777** Nicol. and Burn 615: Virgate of land; a yard land consisting (as some say) of 24 acres, whereof four virgates make an hide, and five hides make a knight´s fee. **1872** Robertson 102: In the fertile lands of Herefordshire, three, two and a half, and sometimes only two hides were held for a fee, when in a different part of the very same county six hides and a half were only counted as a quarter of a fee. **1895** Round 2.293: No fixed number of hides constituted a knight´s fee; ibid 294: In the cartae of 1166 we have fees of 5 hides, of 4, of 6, of 10, of 2 1/2, and even of 2; ibid 294-95: The six fees of St. Albans...being 5 1/2, 7, 8 1/2, 6, 5 1/2, 7 1/2; ibid 295: In the Abingdon Cartulary...we find four fees containing 19 hides, three containing 14, a half-fee 4.

knipperkin. NIPPERKIN

knitch [ME knytche, knucche fr OE gecnycc, bond; akin to MLG knocke, bundle, MHG knock, back of the neck, knoche, bone]. A m-q, a bundle or sheaf, for unbroken straw in northern England and Scotland (c1800-1900) having a circumference of 34 inches (8.636 dm) (Britten 172). See BUNDLE; SHEAF; THRAVE; and other meadow or lea measures

knoggin. NOGGIN

knot [ME knot, knotte fr OE cnotta; so called from the knot tied around a skein of yarn after reeling]. A m-q for wool yarn in Essex (c1800) consisting of 80 turns around a reel (Second Rep. 21).

knyghtes fee, knyghts fee. KNIGHT´S FEE

koome. COOMB

kostorell. COSTREL

krenneke, krennock. CRANNOCK

kylderken, kylderkyn, kylderkyne, kynderkyn, kynerkyn,

kynerkyne. KILDERKIN

kyntal, kyntall, kyntayl. QUINTAL

kynterkyn. KILDERKIN

kype. KIPE

kyppe. KIP

lade. LOAD

lagan. LAGEN

lagen—2-8 L lagena; 5-6 L lagina; 5-9 lagen; 7-9 lagan (OED); ? laggon (OED), L legina (Finchale) [L lagena, a flask]. A m-c for liquid and dry products generally containing 1 gal (c3.78 l).—c**1150** Acts Scotland 1.310: Item lagena debet esse sex pollicium et dimidii in profunditate. In latitudine inferiori debet esse .viij. pollicium et dimidii. cum spissitudine ligni vtriusque partis. In rotunditate superiori debet esse .xxvij. pollicium et dimidii. In rotunditate inferiori debet esse .xxiij. pollicium. **1221** Cur. Reg. 4.74: Et j. dolium de cicera de lx. lagenis. **1256** Burton 376: Vendere in civitatibus duas lagenas ad denarium. **1287** Select Cases 2.19: Goldingus de Gepewyz, lagena falsa, quarta falsa, et quia fregit assisam et vendidit pro xvj. d. **1290** Fleta 118: Item scire debet naturam et originem ponderum et mensurarum vt veraciter et perfecte sciat quantum bladi teneat lagena et quantum bussellus. **1299** Liber 367: Lagena cerevisiæ. c**1300** Brit. Mus. 21.61: Octo lagene faciunt bussellum. **1320** Rot. Parl. 1.375: Ad Petitionem hominum de Com´ Devon´ & Cornub´ conquer´ de Mercatoribus vinorum, qui vendunt vina apud civitatem Exon´ pro vi d. videlicet lagenam...& in partibus London venditur lagena pro IIII denar´. **1373** Gross II.102: Pro vna lagena vini. **1395** York Mem. 2.10: Lagena, potella...pro cervisia. c**1420** Evesham 283: Omni septimana duos prichpottos, octo lagenas cervisiæ continentes. **1440** Scrope 229: 2 other pannys of xvj. lagens. **1443** Brokage II.40: ii barellis olei continente l barello xvi laginas; ibid 139: l barello olei continente xvi lagenas; ibid 156: l jarre olei continente xii lagenas. **1448** Abingdon 126: Et remanet j olla enea continens ij lagenas. c**1461** Hall 7: Et viij libre faciunt lagenam...et viij lagene faciunt busshelum Londonie. **1526** Jacobus 51: Expen´ xxx lagine ceruisie; ibid 81: Item j lagina acetj. **1540** St. Mary´s 58: Quolibet modio continente xij. lagenas. **1607** Cowell sv clerk of the market: Of elns, yards, lagens, as quarts, pottels, gallons, &c. **1678** Du Cange sv lagena: Mensuræ species apud Anglos.... Fuit etiam Lagena non liquidorum dumtaxat, sed et aridorum mensura. **1745** Fleetwood 81: I have observ´d before, that Lagena...holds 4 quarts.

lagena, laggon, lagina. LAGEN

laid. LOAD

langenekre [ME langen, long, + ekre, ACRE]. A m-a for land in Kent (c1400) containing 1 1/2 acres (c0.61 ha) (Prior 147 and Robertson 88, 90).

langhsester [ME langh, long, + SESTER]. A m-c at Glastonbury (c1300) which probably contained 5 to 6 gal (c1.89 to c2.27 dkl) (Prior 154).

lasse. LAST

last—1 hlæst (OED); 3 L lestum; 3-6 L lastum; 3-7 L lastus, L lestus; 3-9 last; 4-6 laste, leste (OED); 4-7 lest; 6 lasse (OED); 7 L lasta [ME last, load, fr OE hlæst, load]. A m-c for dry and liquid products: ashes for soap and barrel fish, 12 bbl (c17.76 hl); beer, 12 bbl (c19.92 hl); bowstaves, 6 C; butter, 12 bbl (c17.76 hl); codfish, 4 C; cork, 12 bbl (c17.76 hl); feathers, 1700 lb (771.103 kg); flax, 6 C bonds or 1700 lb (771.103 kg); grain, generally 10 SEAMS or 80 bu (c28.19 hl before 1824 and 29.09 hl afterward); gunpowder, 24 bbl or 2400 lb (1088.616 kg); herrings, 12,000 in number; iron, 12 bbl (?); hides, 20 DICKERS or 200 in number; oatmeal, 12 bbl (c17.76 hl); oats, Cambridgeshire, 10 1/2 seams (c29.60 hl), Huntingdonshire, 1 1/2 tons (1524.060 kg); olive oil, 4 PIPES (c19.08 hl); orchil, 3 Cwt (152.406 kg); peas, 12 bbl (c17.76 hl); pitch, 12 bbl (c17.76 hl); potash, 12 bbl or 2688 lb (1219.248 kg); quern stones, 12 pair; raisins, 24 bbl or 24 Cwt (1219.248 kg); rapeseeds, Yorkshire, 10 seams (c28.19 hl); salmon, 6 PIPES or 504 gal (c19.08 hl); salt, 10 WEYS or 420 bu (c148.00 hl); seeds, Huntingdonshire, 10 1/2 seams or 84 bu (c29.60 hl); soap, 12 bbl (c17.76 hl); sprats, 10 CADES or 10,000 in number; stones (dog), 3 pair; tar, 12 bbl (c14.28 hl); and wool, 12 SACKS or 4368 lb (1981.290 kg).—**c1243** Select Cases 3.lxxxvi: Et in predicta navi fuerunt vj. lesta correi, ij. dakeres minus. **1249** Gross II.359: Et de quolibet lesto allecium quatuor denarios. **c1253** Hall 11: Et [ii] ways de layne sunt un sac, et xii sacs sunt un last.... Le last de arang´ est de xM., et checun mil est de x cent, et chescun cent de vi [X] xx.... Le last de quir est de xx dakers, et checun dakir de x quirs. **c1272** Ibid 10: Et due waye faciunt unum saccum. Et duodecim sacci continent le last. **1290** Fleta 119: Et due waye lane faciunt vnum saccum, et xij. sacca faciunt vnum lestum.... Lestus autem allecii consistit ex x. miliaribus et quodlibet miliare consistit ex decies centum, et quodlibet centum ex secies viginti.... Item lastus coreorum consistit ex [xx.] dakris, et quodlibet dakrum ex x. coreis. c**1300** Brit. Mus. 21.60v: xij. sacci...le last. c**1330** Gross II.229: Item pour vn last de quirs. **1390** Henry Derby 47: Pro iiij lastes cum di. de bere, xij barellis pro le last. **1402** Gras 1.554: Pro ii lastis sope. **1439** Southampton 2.9: 2 last´ allecii albi; ibid 10: 2 last´ saponis nigri; ibid 12: 1 last´ de pyche; ibid 22: 7 last´ et di. allecii rubii; ibid 27: 2 last´ de tarr. c**1461** Hall 17: Also hyds of bestes, fresh, salt and tannyd be sold by the dyker; and x hydes make a dyker; and xx dyker make a last; ibid 18: And xij barrell Osmond [iron] is a last in byenge and sellynge; ibid 19: Flax, vi [X] c bonds make a last; Bowstavys, vi [X] c make a last. c**1475** Gras 1.193: Of a last wood asshen. **1507** Ibid 696: Corke made in barrelles the laste; ibid 697: Dogestonys the laste...Elys called chaffte elles the last; ibid 698: Fyche barreled the laste; ibid 699: Herynge fulle the laste...Herynge shotton´ the laste. **1545** Rates 1.1: Asshes called woad asshes the laste. **1549** Gras 1.630: Pro uno lasto wheate meale. **1555** York Mer. 155: Item, a last of flax and osemondes, for two shillings and sex pence...a last of ashes, for twentye pence...a last of tarr or pyke, two shillings...a last of rede heringe, for two shillings...a last of stockfyshe, for two shillings and sex pence. c**1590** Hall 21: The last of corne is 80 bushells of corne: 10 quarters makith a last; ibid 22: The last of gunpowder is 24 barills, and euery barill contening a hundred waight...[of] fyve skores poundes waight, haberdepoyse; ibid 23: 26

stonnes of woolle is the sacke of woole and the sacke contenith 364 pound waight: 12 sakes is the last of wool.... The but of salmone ought to be 84 gallons fully packed...the last is 6 buttes conteninge 504 gallons.... The last of woole is 4368 poundes waight of woole haberdepoise.... The last of lether consistith 20 dickers of leather; ibid 28: The last of sault is 420 bushells; the way of sault is 42 bushells: 10 wayes makith a last. **1590** Rates 2.9: Codfish the last containing iiij c...Codfish the c. containing v [X] xx; ibid 10: Corck made for Diers the last containing xij. barrels; ibid 27: Orchall the c. containing v [X] xx [+] xij li...the last waying iij c; ibid 28: Pitch and Tarre the last containing xii. barrels; ibid 30: Quern stones the last containing xii paire of the greatest sort; ibid 34: Sope called flemish sope the last containing xij. barrels; ibid 42: Meale the last being x. quarters; ibid 43: Sprots [sic] the last containing x cades. **1613** Tap 1.61: Salmon & Eels. One Last containeth Buts. 6 Barrells. 12 Firkins. 48 Gallond. 504.... Salt. One Last containeth...Bushel. 420. **1616** Hopton 162: Herring...Last being 10000, euery thousand being 1200, which is 12000 Herrings in the Last, at 120 to the hundred; ibid 164: The Last is 20 Dickers, or 200 hides.... A Last of Barrell-fish is twelve Ale Barrels. **1635** Dalton 149: Leather, the content of the dicker, and the last. **1638** Bolton 271: Ten Quarters of corne is a Last. **1665** Assize 5: The sack of Wool is three hundred twenty eight pounds, and a hundred and twelve pounds to every hundred weight...Two weights of wool make a sack, and 12 sacks make a Last. **1678** Du Cange sv lasta: Lastus, Lestus, Last, Lest, voces Onus, pondus, sarcinam in genere denotantes; sed quæ in specie certis quibusdam mensuris ac ponderibus aptantur. **1682** Hall 29: Barrel fish hath 12 Ale barrels to a Last. **1701** Hatton 3.9: In a Last of Wooll, are...12 Sacks. 24 Weys. 156 Tod. 312 Stone. 624 Cloves. 4368 Pounds. **1704** Mer. Adven. 243: Ffor takeing up and bearing a last of redd herring. **1707** Justice 7: A Last of Gun-Powder, contains 24 Barrels, and the Barrel 100 Pound; ibid 43: In Measure, there is allowed to a Last...12 Barrels of Pease...4 Pipes or Butts of Oyl of Olives. **1708** Chamberlayne 205: Herrings 120 to the C, 12 Hundred to the Thousand, which make a Barrel; and 12 Barrels a Last. **1717** Dict. Rus. sv dry measure: And ten Quarters a Last, which contains 5120 Pints, and so many Pounds Troy-Weight. **1787** Hale 199: Of every last of hides containing 20 dickers, and every dicker ten hides. **1805** Macpherson I.471: 2 weyes (of wool) 1 sack, 12 sacks 1 last. **1820** Second Rep. 21: Last...of ashes, codfish, pitch, tar...12 barrels.... Of butter and soap, 12 ale barrels.... Of corn and seed, 10 quarters.... Of feathers...1,700 lbs.... Of gunpowder and raisins, 24 barrels.... Of oatmeal and potash, 12 barrels.... Cambridgeshire: of oats, 21 comb = 10 1/2 quarters.... Huntingdonshire: of...seeds, 10 1/2 quarters = 84 bushels...of oats, 1 1/2 ton.... Yorkshire, N. R...of rape seed, 10 quarters. **1831** Pope 306: TAR, the last of twelve barrels. **1834** Pasley 115: 1 Last of Feathers and Flax...1,700 [lb]. **1850** Alexander 50: Last; for wool...4368.—pounds. **1882** Jackson 238:

Last of gunpowder = 2400 pounds. **1956** Economist 58: 1 last [of wool] = 4,368 lb. **1966** O'Keefe 671: 1 last = 80 bu. = 29.09 hl.

lasta, laste, lastum, lastus. LAST

lay. LEA

layde. LOAD

lea—4-9 lee; 7-9 lay, lea; 9 ley (OED) [ME lee, perh back-formation fr lees, unit of measure of thread]. A m-l for thread and yarn generally of 300 yd (27.432 dkm), but variations from 80 to 800 yd (7.315 to 73.153 dkm) were sometimes used.—c**1440** Promp. Parv. 291: Lee of threde. **1696** Phillips sv: Every Lea of Yarn at Kidderminster shall contain 200 Threds reel'd on a Reel four yards about. **1776** Act 17 Geo. III, chap. 2,2: Every hank of...yarn shall...contain seven raps or leas, and...every such rap or lea shall...contain eighty threads. **1820** Second Rep. 21: Lay, Lea or Lee...of thread or worsted reeled, 800 yards; 200 threads on a reel of 4 yards.... Hampshire: measured on a reel of 2 yards.... Suffolk: 40 threads of 2 or 3 yards.... Derbyshire: of cotton, a lee is 120 yards. **1887** Bonwick 359: There are 560 yards of worsted to a hank.... The seventh part of a hank is a lea. **1888** Paton 666: Throughout the United Kingdom the standard measure of flax yarn is the 'lea', called also in Scotland the 'cut' of 300 yards. **1956** Economist 58: Linen...1 lea (or cut) = 300 yards.

leag, leage. LEAGUE

league—3 leuce; 3-4 L leuga; 3-7 L leuca; 4 leuk (York Mem. 1), L lewa (Prior), lewge (OED); 4-5 leghe (OED), lywe; 4-6 lege, leuge; 5 leeke

(OED), leuke, lewke, lieke; 5-6 leege (OED); 6 legge (OED), lig (OED); 6-7 leag (OED), leage (OED); 6-9 league; ? L leuua (Maitland) [ME lege fr LL leuga, leuca, of Gaulish origin]. A m-l generally of 15,840 ft (4.827 km) or 3 mi of 5280 ft each. However, various other lengths were occasionally used: a league of 7500 ft (c2.29 km) or 1 1/2 mi of 5000 ft each; of 7680 ft (c2.34 km) or 12 linear FARTHINGDALES of 40 PERCHES each, the perch containing 16 ft; of 7920 ft (c2.41 km) or 12 FURLONGS of 40 perches each, the perch containing 16 1/2 ft; of 8910 ft (c2.72 km) or 13 1/2 furlongs of 40 perches each, the perch containing 16 1/2 ft; of 9375 ft (c2.86 km) or 15 furlongs of 125 PACES each, the pace containing 5 ft; of 10,000 ft (c3.05 km) or 16 furlongs of 125 paces each, the pace containing 5 ft; and of 15,000 ft (c4.57 km) or 3 mi of 5000 ft each. It sometimes was abbreviated l. or lea.—**1227** trans in Cal. Char. 1.17: Five leagues (leuce) from Croyland; ibid 20: Eighteen leagues (leugas) of meadow and a fishery and a manse. c**1289** Bray 10: Ambitus villae de Herleston est ij leucae et quarta pars unius leucae et continet quaelibet leuca xij quadrentenas [farthingdales] et continet quaelibet quadrentena xl perticas et continet quaelibet pertica xvi pedes de pedibus rectis.... Quinque pedes passum faciunt; passus quoque centum viginti quinque stadium [furlong]; si miliare des octo facet stadia; duplicatum dat tibi leucam. c**1300** Hall 7: Unde 5 pedes faciunt passum, et 125 passus faciunt stadium...et 16 stadia faciunt miliare Gallicum, quod vocant Gallici unam leucam. **1302** Rot. Parl. 1.152: Quod nulli Mercatores in Civitate predicta vel Suburbio nec infra septem leucas circumquaque discarcare mercandisas suas presumant, nec emant vel vendant nisi infra portas nundinarum predictarum. c**1325** Rameseia I.76: Pedes quinque passum; passus centum viginti et quinque unum stadium.... Et stadia quindecim unam leucam. c**1350** Higden V.244: Qui centum leugas in longitudine. c**1350** Swithun 66: Per septem leucas in circuitu feriæ illius. **1387** Higden II.11: That is from the Penwythis strete fifte[ne] leges. c**1400** Henley 8: Byen sault ke vne coture deyt estre de quarante perches de long...E la perche le rey est de xvi pez et demi...ceo fet asauoyr ke xii cotures sunt vne lywe. c**1425** Hall 9: Et sexdecim pedes et dimidia faciunt perticatam Regis. Et quadraginta perticate faciunt unum stadium. Et tresdecim stadia et dimidium faciunt leucam. **1430** Rot. Parl. 4.380: Pur l'espace de XII liekes environ le dit Burgh.... De user lour poisure pur XII leuges environ mesme le Burgh. c**1450** Higden II.11: xv. leukes behynde Mochillestowe; ibid V.245: Whiche conteynethe a c. lewkes in lengthe. **1494** Fabyan 63: An Hundreth Legis...whereof euery Lege conteyneth .iii. Englysshe myles. **1561** Eden xviii: Also. 125. Geometricall pases, make a furlong. viii. furlonges one myle, whiche is a thousand pases: And thre myles one league.... Let us gyue to euerye league, thre thousand pases, and to euery pase fiue foote. **1688** Bernardi 202: Pes Anglicus...1/15840...Leucæ maritimæ. **1756** Rolt sv: A sea league is usually reckoned 3000 geometrical paces, or three English miles. **1820** Second Rep. 21: League...3 miles. **1832** Edinburgh XII.569: 3 Miles = 1 League = 4827.9179 [m]. **1878** Wedgwood 380: League...a measure of distances, properly the stone which marked such a distance on the public roads. **1956** Economist 8: League...3 miles. See STADIUM

leap[1]—3-4, 7 lepe; 3-6, 9 lep; 4-5 leep, leepe (OED); 5 leippe (OED); 6-7 leape (OED); 7 L lepa; 7-8 lib; 7-9 leap; 8 lip (OED) [ME leep, basket, fr OE lēap]. A m-c for grain in Sussex, Norfolk, etc. generally containing 1/2 bu (cl.76 dkl).—**c1440** Promp. Parv. 296: Leep, or baskett...Sporta, calathus. **1674** Ray 70: A Leap or Lib; Suss. Half a bushel. **1678** Du Cange sv lepa: Mensuræ species apud Anglos. Vox formata a Saxonico Leap, Calathus, corbis. **1695** Kennett Glossary sv seed: The Saxon Leap was properly a basket or pannier made of Osiers.... From this Continent they borrowed the Latin word Lepa, a Lepe, or measure of about five gallons.... The words Leap and Lib in Sussex do now signifie the measure of half a bushel, or four gallons. **1853** Cooper 58: Leap...Half a bushel. **1880** Britten 151: Lep (Norf., &c.), a large wicker-basket.

leap[2]—3 leep (OED), leepe (OED); 4-6 lepe (OED); 6-7 leape (OED); 6-9 leap [ME leep fr OE hlȳp; akin to OE hlēapan, to run, leap]. A m-l of 6 ft 9 inches (2.057 m) in Wales.—**1820** Second Rep. 21: Leap, Wales: formerly 6 feet 9 inches.

leape. LEAP[1]; LEAP[2]

lee. LEA

leege, leeke. LEAGUE

leep, leepe. LEAP[1]; LEAP[2]

lege, legge, leghe. LEAGUE

legina. LAGEN

leippe. LEAP[1]

leippie. LIPPY

leispound. LISPOUND

lep, lepa. LEAP[1]

lepe. LEAP[1]; LEAP[2]

leshpund, lespund. LISPOUND

lest, leste, lestum, lestus. LAST

leuca, leuce, leuga, leuge, leuk, leuke, lewa, lewge, lewke. LEAGUE

ley. LEA

lib. LEAP[1]

liber, libra. POUND

librat, librata. LIBRATE

librate—2-7 L librata; 6-? librate; 7 librat (OED) [MedL librata fr MedL libra, English pound]. A m-a for an amount of land worth 1 pound a year. Its total acreage depended on local soil conditions and on the value of the pound, and it seems to have varied from several BOVATES or OXGANGS (often 4) to as much as 1/2 KNIGHT'S FEE.—c**1139** Malcolm 139: Preter .x...libratas terre quas priusquam Roberto Foliot dederam. **1163** St. Edmunds 98: Sciatis me concessisse et presenti carta confirmasse ecclesie sancti Ædmundi et monachis ibidem deo seruientibus VI libratas terre quas Warinus filius Geroldi eis dedit in Sabrichtesuuorde et carta sua confirmauit. **1200** Cur. Reg. 8.145: Et per finem concordie dedit idem Willelmus eidem Frarico terram illam pro clamio x. libratarum terre quas clamavit versus eum. c**1230** Red Book 356: Sed dominus Gwarinus, frater meus, dedit Sancto Edmundo pro anima sua vj libratas terræ in eadem villa quietas. c**1260** Bracton IV.242: Per carucatas, vel libratas, vel virgatas. c**1280** Cal. Char. 1.307: Quinque libratas terre in esterlinggis. **1607** Cowell sv farding deale: You haue also Denariata & obolata, solidata, & librata terræ, which by probabilitie must rise in proportion of quantitie...as an halfepeny, peny, shilling, or pound rise in valew and estimation. **1665** Sheppard 24: And that Librata terræ, some say, containeth four Oxgangs, and every Oxgang thirteen acres. **1777** Nicol. and Burn 612: Librate of land, is a quantity containing four bovates or oxgangs. **1780** Paucton 793: Librata terræ = 240 acres. **1867** C. I. Elton 71: There are, however, good reasons for supposing that the librate varied according to the quality of the land from twenty to forty acres.

libre. POUND

lieke, lig. LEAGUE

linck, lincke. LINK

line—7 L linea; 7-9 line [F ligne, line, fr L linea]. A m-l equal to 1/12 inch (2.12 mm). Lines were sometimes referred to as "parts" of an inch.—**1678** Du Cange sv alna: Pes Regius est 12. pollicum; pollex 12. linearum. **1855** Jessop 14: The line = 1/12 inch. **1880** Courtney 167: 12 lines or 3 barley-corns 1 inch. **1894** Francis 34: 12 parts = 1 inch. **1956** Economist 8: Line...1/12 inch.

linea. LINE

link—7 linck, lincke; 7-9 link [ME link, of Scand origin; see WNID3]. A m-l for land surveying: generally 1/100th of a Rathborn Chain or 1.98 inches (5.029 cm); 1/100th of an Engineer´s Chain or 1 ft (0.3048 m); and 1/100th of a Gunter´s or an Imperial Chain or 7.92 inches (0.2012 m). In Scotland the link for Gunter´s Chain was 8.928 inches (0.2268 m), while in Ireland it was 10.08 inches (0.2560 m). It is sometimes abbreviated li.—**1610** Folkingham 52-53: The Prime of the Chaine into two lincks, with three rings betweene euery lincke to keepe it from crossing. **1651** Jager 77: One Land measurer hath a chaine 4 perches long, consisting of 100 lincks. **1653** Leybourn 1.46: Equall parts or Links, called Seconds; ibid 47: As every Pole of Master Rathborns Chain was divided into 100 Links, so Master Gunters whole Chain (which is alwayes made to contain four Poles) is divided into 100 Links.... Each Link of this Chain will contain 7 Inches and 92/100 of an Inch. **1677** H. Coggeshall 1.31: To take off ten Chains and 46 Links or 10.46 Chains. **1779** Swinton 23: Gunter Link...7.92 [inches]. **1814** Brown 187: To Reduce Square Links of the Scotch Chain into Scotch Ells, Multiply by... .0576. To Reduce Square Links of the Scotch Chain into English Yards, Multiply by... .06125625. **1834** Pasley 3: 100 Links...1 Chain of 22 yards. **1843** Strachan 37: Links 1...7.92 [inches]. **1862** Ewart 21: Each link is .66 foot. **1889** Francis 12: The length of each link, together with half the rings connecting it with the adjoining links, is consequently 66/100 of a foot, or...7.92 inches. **1899** Browne

118: 4 Poles or 100 Links, or 22 Yards = 1 Chain = 20.12 [m]. **1907** Hatch 22: 1 link = 7.92 inches = 0.66 foot. **1956** Economist 7: 7.92 inches = 1 link. **1969** And. & Bigg 18: 1 link (0.66 ft) = 0.201168 m.

lip. LEAP[1]

lippie. LIPPY

lippy—7 leippie (OED); 7-9 lippy; 8-9 lippie [dim of LEAP[1]]. A m-c in Scotland for dry products: wheat, peas, beans, rye, and white salt, 137.333 cu inches (2.251 l) or 1.3281 Scots pt and equal to 0.063863 Winchester bu; oats, barley, and malt, 200.345 cu inches (3.284 l) or 1.9375 Scots pt and equal to 0.093166 Winchester bu (Swinton 32).—**1779** Swinton 32: Lippie or Forpet. **1813** Cooke 103: Linlithgow Bear Measure. A Lippie...200.345 [cu inches]. **1816** Kelly 93: 4 Lippies...1 Peck. **1820** J. Sheppard 91: WHEAT, BEANS, PEAS, RYE, SALT, AND GRASS SEEDS. 4 lippies make 1 peck. **1820** Second Rep. 17: Forpet or Forpit...Scotland: the fourth part of a peck, otherwise called a lippie; _ibid_ 22: Lippie, Scotland: a quarter of a peck = .0932 Winchester bushel. **1860** Britannia 809: 64 lippies or forpets = 16 pecks = 4 firlots = 1 boll. **1883** Simmonds sv: Lippy, a term in Scotland for the fourth part of a peck, also called a forpet. _See_ FORPIT

The lippy (_c_1600-1800), however, had many regional variations (Swinton 53-130). _North_—Nairnshire: wheat, peas, beans, rye, ryegrass-seed, oatmeal, and barleymeal, 167.514 cu inches (2.745 l); barley and oats, 223.352 cu inches (3.660 l). Sutherlandshire: peas, rye, and beans,

161.569 cu inches (2.647 l); oats, barley, and malt, 221.670 cu inches (3.632 l). Northwest--Inverness: wheat, peas, beans, rye, ryegrass-seed, and meal, 157.185 cu inches (2.575 l); oats, barley, and malt, 219.951 cu inches (3.605 l). Ross and Cromarty: wheat, rye, peas,, beans, and lime, 155.106 cu inches (2.542 l); oats, barley, and malt, 206.808 cu inches (3.390 l). Northeast—Aberdeenshire: wheat, rye, peas, beans, meal, and seeds, 168.031 cu inches (2.752 l); oats, barley, and malt, 219.733 cu inches (3.600 l). Banffshire: wheat, beans, peas, rye, and white salt, 144.765 cu inches (2.310 l); oats, barley, and malt, 210.568 cu inches (3.450 l). Caithness: oats and barley, 212.867 cu inches (3.487 l). Moray (Elgin): wheat, rye, peas, and beans, 146.625 cu inches (2.402 l); barley and oats, 210.877 cu inches (3.455 l). Central—Perthshire: wheat, peas, rye, and beans, 141.435 cu inches (2.317 l); oats, barley, and malt, 208.687 cu inches (3.420 l). Stirlingshire: wheat, peas, beans, and rye, 148.643 cu inches (2.435 l); oats, barley, and malt, 214.886 cu inches (3.522 l). West central—Dumbartonshire: wheat, peas, beans, and meal, 160.172 cu inches (2.625 l); oats, barley, and malt, 213.562 cu inches (3.500 l). West—Argyllshire: wheat, rye, beans, and peas, 159.650 cu inches (2.617 l); oats, barley, and malt, 214.886 cu inches (3.522 l). East—Angus: wheat, peas, and beans, 142.180 cu inches (2.330 l); oats, barley, and malt, 207.616 cu inches (3.402 l)—both lippies average of Montrose, Forfar, Brechin, Dundee, and Arbroath lippies. Fifeshire: wheat, peas, and beans, 142.180 cu inches (2.330 l); oats, barley, and malt, 206.808

cu inches (3.390 l). Kincardineshire: wheat, rye, and peas, 155.106 cu inches (2.540 l); oats and barley, 213.271 cu inches (3.495 l). Kinrossshire: wheat, peas, and beans, 140.969 cu inches (2.310 l); oats, barley, and malt, 206.404 cu inches (3.382 l). South—Lanarkshire, Glasgow and Lower Ward: wheat, 144.637 cu inches (2.370 l); peas and beans, 204.45 cu inches (3.350 l): oats and barley, 208.712 cu inches (3.420 l). Peeblesshire: wheat, peas, beans, and rye, 147.15 cu inches (2.412 l); oats, barley, and malt, 209.28 cu inches (3.430 l). Southwest—Ayrshire: wheat, rye, peas, and beans, 153.6 cu inches (2.517 l) in Kyle and Carrick, 157.064 cu inches (2.575 l) in Cunningham; oats, barley, and malt, 226.36 cu inches (3.710 l) and 252.002 cu inches (4.130 l) in Kyle and Carrick. Southeast—Berwickshire: all grain, 210.039 cu inches (3.442 l). East Lothian: wheat, peas, and beans, 141.372 cu inches (2.315 l); oats, barley, and malt, 206.404 cu inches (3.382 l). Midlothian: wheat, peas, and beans, 139.757 cu inches (2.290 l); oats, barley, and malt, 203.576 cu inches (3.337 l). Roxburghshire: wheat, peas, and beans, 142.180 cu inches (2.330 l); oats, barley, and malt, 213.271 cu inches (3.495 l). Selkirkshire: wheat, rye, beans, and peas, 142.522 cu inches (2.336 l) or 1/8 pk of 1140.675 cu inches.

lispond, lispondt. LISPOUND

lispound—6 leshpund, lespund (OED); 7-8 leispound (OED), lispond; 8 lispondt; 8-9 lispound, lispund (OED); 9 lyspund [LG lispund fr MLG lispunt, livespunt fr lis, lives, Livonian, + punt, pound]. A wt in the

Orkney and Shetland islands varying from 12 to 32 lb (5.443 to 14.515 kg) and used principally for barley, butter, malt, meal, oats, and oil.—**1597** Skene 1. sv serplaith: Ane stane and twa pound Scottish makis ane leshpund. **1677** Roberts 32: Thousands, Weighs...Lisponds.... The Lispond also is found to consist sometimes of 15 l. of 16 l. and 20 l. to the Lispond. **1707** Justice 58: The Lispondt, 15 Pound, more or less. **1716** Harris 2. sv weight: Weights of great Content; as Hundreds, Kintals, Centeners, Talents, Thousands, Weighs, Skippounds, Charges, Lispounds, Rooves, &c.... Lispounds, of 15, 16, and sometimes 20 Pound to the Lispound. **1779** Swinton 104: ORKNEYS. In buying and selling foreign goods, and goods from other parts of Scotland, the ordinary measures and weights of Scotland and England are used. But the WEIGHTS For Barley, Oats, Malt, Meal, Butter, and Oil, payable by vassals to their superiors, and by tenants to their landlords, or delivered by these persons to merchants, are quite different from the weights used in any other part of the kingdom. They are originally from Norway. The...weights are called Marks, Setteens or Lyspunds, and Meils. **1820** Second Rep. 22: Lispound...Shetland: 32 lbs English; formerly 24 lbs Dutch = 26 1/4 E. **1820** J. Sheppard 134: 24 marks make 1 seteen or lyspund. See SETTEEN

lispund. LISPOUND

liver, livere, livre. POUND

llath [W llath, a rod, staff, yard]. A m-a for land in South Wales (c1800-1900) varying between 11 1/2 and 24 sq ft (1.068 to 2.230 sq m)

(Second Rep. 22 and Donisthorpe 212, 214).

llathen [see LLATH]. A m-l of 9 ft (2.743 m) for cloth in Wales (c1800-1900) (Second Rep. 22 and Donisthorpe 212).

llestraid [*]. A m-c for grain in Wales (c1800-1900) containing 20 gal (c8.81 dkl) in Cardiff and 22 or 24 gal (c9.69 or c10.57 dkl) in Neath and Swansea (Second Rep. 22 and Donisthorpe 212).

load—3-6 lode; 4-9 lade (OED); 5 layde (OED), lod (OED); 5-6 lood (OED), loode (OED); 5-9 laid (OED); 6-7 loade; 6-9 load [ME lod, lode, load, fr OE lād, course, way, carrying, support]. A m-c originally referring to the amount of goods loaded on a cart or wain, the exact amount varying in relation to the quality of the goods, the strength of the wheels, the condition of the roads, and the distance traveled.

By the late Middle Ages and early modern period, however, standard loads were generally used for the following items: birch brooms, 60 BUNDLES; bricks, 500 in number; bulrushes, 63 bundles; earth or gravel, 27 cu ft (0.764 cu m); hay, 18 Cwt or 2016 lb (914.436 kg) or 36 TRUSSES of 56 lb each; lime, 32 bu (c11.28 hl); oak bark, 45 Cwt (2286.090 kg); sand, 36 bu (c12.69 hl); Scots coal, 1 Cwt (50.802 kg); straw, 36 trusses of 36 lb each, or 1296 lb (587.853 kg); tiles, 1000 in number; wheat, 5 SEAMS (c14.09 hl); and wood or timber, 20 Cwt (1016.040 kg) or 40 cu ft (1.133 cu m) for rough timber and 50 cu ft (1.416 cu m) for hewn.—**1440** Scrope 230: 1 lode hey. c**1590** Hall 27: The load of hay is but 18 hundredes to the loade; and euery hundred 112 poundes waight: 36 trusses makith a loade of haye, and euery trusse is 56 poundes waight haberdepoyse; ibid 28: Every loade of wood ought to be 20 hundred waight, which is a tunne. **1595** Powell F2: The Assise of Heye. First the lode.... They do reckon and account eighteene hundreth weight of Auoirdupois weight, to be the common loade of hey.... The Assise of Timber...Fiftie foot of Timber for the lode. **1635** Dalton 149: Timber well-hewen, and perfectly squared, fifty foot thereof the load. **1660** Bridges 32: A Load is 50 foot of square Timber. **1665** Assize 13: They do reckon and account eighteen hundred weight of Avoirdupois weight to be the common load of Hay. **1701** Hatton 3.17: Of Timber...50 Foot unhew'd 1 Load. **1779** Swinton 37: Hay is sold by the load of 18 cwt. nearly = 93 Scotch stones Trone weight. Straw, by the load of 36 trusses, each truss weighing 36 lb. Avoirdupois, nearly = 59 1/2 Scotch stones Trone weight. **1816** Kelly 85: Hay and Straw are sold by the Load of 36 Trusses. The Truss of Hay weighs 56 lb. and of Straw 36 lb; ibid 87: 40 Feet of rough Timber, or 50 Feet of hewn...1 Load or Ton. **1820** Second Rep. 22: Load of bullrushes, 63 bundles...of hay, 36 trusses of 56 lb each...of wheat is properly 5 quarters...of earth or gravel, 1 cubic yard = 27 cubic feet...of lime, 32 bushels...of oak bark, 45 cwt...of timber, round, 50 cubic feet...of sand, 36 bushels...of Scotch coals, 1 cwt...of birch brooms, 60 bundles. **1840** Waterston 147: Bricks, load...no. 500; ibid 148: Tiles, load...no. 1000. **1855** Jessop 41: A load of earth = 27 cubic feet. **1882** Steele 41: In London, and for a distance of 25 miles round, hay is generally bought and sold by the load of 18 cwt. or 36 trusses, weighing 56 lbs.

each truss. **1931** Naft 196: Great Britain: 1 load of hay = 2016 lbs. or 18 cwts. **1956** Economist 67: Load = 40 cubic feet of roundwood...50 cubic feet of sawnwood.

But these and many other items (c1800-1900) had local variations (Second Rep. 22-23, Britten 172-73, and Pasley 114). North—Durham: lime, 27 bu (c9.51 hl). Northwest—Cheshire: oatmeal, 240 lb (108.862 kg). Lancashire: barley, 6 bu (c2.11 hl); beans, 4 1/2 or 5 bu (c1.59 or c1.76 hl); peas, 4 1/2 bu (c1.59 hl); oats, 7 1/2 or 9 bu (c2.64 or c3.17 hl); potatoes, generally 2 Cwt (101.604 kg), except at Ulverstone, 250 lb (113.397 kg) for washed potatoes and 260 lb (117.933 kg) for unwashed; and wheat, 4 1/2 bu (c1.59 hl). Westmorland: potatoes, 4 1/2 heaped bu (c2.03 hl). North central—Derbyshire: charcoal, 144 level bu (c50.74 hl); lead ore, 9 DISHES (9.909 to 11.322 dkl). Central—Bedfordshire: wheat, 5 bu (c1.82 hl). Oxfordshire: straw, 22 1/2 Cwt (1143.045 kg). East—Cambridgeshire: osiers, 80 BUNCHES; and wheat, 5 bu (c1.76 hl). Suffolk: carrots and turnips, 40 bu (c14.09 hl). Southeast—Buckinghamshire: chalk, 16 BUCKETS of 1 1/2 bu each equal to 24 bu (c8.46 hl); straw, 11 1/2 Cwt (584.223 kg); and wheat, 5 bu (c1.76 hl). Essex: chalk, 90 bu (c31.71 hl); clay, 40 bu (c14.09 hl); osiers, 80 BOLTS; and shingle, 24 bu (c8.46 hl). Hertfordshire: chalk, in some places, 22 buckets equal to 33 bu (c11.63 hl); and wheat, 5 bu (c1.76 hl). Middlesex: new hay, 2160 lb (979.754 kg) or 36 TRUSSES of 60 lb each; and old hay, 2016 lb (914.436 kg) or 36 trusses of 56 lb each. South—Hampshire: rafter poles, 30 bundles. Surrey:

chalk, 30 to 35 bu (c10.57 to c12.33 hl); hoops, 30 bundles or 1800 in number; and limestone, 40 bu (c14.09 hl). Sussex: faggots, 100; limestone, 12 bu (c4.23 hl); oats, 80 bu (c28.19 hl); and wheat, 40 bu (c14.09 hl). See CARTLOAD; FOTHER; HUNDRED

loade, lod, lode. LOAD

log. LUG

loggin [*]. A m-q, a bundle, for straw in Yorkshire (c1800-1900) weighing 14 lb (6.350 kg) (Britten 151).

lood, loode. LOAD

lug—3-7 lugge; 7 log; 7-9 lug, lugg [ME lugge, of obscure origin]. A m-l generally varying from 15 to 20 ft (4.575 to 6.100 m) with 16 1/2 ft (5.029 m) being the most common. It occasionally was equivalent to the GOAD, PERCH, POLE, and ROD. In Herefordshire (c1800) a lug was a m-a of 49 sq yd (40.969 sq m) for coppice wood.—**1607** Cowell sv furlong: Twenty lugs or poles in length, and euery pole 16 foote and a halfe; ibid sv mile: Euery lugge or pole to containe 16. foote and a halfe. **1639** Bedwell B2: And every Lugge or Poale, to containe 26 [sic] foot and an halfe. **1665** Sheppard 25: 40 Luggs, Perches, or Poles in length. **1669** Worlidge 330: A Perch, or Lug is sixteen foot and a half Land-measure, but is usually eighteen foot to measure Coppicewoods withal. **1695** Kennett Glossary sv pertica: But now commonly a Perch, a Rod, or Pole, in Wilshire a Log, is sixteen foot and a half in length. **1696** Phillips sv pole: In measuring, it is the same with Pearch or Rod, or as some call it Lugg. **1717** Dict. Rus. sv mile: Every Furlong containing 40 Lugs or Poles; and every Lug or Pole 16 Foot and a half. **1725** Bradley sv mile: Every Furlong forty Lugs or Poles...every Pole sixteen Foot and a Half. **1820** Second Rep. 23: Lug or Lugg, Dorsetshire...15 feet and an inch; called also Goad, used instead of a pole of 16 1/2...Hertfordshire: 20 feet...Wiltshire: a pole or rod of 15, 16 1/2 or 18 feet. **1842** Akerman 33: Lug...A pole in land measure, 5 1/2 yards. **1868** Huntley 49: Lug...A measure of land, a perch. **1880** Britten 173: Lug (Dors.), of land, 15 feet 1 inch; called also Goad, used instead of a pole of 16 1/2.... (Herts.), 20 feet. (Wilts.), a pole or rod of 15, 16 1/2, or 18 feet.

lugg, lugge. LUG

lyspund. LISPOUND

lywe. LEAGUE

M

M. MIL

maand. MAUND

mace. MEASE

maen [W maen, stone]. A wt of 26 lb (12.700 kg) or 4 TOPSTONS for wool (c1800-1900) in South Wales (Second Rep. 24 and Donisthorpe 213).

maenol [W maenol, stony]. A m-a for land in Wales (c1300) containing 4 TREVS or 1024 ERWS (c369.66 ha) (Laws Wales 1002).

mainard—3 L mainardus (Bish. Winch.); 5 mainard (Gras 2); ? maynarde (Prior) [*]. A wt of 32 lb (14.515 kg) for cheese and wool (c1200-1400) in southern England (Gras 2.718 and Bish. Winch. 1).

mainardus. MAINARD

mais, maise, maize. MEASE

mand, mande. MAUND

mark [ME mark fr OE marc, prob of Scand origin; akin to ON mark, mörk, mark (weight), OE mearc, mark, sign]. A Scots wt (c1600-1800) for barley, butter, oats, malt, meal, and oil: Orkney Islands, 1.3596 avdp lb (616.699 g) or 1/24 LISPOUND of 32.6306 avdp lb; Shetland Islands, 1.2689 avdp lb (575.563 g) or 1/24 lispound of 30.4553 avdp lb (Swinton 104-07 and J. Sheppard 1340).

mase. MEASE

mast [perh fr mass; see OED sv mast]. A wt of 2 1/2 lb (1.134 kg) for amber, dyes, gold, silver, and other valuable products (c1600-1800) (Second Rep. 24, Rates 2.2ff, Hatton 3.230, and Donisthorpe 213).

math—6-7 mathe (OED); 6-9 math [OE mǣth, a mowing]. A m-a in

Herefordshire equal to approximately 1 acre (0.405 ha) or to the amount of land that a man could mow in a day.—**1819** Cyclopædia sv weights: 1 Day´s math...About a statute acre of meadow or grass land, being the quantity usually mown by one man in one day. **1820** Second Rep. 24: Math, Herefordshire: mowing; a day´s math is about an acre, or a day´s work for a mower. **1888** Round 3.219: Mr. Palmer proceeds to note that in common parlance a meadow is still spoken of...as containing so many ´days´ math´.

mathe. MATH

matt [var of mat fr ME mat fr OE matt, matte, fr LL matta, of Sem origin]. A wt for cloves (c1850) of 80 lb (36.287 kg) (Waterston 147). The name probably originated from the manner in which the cloves were tied or bundled together to give the appearance of a "mat" or "rug."

maun. MAUND

maund—5 mawnde; 5-6 mande; 5-7 maunde, mawnd (OED); 5-9 mand (OED), maund; 7 moane (OED); 8 maand (OED); 9 maun (OED), mawn (OED), mound (OED) [ME maund, hand basket, fr MF mande; akin to OE mand, MLG mande]. A m-c containing perhaps 2 or 3 pk (c1.76 or c2.64 dkl) for most goods and 2 FATTS or 8 BALES or 40 REAMS for unbound books. It was a wicker-type basket with handles.—**1420** Gras 1.472: Pro 1 fatt´ vi pokis i maunde. **1439** Southampton 2.9: 1 maund calcarium et panni picti...1 maunde de wastyng paper; ibid 41: 12 maundys pomarum; ibid 49: 1 mawnde panni picti. **1443** Brokage II.81: ii maundes orenges; ibid 111: i maunde patellarum ferrearum; ibid 116: i mawnde cum diversis haberdasshe et grocer; ibid 135: ii maundes skowryngstonys. **1509** Gras 1.563: i maunde cum xii dossenis pannorum depictorum; ibid 571: l maunde i fat cum iiii grosses gloves; ibid 572: ii fardelli i cista i mande i barel trane...ii mandes cum ii mastis ambr[e]. **1545** Rates 1.7: Bokes unbounde the basket or mande. **1590** Rates 2.4: Bookes unbounde the whole maund fortie remes; ibid 42: Glouers clippings the maund or fat. **1603** Hostmen 36: Not to exceed in any one Keell or Lighter above two smale maunds or pannyers full, holdinge two or three pecks apeece. c**1610** Lingelbach 113: Baste or Strawe Hattes by the Maund; ibid 114: Turnout Tykes by the Maund. **1664** Gouldman sv: A maund or basket. Alveolus, cophinus, sporta. **1701** Hatton 3.230: Maund...(of unbound Books is) 8 Bales, each 100 l. weight or 2 Fats. **1710** Harris l.sv: Maund, was anciently a Measure of Capacity with us, being a kind of great Basket or Hamper containing 8 Bales, or 2 Fatts. **1717** Dict. Rus. sv fat: Fat...of unbound Books half a Maund or four Bales. **1840** Waterston 148: Indigo (E.I.), chest, about 3 1/2 maunds, or...lbs. 260.... Opium,, chest (E.I.), 2 maunds, or...lbs. 149 1/3.

maunde, mawn, mawnd, mawnde. MAUND

maynarde. MAINARD

mayse, maze. MEASE

meal [ME mel, mele fr OE mǣl, measure, mark, sign, fixed time]. A m-c for milk in Suffolk and Sussex (c1850-1900) equal to the quantity taken from a cow at one milking (Cooper 60 and Britten 151).

meas. MEASE

mease—5 meyse; 5-6 mayse (OED); 5-9 meise; 6 meaz; 6-8 mese; 6-9 maise, meaze (OED); 7 maze (OED), mes (OED); 7-9 mease, mesh (OED); 9 mace, mais (OED), maize, mase (OED), meas (OED), meash (OED) [ME meise fr MF maise, a receptacle for herrings, ft MLG meise, mēse, barrel]. A m-q for herrings, varying in number from 500 to 630, equal to 1/20 LAST.—**c1400** Hall 41: In uno meyse allecis sunt v [X] c et xx: Item xx meyses faciunt lastall. **c1550** Welsh 58: 1 maises red herring; ibid 201: 24 meaz of shotten herrings. **1597** Skene 1. sv mese: Of herring, conteinis fiue hundreth. **1603** Henllys 139: The meise consisteth of XXXI [X] xx of herringes. **1701** Hatton 3.230: Mease...Five hundred Herrings. **1820** Second Rep.. 24: Maise...of herrings, 30 score of 21 each = 630. **1883** Simmonds sv: Mease, 500 herrings; sv mace: a mace or maize of herrings being 500 in number. **1895** MAISE: South Wales: of herrings, 30 score of 21 each.

meash. MEASE

measure [ME mesure fr OF mesure fr L mensura fr metiri, mensus, to measure]. A m-c (c1800-1900) for several products (Second Rep. 24 and Britten 173): apples, Guernsey and Jersey, 3 Winchester bu (1.057 hl); barley and oats, Cheshire, 38 qt (c4.18 dkl); coal, Kincardineshire, 48 Scots pt (8.196 dkl); lime, Kincardineshire, 64 Scots pt (1.093 hl); malt, Cheshire, 32 to 36 qt (c3.52 to c3.96 dkl); oatmeal, Westmorland, 16 qt (c1.76 dkl); potatoes, Lancashire, 90 lb (40.823 kg), and Guernsey and Jersey, 7 gal (c3.08 dkl); and wheat, Cheshire, 38 qt = 75 lb (34.019 kg).

meaz, meaze. MEASE

meel. MELL

meiliaid [*]. A m-c for grain in Llandovery (c1800-1900) equal to 1/4 bu (c8.81 l) (Second Rep. 24).

meise. MEASE

mel, mele. MELL

mell—5 mel (Fab. Rolls), mell (Prior), miell (Fab. Rolls); ? meel (Prior), mele (Prior) [see MEAL]. A m-c for lime and other dry products (c1400) generally equal to 2 SEAMS (c5.64 hl) (Prior 167).

merk [prob var of mark, boundary, limit, border, fr ME marke, merke fr OE mearc, merc]. A m-a for land in Shetland (c1800-1900) varying from 1/2 to 2 acres (c0.20 to c0.81 ha) (Second Rep. 24 and Donisthorpe 213).

mes, mese, mesh. MEASE

met—3-7 L mitta; 3-9 met; 6 mett; 7 mette (Best); ? L metra (Prior) [OE gemet, a measure]. A m-c for grain and other dry products generally containing 2 bu (c7.05 dkl), but variations from 1/2 bu (c1.76 dkl) to 1 SEAM (c2.82 hl) were not uncommon.—**c1200** Rameseia III.158: Et quatuor communes ringæ, duo busselli, faciunt mittam gruti.... Et quinque communes ringæ brasei et præbendæ faciunt mittam; ibid 159: Et hoc facto, continet mitta gruti quatuor communes ringas, duos bussellos. **1297** Neilson 11: Item in missione apud Rameseiam xv ring. quæ fecerunt iii mittas de novo grano. **1587** Acts Scotland 3.521: The boll/mett/firlot/and peck. **1674** Ray 48: A Met: a Strike or four Pecks...in York-sh. two Strike. **1678** Du Cange sv mitta: Mensuræ salariæ et frumentariæ species, a Saxonico mitten, mensura. **1819** Cyclopædia sv weights: At Lancaster, and the neighbourhood, they have several different weights, as the Lancaster peck, of twenty-four quarts...the met, of fifty-six quarts. **1872** Robertson 68-69: The Mitta contained two Ambers, answering by the London standard to a quarter, or the fourth of a chaldron.

metra, mett, mette. MET

meyse. MEASE

miell. MELL

mil—1-3 L milia, mille, L millia; 1-9 L M; 2 L millerium; 3 L miliare, L millarium; 3-9 mil; 4 L milliare; 5, 8 mill; 7 L miliarium [L mille, thousand]. Equivalent to THOUSAND. Occasionally in the early modern period it was abbreviated m.—c**750** Brit. Mus. 11.105: M...mille. **1086** Sussex 14: In Lewes x burgenses de lii denariis et de villanis xxxviii. milia allecium et quingenti; ibid 84: In LEWES xliiii hagæ de xxii solidis et iiii milia allecium. c**1195** Devizes 387: Willelmus Eliensis electus, datis tribus millibus librarum argenti, sigillum regis sibi retinuit, licet Reginaldus Italus quartum millerium superobtulerit. **1202** Feet 3.196: Et sex millariorum et duodecim strikarum anguillarum. c**1205** Hoveden IV.81: xx. millia marcarum argenti. c**1225** Coggeshall 101: Triginta millia marcarum; ibid 166: Et præterea mille marcas sterlingorum. c**1253** Hall 11: Le last de arang´ est de xM., et checun mil est de x cent, et chescun cent de vi [X] xx. **1290** Fleta 119: Lestus autem allecii consistit ex x. miliaribus et quodlibet miliare consistit ex decies centum, et quodlibet centum ex secies viginti. **1303** Gras 1.160: M ceparum.... M de stagno; ibid 161: M de cupro; ibid 162: M pellium squirellorum; ibid 166: Milliare ceparum. **1304** Ibid 168: Pro xM ferri. **1396** Ibid 437: MMMM tunholt. **1409** Rot. Parl. 3.626: & achatent ascun foitz come en un an mill ou deux mill Draps de blanket fyne ou pluys. **1410** Ibid 642: Les parties pleintiefs mettre loure damages a deux centz, trois centz, ou mill li. c**1461** Hall 13: Also of this Weyght there goo v [X] xx [+] xii lb. to the C; and x [X] c make a M of ony weyght.... xxviij lb. [make a quarter Cwt]; lvj lb. make half a C; v [X] xx [+] xij lb. make a C...and x [X] c make a M; ibid 17: Also stocke fyssche ys sold by vi [X] xx and a M fysschys make a last. **1507** Gras 1.695: All blades for shomakrs the M...Bodkyns the M...Ballys the M. **1524** Ibid 196: Pro uno M hoopis. **1549** Ibid 627: Pro M waight rosen. **1590** Rates 2.23: Lemmons the M; ibid 27: Orenges the M; ibid 28: Pinnes the dosen M; ibid 36: Swan quilles the M; ibid 43: Tips of hornes the M. **1678** Du Cange sv miliarium: Mille pondo librarum. **1721** King 294: Oxbones...per Mill; ibid 303: Hilling Stones...per m. **1732** J. Owen 115: M. in number 1000. **1854** Bowring 12: M the initial of mille.

mile—1 L miliaria; 2-5 L miliarium; 3 L miliarius; 3, 7 L milliarus; 3-7 myle; 4 L mileare; 4-5 L miliare; 4-6 myl (OED); 5 myill (OED); 5-7 mylle; 7 mille (OED); 7-9 mile [ME myle fr OE mīl fr L milia, miles (fr milia passuum, thousands of paces), pl of mille, mile, fr mille passus, thousand paces, fr mille, thousand]. A m-1, standardized under

Elizabeth I at 5280 ft (1.609 km) or 1760 yd, equal to 8 FURLONGS of 40 PERCHES each, the perch containing 16 1/2 ft. Prior to standardization several other lengths for the mi were common: 5000 ft (c1.52 km) or 1000 paces of 5 ft each; 5000 ft (c1.52 km) or 8 furlongs of 125 paces each, the pace containing 5 ft; 6600 ft (c2.01 km) or 10 furlongs of 220 ft each; and for the Old English mi, 1500 paces, the pace varying in size from one region to another. The Scots mi contained 320 FALLS or 1920 ELLS and was equal to 1984 English yd or 5952 English ft (1814.170 m). The Irish mi contained 2240 yd or 6720 ft (2.048 km). Since the establishment of the Imperial system in 1824, the mi has assumed two standards: 5280 ft (1.609 km) for the statute mi and 6080 ft (1.853 km) for the nautical, geographical, or sea mi.—**c1075** Hall 2: Digitus, uncia, palmus, sextas, pes, cubitus...passus...stadium, miliaria. **c1260** Bracton I.58: Stadium vero dicitur octava pars milliarii. **c1300** Hall 7: Unde 5 pedes faciunt passum, et 125 passus faciunt stadium [furlong]; et 8 stadia faciunt mileare Anglicum. **c1325** Rameseia I.76: Pedes quinque passum; passus centum viginti et quinque unum stadium. Stadio octo unum miliarium. **1395** York Mem. 1.142: Item, quinque pedes faciunt passum; centum et triginta [sic] quinque passus faciunt stadium; octo stadia faciunt miliare Anglie. **c1400** Brit. Mus. 20.1v: viij stadia faciunt miliare Anglie. **c1400** Hall 5: Stadium passus 125 constat. Miliarium 8 stadia, i.[e.] passus mille continet. **c1461** Ibid 14: And there go viij forelonges to a myle, in Yngland. **1502** Arnold 204: V fote make a pace...CXXV pace make a furlong and VIII furlong make an English myle. c**1536** Leland 125: It semid to me a veri hy montaine, and was distant by gesse a vi. miles.... Going from Montgomerik to the Walsche Poole a v. myles of I passid over a forde of Severn. **1592** Berriman 170: A myle to conteyne eight furlongs and every furlong to conteyne fortie luggs or poles and ev´y lugg or pole to conteyne sixteen foot and half. c**1600** Brit. Mus. 16.70: And .viij. furlonges maketh an Englishe mylle. **1616** Hopton 165: Also an English mile is 8 Furlong, 88 scores, 320 pearches, 1056 paces...1760 yards, 5280 feét, 63360 Inches. **1624** Huntar 10: OF THE ENGLISH MYLE. They compt 40 pearches to a furlong, and 8 furlongs to a myle which is 320 pearchs or 176[0] yards, & containeth of Paces 1056. **1635** Dalton 150: Note that our English mile containes 280 foot more than the Italian mile...of 1000 paces, and five foot to a pace. **1665** Sheppard 25: Forty Pole in length make a Furlong; eight Furlongs or 320 Pole, an English Mile. **1682** Hall 29: A Myle is 8 furlongs, or 320 pearches. **1688** Bernardi 202: Pes Anglicus...1/660 Stadii aut Furlongi, et 1/5280 Milliaris Anglici. **1708** Chamberlayne 207: 8 Furlong or 320 Perch make an English Mile, which...ought to be 1760 Yards, 5280 Foot, that is 280 Foot more than the Italian Mile. **1717** Dict. Rus. sv furlong: Furlong...contains 40 Poles...in length, being the eighth part of a Mile. **1761** Thomson viii: The Scots mile contains 5952 [English ft]. **1779** Swinton 24: Mile. = 1984 [English yd].... [This] is the computed Scotch mile, although by act 44. James VII. in 1685, the Scotch mile is ordained like the English, to contain 1760 yards of 36 inches each.

1805 Macpherson II.203: That the length of a statute mile...should be eight furlongs, each furlong containing forty poles or perches, and every pole to contain sixteen feet and a half in length; so that an English mile was hereby to contain 1760 yards in length. **1816** Kelly 94: 1 [Scots] Mile = 5952 English Feet.... 80 Scotch miles = 91 English miles. **1832** Edinburgh XII.569: 8 Furlongs = 1 Mile = 1609.3059 [m]; ibid 571: On Irish Measures.... 7 Yards = 1 Perch...2240 Yards = 1 Mile...11 Ir. miles = 14 E. miles. **1872** Herschel 424: The Romans reckoned their distances by intervals of 1000 paces (millia passuum) whence our name for a mile, though differing widely in reality. **1883** Thurston 22: 60 Geographical Miles make one Degree. **1907** Hatch 37: 1 mile (8 furlongs) = 1.6093 kilometres. **1951** Trade 27: Mile = 1,760 yards. **1956** Economist 4: Mile: (a) United Kingdom—Statute = 5,280 feet.... Geographical = 6,080 feet (in practice often 6,000 feet); ibid 7: 1 nautical mile = 1.15152 land miles. **1969** And. & Bigg 11: 1 (UK) nautical mile = 1.85318 km. See STADIUM

mileare. MILE

milia. MIL

miliare. MIL; MILE

miliaria. MILE

miliarium. MIL; MILE

miliarius. MILE

mill, millarium. MIL

mille. MIL; MILE

millerium, millia, milliare. MIL

milliarus. MILE

mina [L mina fr Gr mna, of Sem origin]. A m-c for dry products sometimes considered equal to 4 1/2 bu (c1.58 hl) (St. Paul's lxxxii), but more commonly defined as a vessel containing 3 to 7 SKEPS.—**c1320** Du Cange sv: Mensura...ad frumentum, et ad bladum, et ad pisa, quæ alio nomine Mina vocatur, continet 5. eskippas de duro blado; et istæ 4. Minæ, cum gata quæ dicitur Gundulfi, faciunt 3. sumas...unde Mina et gata faciunt 3. quarteria. Mina ad grutdum recipiendum continet 7. eskippas. Mina ad brazium continet 3. eskippas de duro blado. Mina ad farinam in pistrino continet largiter 7. eskippas, et debet mensurari sicut sal, et radi.

minim [E minim fr L minimus, the least, smallest, a superlative fr root of L minor]. The smallest Imperial apothecaries' liquid measure, approximately a drop, defined as the volume of 0.9114583 gr (0.059 g) of distilled water at 62° F, and equal to 1/60 ap fluid dr or 0.059192 ml (Hatch 24, 36, 38, Stevens 2-3, and Economist 7).

mite—7 myte; 7-9 mite [ME mite fr MF or MDu mite, small copper coin; see OED and WNID3]. A moneyer's unit of wt equal to 1/20 t gr (0.00324 g) or 24 DROITS or 480 PERITS or 11,520 BLANKS. It belonged to a series of imaginary wt used to compute exact coin wt by alternate subdivisions of 20 and 24.—**1665** Sheppard 15: 20 mites make a grain; 24 Droits make a Myte. **1707** Justice 4: One Grain into 24 Mites. **1725** Bradley sv weights: The Moneyers subdivide the grain thus: 24 Blanks make 1

Perrot; 20 Perrots 1 Dwit; 24 Dwits 1 Mite; 20 Mites 1 grain. **1727** Arbuthnot 109: 21 Grains and 15 Mites (of which there are 20 in the Grain) of Sterling Silver. **1784** Ricard II.151: On divide le grain en 20 mites. **1819** Cyclopædia sv weight: The grain troy is divided into 20 mites, the mite into 24 doits, the doit into 20 periots, and the periot into 24 blanks. **1840** Ruding I.411: Twenty pennyweights an ounce, twenty-four grains a pennyweight, twenty mites a grain, twenty-four droits a mite, twenty perits a droit, twenty-four blanks a perit. **1868** Eng. Cyclo. 822: In some old books a grain is 20 mites. **1896** Klimpert 228: Mite...3,24 mg.

mitta. MET

moane. MAUND

mogg [perh var of MUG]. A m-c or m-q of uncertain size or wt for salt.—**1400** trans in Cal. Close 17.149: 58 bundles of leather, one runlet of grain and about 180 ´mogges´ of salt.

molley, mollie. MOLLY

molly—8 molley, mollie (OED); 8-9 molly [*]. A m-c, a large basket, for vegetables: Chester, 3 1/2 gal (c15.91 l); Northamptonshire, 12 to 40 lb (5.443 to 18.144 kg).—**1896** Wagstaff 36: A ´molley,´ for vegetables, in Northamptonshire varies from 12 to 40 lbs., there being large and small molleys. A ´molley,´ for vegetables, in Chester = 3 1/2 gallons.

mound. MAUND

mount [prob mount, a high hill (here, of material), fr ME mount, munt, mont fr OE munt and OF mont, both fr L mons, montis]. A wt of 3 M

(1524.060 kg) for plaster of Paris (c1600-1800) (Rates 2.29, Hatton 3.230, and Second Rep. 24). See THOUSAND

mow—5-7 L muwes; 7 mow [ME mowe fr OE mūga, mūha, mūwa, mow, heap]. A m-c for grain and other dry products.—**1678** Du Cange sv muwes: Mensuræ species, nisi ab Angl. Mow, quod acervum, cumulum sonat, accersas.

muchekyn, muchkin. MUTCHKIN

mug—6 mugge (OED); 6-9 mug; 7-8 mugg (OED) [cf Sw mugg, Nor mugge, mugga, an open can or jug]. A m-c for ale in Bedfordshire containing 1 pt (c0.55 l). It was generally a cylindrical earthenware vessel, often having a handle.—**1820** Second Rep. 24: Mug, Bedfordshire: of ale, a pint. **1895** Donisthorpe 87: In Bedfordshire a pint was usually called a mug.

mugg, mugge. MUG

musking, mutchen, mutchin, mutchken. MUTCHKIN

mutchkin—5 muchekyn; 6 musking (OED), mutskin (OED), mychkin (OED); 6-9 mutchkin; 7 mutchen, mutchin (OED), mwching (OED); 7-8 muchkin, mutchkine; 8 mutchken (OED) [ME (Sc) muchekyn; cf Du mudseken, a liquid measure]. A m-c for liquids in Scotland containing 4 GILLS or 25.851 cu inches (c0.42 l) and equal to 1/4 Scots pt or 1/2 CHOPPIN.—**1425** Acts Scotland 2.12: Now ordanit ix pyntes & thre muchekynis. **1618** Ibid 4.588: Quart/Chopin/Mutchkin/and halfe mutchkine. **1624** Huntar 4: Everie pinte is devided in 2 choppins and 4 muchkins.... The pinte doth weigh 55 ounces...the muchkin full, 13. vnces 12 drop. **1681** Acts Scotland 8.400: Pynt choppin and mutchen stoups. **1779** Swinton 29:

Mutchkin...25.851 [cu inches]. **1816** Kelly 93: 4 Gills...1 mutchkin. 2 Mutchkins...1 Chopin. **1820** Second Rep. 24: Mutchkin, Scotland: 1/4 pint = 1/2 a chopin = 4 gills. **1883** Simmonds sv: Mutchkin, a Scotch liquid measure of 4 gills, = 25.851 cubic inches; the fourth of the Scotch pint.

mutchkine, mutskin. MUTCHKIN

muwes. MOW

mwching, mychkin. MUTCHKIN

myill, myl, myle, mylle. MILE

N

naggin. NOGGIN

nail—3 neil (OED); 3, 7 neile; 3-7 naile, nayle; 4-5 naille (OED), nayll (OED), naylle (OED); 4-5, 7 naill; 4-6 nale (OED); 4-8 nayl; 4-9 nail; 6 neayle (OED), neyll; 6-7 nall (Halyburton), neale (OED); 7 neyle [ME *nail* fr OE *nægl*; *see* WNID3]. A m-l for cloth, originally a unit of body measurement referring either to the distance from the end of the thumb nail to the joint at the base of the thumb, or to the last two joints of the middle finger, and taken equal to 1/2 FINGER, 1/4 SPAN, and 1/8 CUBIT. Based on the ft of 12 inches, it was made equal to 2 1/4 inches (5.715 cm) or 1/4 quarter of cloth measure or 1/16 yd. It was also a wt synonymous with the CLOVE and was sometimes abbreviated *na.*—*c***1461** Hall 13: Also woll is weyd by this weyght, butt itt is nott rekynnyd soo, for ytt is bowght odyr by the nayle.... vij lb. make a nayle; *ibid* 19: For thai use to by or sell most comynly odyr by the Clawe, the Nayle. **1569** Remembrance 109: ii ledyn waytts of xiii neyll apes. **1577** D. Gray 8: The yarde is diuided into 4 quarters, and euery quarter into 4 nayles. *c***1590** Hall 23: 7 poundes waight haberdepoyse is the halfe stonne or clave of woole, or nayle.... 7 pounds daberdepoyse [*sic*] is the claue or nayle of woole. **1600** Hill 66: 8. Pounds hauerdepoise weight maketh 1. Naile. **1607** B. J. 19: About an Ell lesse a naile of our English measure. **1624** Huntar 40: 4 Acres, 1 Roode, 16 Falles, 2 Ells, 3 quarter an of Ell and a Naill. **1628** Hunt B: 20 Neyles the English Ell; *ibid* D2: The proportion of price betweene the Ell and parts stands thus; one q; the neile, is 1 d. **1660** Bridges 24: In

256]

adding of the Measures of Cloth sold by the Yard, Quarter, and Nail, because 4 Nails make a quarter, and 4 quarters a yard, for every 4 nails carry a quarter, and for every 4 quarters a yard. **1696** Jeake 80: Beef, in 1 Nail, 8 Pounds of common use. **1784** Ency. meth. 137: Cloves ou Nayls. **1820** Second Rep. 24: Nail of cloth: 1/16 yard = 2 1/4 inches. **1829** Palethorpe sv: NAIL, a long measure used for measuring linens, silks, &c. and is 2 1/4 inches long. 4 nails make 1 quarter, and 4 quarters 1 yard, cloth measure. **1850** Alexander 73: Nail...2,25 inches. **1956** Economist 8: Nail...2 1/4 inches.

naile, naill, naille, nale, nall, nayl, nayle, nayll, naylle, neale, neayle, neil, neile. NAIL

nest—6 neste; 6-8 nest [ME nest fr OE nest; akin to OHG nest, L nidus, OIr net, nest, Skr nīda, resting place, nest]. A m-q for any item or sets of items, consisting of 3 in number.—**1545** Rates 1.6: Boxes the neste; ibid 11: Counters the neste; ibid 20: Hampers the neste; ibid 32: Painted coffers the neste. **1590** Ibid 2.9: Cofers with iron barres the nest containing three; ibid 11: Counters the nest containing three in one; ibid 18: Hampers the nest containing three; ibid 27: Painted cofers, the nest containing three to the nest; ibid 34: Sipers Chests the nest. **1609** Clode 97: One nest of Bowles with a cover. **1701** Hatton 3.231: Nest...of Chests of Coffers...3.

neste. NEST

neyle, neyll. NAIL

nip [shortened form of NIPPERKIN]. A United Kingdom m-c for beer containing 2/3 pt (0.379 l) (Economist 54).

niperkin. NIPPERKIN

nipper [prob abbreviation of NIPPERKIN]. A m-c for vegetables (c1895) in Middlesex totaling 12 lb (5.443 kg) (Wagstaff 36).

nipperkin—7 niperkin (OED); 7-8 knipperkin (OED); 7-9 nipperkin [cf Du nippertje, a small measure for liquor, and Du nippen, to sip]. A m-c for liquor (c1600-1800) containing no more than 1/2 pt (c0.24 l) (Rolt sv and Shipley 455).

nive [perh fr MF niveau, nivel, alter of livel, level]. A m-c for salt (c1550) containing 7 bbl (c10.36 hl) (Welsh 178).

noggan. NOGGIN

noggin—7 nogging (OED); 7-9 noggin; 8 knoggin (OED), noggan (OED); 8-9 naggin [cf E nog, ale]. A m-c for liquids (c1600) generally containing 1/2 pt (c0.24 l) and sometimes synonymous with the GILL (Jones 90). In Ireland, the "naggin" (c1800) contained 6.8 cu inches (0.111 l) and equaled 1/4 Irish pt of 27.2 cu inches (Edinburgh XII.572 and Skilling 190-193). Since the establishment of the Imperial system the noggin for wine or spirits has been reckoned as 5 fluid oz (14.206 cl) and equal to 1 GILL, 1/4 pt, or 1/32 gal (Economist 55).

nogging. NOGGIN

nok, noka, noke. NOOK

nook—3 L noka; 3-4 nok (OED); 4-6 noke (OED); 4, 6-7 nouke (OED); 5-7 nooke; 6 noque (OED); 6-9 nook [ME nok, noke; cf Nor dial nok, hook, bent figure]. A m-a for land in northern England and Scotland containing 20 acres (c8.10 ha) and equal to 2 FARTHINGDALES of 10 acres each.—c**1290** Worcester 41b: Villani tenent de dominico xlviij. Nokas; ibid 43a: Et una noka...et dimidia noka; ibid 56a: Nicholas Frewin tenet unam nokam. Cristina vidua tenet aliam nokam. Iste duo faciunt quantum dimid. virg. **1634** Noy 57: You must note, that two Fardells of Land make a Nooke of Land, and two Nookes make halfe a yard of Land. **1874** Hazlitt 434: A nook of land.... Noy, in his Complete Laywer, p. 57, says, two fardels of land make a nook, and four nooks make a yard-land.

nooke, noque, nouke. NOOK

oenophorum—3 L anaphorum (Chron. Abing.), L enoforium (Chron. Abing.), L oenophorum, L onophorium [L oeno fr Gr oino fr oinos, wine, + L phorum fr Gr phoros, bearer]. A m-c for wine equal to a gal (c3.78 l).—c**1275** Chron. Abing. II.339: Primo die admissionis abbatis Abbendonæ debet in refectorio discumbere; conventui necessariæ in cibis et potibus honorifice invenire; scilicet, onophorium, id est galonem vini, unicuique placentam integram, tria fercula piscium honorabilia, exceptis ferculis de consistorio per tabulas in invicem succedentibus; ibid 394: In duobus anniversariis, scilicet, Faricii, Vincentii inveniet in refectorio, unicuique monacho oenophorum, id est, galonem vini, et his fercula piscium honorabilia, excepto generali et aliis ferculis consuetudinariis; ibid 400: Quoties conventus oenophorum, id est, galonem, habuerit, refectorarius, excepto communi, obbatam vini habebit, obbaque prioris implebitur.

oince. OUNCE

omber, ombor, ombra. AMBER

once. OUNCE

onophorium. OENOPHORUM

oonce. OUNCE

ordeum. BARLEYCORN

osken, oskin. OXGANG

ounc. OUNCE

ounce—1-7 L uncia; 3-6 unce; 5 ouns (OED), oyns (OED), unc, unch (OED), vunce (OED); 5-7 once, owns (Halyburton); 5-9 ounce; 6 oince (OED), oonce (OED), ownche (OED), wnce; 6-7 ounc, ownce [ME unce fr MF unce fr L uncia, a twelfth, the twelfth part, ounce, inch]. A unit of wt in the ap, avdp, merc, tow, English t, and Scots t systems.

The ap oz contained 24 s, or 8 dr, or 480 t gr (31.103 g), and was equal to 1/12 ap lb of 5760 gr (373.242 g). In the Imperial system an ap fluid oz is a m-c containing 8 ap fluid dr or 1.733875 cu inches (2.84123 cl) or 480 MINIMS or the volume of 437.5 gr (28.350 g) of distilled water at 62° F and equal to 1/20 ap pt of 34.6775 cu inches.—c**1450** Hall 34: Uncia pars libre duodena, quis ambigit, inde; ibid 35: Et quelibet uncia constat ex octo dragmis. c**1600** Ibid 36: Scrupuli is 20 barley cornes...3 scruples contain a drachme...8 drachmes, an ounce. **1628** Young II.49: Put in an ounc of nutmegs & a ounc of acorns. **1660** Bridges 28: 20 Grains make 1 Scruple. 3 Scruples make 1 Dram. 8 Drams make 1 Ounce. **1688** Bernardi 137: Vel more Pharmacopolarum: Libra de Troy, 12 Unciæ...96 = 12 X 8 drachmæ ʒ: Scripuli ℈. 288 = 96 X 3: grana monetaria rursus 5760 = 288 X 20. **1696** Cocker 108: (℥) an ounce. **1708** Chamberlayne 205: The Apothecaries reckon 20 Grains Gr. make a Scruple ℈, 3 scruples 1 Drachm ʒ, 8 Drachm 1 Ounce ℥, 12 Ounces 1 Pound ℔. **1728** Chambers 1.360: The Apothecaries also use the Troy Pound, Ounce, and Grain; but they differ from the rest, in the intermediate Divisions.—They divide the Ounce into 8 Drachms; the Drachm into 3 Scruples; and the Scruple into 20 Grains. **1829** Palethorpe sv: OUNCE, in commerce, a small weight, being...the 12th part of the apothecaries´ and troy lb. **1907** Hatch 24: 1 fluid ounce (fl. oz.) = 8 fluid dracms = 1.733875 cubic inches.... 1 fluid ounce (f. ℥) is the volume of 437.5 grains of distilled water at 62° F. **1920** Stevens 2: 1 Pint, O. = 20 Fluidounces. 1 Fluidounce, fl. oz. = 8 Fluidrams, fl. dr. = 480 Minims, ♏. **1951** Trade 22: 8 drachms...1 apothecaries ounce.... 1 apothecaries ounce = 480 grains. **1966** O'Keefe 671: 1 fl. ounce = 2.84123 cl.... 1 fl. ounce = 1.7339 cu. in; ibid 673: 1 apothecaries' fluid ounce = the volume at 16.7° C (62° F) of 1 ounce avoirdupois of water.

The avdp oz contained 16 avdp dr or 437 1/2 gr (28.350 g) and was equal to 1/16 avdp lb of 7000 gr (453.592 g). It was sometimes erroneously described as the equivalent of the t oz of 480 gr (31.103 g), thereby making it equal to 1/16 lb of 7680 gr (497.664 g). Actually the avdp oz was 42 1/2 gr (2.754 g) lighter than the t oz. Nonetheless, because of its greater number of the smaller oz, the avdp lb was 1240 gr heavier than the t lb. Comparatively the avdp oz was 0.075976 English t lb, 0.057461 Scots t lb of 7616 t gr or 14.71 Scots t DROPS, and 0.045969 Scots tron lb.—**c1400** Hall 37: 16 uncie...faciunt libram. **1474** Cov. Leet 396: & xx sterling makith a Ounce of haburdepeyse; and xvj Ouncez makith a li. **1496** Seventh Rep. 29: The same tyme ordeined that xvi uncs of Troie maketh the Haberty poie. **1517** Hall 48: So makyth the whete afore namyd the Habar de Poyse once.... And xvi of that onces the trewe habar de poix lib. **c1600** Brit. Mus. 31.213: 1 once wryte 1/16 of a pound...10 onces write 5/8 part of a pound. **1635** Dalton 143: And this hath to the pound xvi ounces. **1682** Hall 29:

Aver-du-pois conteynes: every pound, 16 ounces; every ounce, 8 drgmes [sic]; every dragme, 3 scruples; every scruple, 20 graines.... But the ounce Troy is greater than the ounce Averd.; for 73 ounces Troy are equall to 80 ounces Aver-du-pois. **1688** Fox 102: 1 pece of plate 11 ounc. **1688** Bernardi 135: Insuper uncia Avoirdupois pro mercibus caducis explicat 8 drachmas aut 3 X 8 = 24 scripulos Avoirdupois...et vero 1/16 libræ suæ unciæ Romanæ prorsus æqualis ideoque 1/12 libræ Romanæ. **1717** Dict. Rus. sv: Ounce...the Sixteenth part of a Pound Avoir-du-Pois. **1794** Martin 20: The Avoirdupoise ounce is equal to 437.5 grains. **1820** Second Rep. 24: Ounce...Avoirdupois, 1/16 lb. = 7000/16 = 437 1/2 grains troy. **1834** Pasley 111: 1 Ounce Avoirdupois...437.5 [gr]. **1907** Hatch 20: 16 drachms = 1 ounce (oz.) = 437.5 grains; ibid 38: 1 ounce (16 drams) = 28.350 grammes. **1951** Trade 22: 1 avoirdupois ounce = 437.5 grains.

The merc and tow oz contained 450 gr (29.160 g), but the merc oz equaled 1/15 merc lb of 6750 gr (437.400 g), while the tow oz equaled 1/12 tow lb of 5400 gr (349.920 g). Both oz were determined as 20 dwt, the dwt being 32 wheat gr (22 1/2 t gr or barleycorns), and hence each merc or tow oz equaled 640 wheat gr or 450 t gr or barleycorns.—c**1253** Hall 11: En lituaris e confeciuns la liver est de xii uncis; en tutes autre chosis la li. est de xv uncis. **1290** Fleta 119: Item denarius sterlingus, sicut dictum est, ponderat xxxij. d. facit vnciam, et quindecim vncie faciunt libram mercatoriam.... Sterlingus...debet ponderare xxxij. grana frumenti mediocra. Et vnde xx. d. faciunt vnciam et xij. vncie faciunt libram xx. s. in pondere et numero. c**1400** Hall 7: Le denier d´Engleterre round et sanz tonsure poisera xxxii greins de froument en my le spic. Et xx d. font la unce. **1607** Cowell sv weights: 15. ownces make the Merchants pound. **1840** Ruding I.102: The old Tower, or Saxon ounce, as taken from the accounts in our exchequer, A.D. 1527...Troy Grains. 450.

The Scots t oz had two variations: for gold and silver, 480 gr (31.103 g) or 16 DROPS of 30 gr each (1.944 g) and equal to 1/12 t lb of 5760 gr (373.242 g); for meal, meat, hemp, and iron, 476 gr (30.845 g) or 16 drops of 29.75 gr each (1.928 g) and equal to 1/16 t lb of 7616 gr (493.517 g). The Scots tron oz contained 16 drops of 29.75 gr each (1.928 g) or 476 gr (30.845 g) in all, but equaled 1/20 tron lb of 9520 gr (616.896 g) (Swinton 36-38). Comparatively the Scots t oz of 476 gr was 0.082638 English t lb and 0.06798 avdp lb. The English t oz contained 480 gr (31.103 g) and was equal to 1/12 t lb of 5760 gr (373.242 g). Consisting of 20 dwt of 24 gr each, the English t oz was 30 gr (1.944 g) heavier than the merc and tow oz and 42 1/2 gr (2.754 g) heavier than the avdp oz. It was also the standard for the ap oz of 480 gr (31.103 g), the only difference being that the ap oz was divided into 8 dr (3.888 g) of 60 gr each or 24 s (1.296 g) of 20 gr each, while the t oz was divided into 20 dwt (31.103 g) of 24 gr each. Comparatively the t oz was 0.068552 avdp lb or 1 avdp oz and 1.549 avdp dr, 0.063025 Scots t lb of 7616 t gr or 1 Scots t oz and 0.134 Scots t drop, and 0.05042 Scots tron lb.—**1496** Keith 1.23: And xx starling maketh an once, and xii onces maketh a li...of Troy weight. **1587** Acts Scotland 3.521: Ilk trois pund contening sextene wnce. c**1590** Hall 22: The coyners in the Towre allowith but 24 grayns to a peny sterlinge waight [= 1/20 t oz]. **1606** Ibid 38: Ffor 24 graines or barleycornes, drie, out of the middest of the eare, doe make a 1 d. [wt]...Soe the pounde waight is 12 oz...or 5760 graines. **1607** Cowell sv weights: The pound of 12. ownces. **1616** Hopton 159: Euery ounce, 20 peny weight, euery peny weight 24 graines. **1635** Dalton 143: Troy weight...hath to the pound xii. ounces. **1665** Sheppard 15: Twenty penny weight make an ounce; 24 grains make a penny weight. **1682** Hall 29: Troy weight conteynes: every pound, 12 ounces; every ounce, 20 penny weight; every penny weight, 24 graines. **1688** Bernardi 134-35: Uncia Anglica de Troy...480 grana argenti triticive; ibid 137: Libra Anglica de Troy. 12 Unciæ...12 X 20 = 240 p. w. **1696** Oldfield 2: That in all Silver Weight, 12 Ounces make one Pound; 20 penny Weight makes one Ounce. **1708** Chamberlayne 205: 20 Pennyweight make one Ounce. **1717** Dict. Rus. sv troy-weight make one Ounce; and twelve Ounces one Pound. **1761** Thomson viii: The Scots ounce is equal to 476 [t gr]. **1805** Macpherson I.471: 20 pennies (of money) 1 ounce, 12 ounces 1 pound of London. **1816** Kelly 92: The Scotch jewellers divide the troy ounce into 16 drops, each drop being 30 troy grains. **1883** Simmonds sv: The troy ounce in England weighs 480 grains. **1907** Hatch 35: 1 ounce = 31.1034807566 grammes. **1951** Trade 25: 24 grains = 1 pennyweight...20 pennyweights = 1 ounce troy.

ouns, ownce, ownche, owns. OUNCE

oxegang, oxegange. OXGANG

oxeland, oxelande. OXLAND

oxengate, oxgait. OXGATE

oxgang—5 oxingang; 5-9 oxgang; 7 oxegang, oxegange, oxgange; 7-9 oskin, ox-going (OED); 7-9 osken [ox fr OE oxa + gang fr OE gang; akin to Du and G gang, a going, ON gangr, Goth gaggs, street, way]. A m-a for land generally synonymous with the BOVATE but occasionally described as the equivalent of either the VIRGATE or the HIDE. Like the acreage of other superficial measures, the total acreage of the oxgang depended on local soil conditions, but oxgangs of 4, 5, 6, 7, 7 1/2, 8, 9, 10, 12, 12 1/2, 13, 13 1/2, 15, 16, 18, 20, 24, 30, 32, 36, and 50 acres (c1.62 to c20.25 ha) seem to have been the most common. It was sometimes called OXGATE and OXLAND.—**c1400** Acts Scotland 1.387: The plew land thai ordanit to contene .viij. oxingang/the oxgang sall contene .xiij. akeris. **1607** Cowell sv librata terræ: Foure oxegangs, and euery oxegange 13. acres. **1610** Norden 59: In the North parts called an Oxe gange; ibid 99: Whether by the yard land, plow-land, oxegang, acres. **1664** Spelman 442: An Oxgang of Land.... Scotis ane Oxengate...quantum sufficit ad iter vel actum unius bovis. Ox enim est bos: gang vel gate, iter. **1665** Sheppard 23-24: An Oxgange of Land (in Latine Bovata terræ) is not a certain quantity of Land, as Fifteen acres; whereof 8 acres make a Plough Land. But (as some say) six Oxganges of Land seem to be as much as six Oxen will plow.... And some would say it alwayes to contain 13 Acres, and that four Oxengates is a pound Land of old extent. **1777** Nicol. and Burn 613: Oxgang of land, as much as one yoke of oxen can plough in a year. **1824** Hunter 118: An Oskin of land; an oxgang contains ten acres in some places, in others, sixteen, eighteen, twenty-four, and fifty in some part of the Bradford parish. **1829** Brockett 222: Osken...an oxgang of land...varying in quantity in different townships, according to the extent of ground.... In our old laws it meant as much as an ox-team could plough in a year. **1872** Robertson 97: In the oldest examples of customary tenure in the Boldon Buke, the oxgang is always reckoned at 30 acres; ibid 100: In the Boldon Buke and the Black Book of Hexham, compiled respectively in the thirteenth and fifteenth centuries, the oxgang by no means appears invariably as a measure of fifteen acres, but varies in extent between seven and a half and thirty-six...though thirty and fifteen are the ordinary amounts in the Palatinate, and twelve in Northumberland. **1883** Simmonds sv: Ox-gang, a parcel of land of about 15 acres, on the average. **1888** Taylor 147: In 1766, at Elsternwick, the oxgangs contained 12 acres in each arable field.... At Keyingham each oxgang consisted of 10 acres in each arable field; ibid 183: At Kirby Underdale, Domesday gives 6 carucates, which would be 48 oxgangs.

oxgange. OXGANG

oxgate—6 oxgait (Robertson); 7 oxengate; 9 oxgate [ox + gate; see OXGANG]. Equivalent to OXGANG.—**1610** Folkingham 59: The Oxe-gang, or Oxengate...called Bouata terræ containes after the originall repute 13.

acres but we find it more or lesse as the custom of the place inures. **1624** Huntar 7: 13. Acres is compted an Oxen-gate. 4 Oxen-gate is esteemed a pund land of old extent. **1665** Sheppard 23-24: And some would say it alwayes to contain 13 Acres, and that four Oxengates is a pound Land of old extent. **1829** Palethorpe sv oxgang: OXGANG, or OXGATE...is generally taken for 15 acres. **1872** Robertson 135: The Ploughgate, or carucate of 104 acres. It was divided [in Scotland], as in northern England, into eight oxgates.

ox-going. OXGANG

oxland—4, 7 oxeland; 7 oxelande; 9 oxland [ox + land; see OXGANG]. Equivalent to OXGANG.—**1387** Higden II.97: Of eueriche bouata terræ, that is, of eueriche oxeland. **1603** Henllys 135: viii acres make an Oxelande...viii oxelandes make a ploweland being...64 acr. **1820** Second Rep. 24: Ox-Land, Glamorganshire and Pembrokeshire: 8 customary acres.

oyns. OUNCE

P

paame. PALM

paas. PACE

pace—1-7 L passus; 3-5 pas (OED); 4-5 paas, pass (OED); 4-5, 7 passe; 4-7 pase; 4-9 pace; 5 pasce (OED); 5-6 pais (OED), paiss (OED); 6 paice (OED) [ME pace, pas fr OF pas fr L passus, a step, pace]. A m-l generally equal to 2 STEPS or approximately 5 ft (c1.52 m).—**c1100** Hall 3: Passus v pedes habet. **c1289** Bray 10: Quinque pedes passum faciunt. **c1300** Hall 7: Unde 5 pedes faciunt passum, et 125 passus faciunt stadium. **c1325** Rameseia I.76: Passus pedes quinque. **1387** Higden I.49: Fiftene thowsand paas in lengthe, and fyue thowsand paas in brede. **1395** York Mem. 1.142: Item, quinque pedes faciunt passum. **c1400** Hall 5: Passus v pedes habet. **c1460** Capgrave 16: In length half a mile and XL. passes. **c1461** Hall 14: And also V fote make a pase. **1561** Eden xviii: Fyue feete a geometrical pase. **c1600** Brit. Mus. 16.70: And .5. foote maketh a pase. **1616** Hopton 165: Also an English mile is...1056 paces. **1635** Dalton 150: Five foot doe make a Geometricall Pace. **1639** Bedwell Bl: One hand breadth, one foote, one passe. **1665** Sheppard 16: 5 foot a Geometrical pace. **1688** Bernardi 202: Pes Anglicus...1/5 Passus Geometrici aut Agri mensorii. **1708** Chamberlayne 207: 5 Foot make a Geometrical Pace. **1717** Dict. Rus. sv: A Geometrical Pace consists of Five Foot, and a thousand such Paces, make up a Mile. **1850** Alexander 78: Pace...60.—inches. **1868** Eng. Cyclo. 817: Two feet and a half, are a step...two steps, or five feet, are a pace. **1956** Economist 8: Pace, geometrical...5 feet.

pack—3-6 pak; 3-7 packe; 4-5 pakke; 4-9 pack; 5 pakk; 5-6 pake [ME pak, packe, pakke, of LG origin; see OED]. A m-c and m-q for many products: cloth, generally 10 PIECES; flax or flour, 240 lb (108.862 kg); teasels, generally 9000 heads for kings and 20,000 heads for middlings, except in Gloucestershire, 40 STAFFS or 1000 GLEANS or 20,000 heads for middlings and 30 staffs or 900 gleans or 9000 heads for kings, and in Yorkshire, 1350 bunches of 10 heads each or 13,500 in all; vegetables, Huddersfield, 240 lb (108.862 kg); wool, 240 lb (108.862 kg), except lamb´s wool in Yorkshire and Lancashire, 44 lb (19.958 kg); and yarn, 4 Cwt or 480 lb (217.724 kg).—**1228** Gras 1.157: 1 pak mailede. **1439** Southampton 2.81: 1 pak de canevas. **1443** Brokage II.27: Flaxe the pack untrussed.... Cum iiii pakkes de pannys; ibid 119: 1 pak straytes; ibid 158: Cum ii pakkys cerseyse. **c1461** Hall 16: Also clothe is sold by numbyr, for x hole clothys make a pak. **1466** Gras 1.614: Pro i pakke lewent. **c1475** Ibid 192: Of a pakke of wulle cloth. **1507** Ibid 695: Brusshys the packe; ibid 698: Flexe the pake containing xx [X] c lbs.; ibid 699: Hather the packe that contains as moche as a packe of wolle; ibid 704: Torche waxe the pack. **1509** Ibid 562: ii packes canvas continent´ iii [X] m ulnarum; ibid 566: i packe cum ii bages ginger continent iii [X] c libras; ibid 590: Pro ii packes cum xii [X] c goodes cotonrusset. **1555** York Mer. 156: A pake of clothe, sixtene pence; a small trusse, as the parties canne agree, so that it excede not the price of the pake, to be rated after the qualitie thereof. **1562** Ibid 168-69: A packe of clothe, xx d. **1607** Cowell sv sarpler: Further that a packe of wolle is a horse loade, which consisteth of 17. stone. two pounds. **1665** Sheppard 66: And further, That a Pack of Wooll is a horseload, consisting of 17 stone and two pounds. **1717** Dict. Rus. sv: Pack of Wooll, is 17 Stone and 2 Pounds, or 240 Pound weight. **1756** Rolt sv: PACK of wool, in commerce, is a horse´s load; containing 17 stone and 2 pounds, or 240 pounds weight. **1820** Second Rep. 25: Pack of yarn, 4 hundred weight, each of 120 lb...of teazles, 9000 heads of kings; 20000 of middlings...Huntingdonshire: of wool, 240 lb...Kent: of flax, 240 lb...Yorkshire, N. R. of teazles, 1350 bunches of 10 each = 13500. **1834** Pasley 113: 1 Pack of Lamb´s Wool in Yorkshire and Lancashire...44 [lb]; ibid 114: 1 Pack of Wool in Huntingdonshire...of Lamb´s Wool in North Wales...of Flax in Kent...240 [lb]. **1850** Alexander 78: Pack; of wool...240.—pounds. **1880** Courtney 154: A pack of wool is 17 stone 2 lbs. = 240 lbs. **1880** Britten 173: Pack, of teazles, 9000 heads of kings; 20,000 of middlings. (Glouc.), of teazles, 40 staffs = 1000 glens = 20,000 of kings, 30 staffs = 900 glens = 9000. **1883** McConnell 15: 20 lbs. = 1 score, and 240 lbs. or 12 scores = 1 pack. **1896** Wagstaff 36: A pack of vegetables in Huddersfield = 240 lbs. **1956** Economist 50: Pack: Flour = 240 lb; ibid 58: 1 pack [of wool] = 240 lb.

packe. PACK

packet—4 pakett; 6-9 packet, pacquet (OED); 7 paquette (OED); 8-9 paquet (OED) [ME pakett, dim of ME pak; see PACK]. A m-c and m-q probably equal to a small PACK or BUNDLE.—**1304** Gras 1.168: Pro i pakett´ canabi. **1820** Second Rep. 25: Packet of leaf metal, 250 leaves. **1831** Pope 175: LEAF METAL (except Leaf Gold) the packet containing 250 leaves.

pacquet. PACKET

pad [alter of ped fr ME pedde, basket]. A m-c for potatoes (c1895) in Middlesex totaling 112 lb (50.802 kg) (Wagstaff 36). It was an open PANNIER, usually made of osiers.

paer. PAIR

paice. PACE

paier. PAIR

pair—3-5 peyre; 3-6 peire; 4-6 payr; 4-7 paire, pare, payre; 4-9 pair; 5 payir (OED), peyer (OED), peyr; 5-6 par, payer; 6 paer, paier (OED), parre (OED), peare, per, pere (OED) [ME peire, paire fr OF paire fr L paria, neut pl of par, equal]. A m-q consisting of 2 of the same item or sets of items. It sometimes is abbreviated pr.—c**1435** Amundesham II.214: Unum par furcarum. c**1440** Promp. Parv. 391: Payr, or a peyr, of tweyne thyngys. **1443** Brokage II.293: Et xvii payr shetes. c**1475** Stonor I.146: Item, j peyre canstyckes; ibid 153: It., for a peyr schone, v. d.;ibid 154: It., a payre hosen off russet, the price, iij. d. **1500** Relation 126: A payre of hosyn of skarlet.... Two payre of hosyn, skarlet, garded with crymsyn velvet; ibid 127: A paire of stirropes.... A paire of buskyns of blacke velvet.... A payre of arminge spores; ibid 128: iij. payre of shoes of whyte clothe.... iij.

paire of yellow clothe.... A payre of arminge shoes.... A payre of slippers of redd letter; ibid 129: A payre of slippers of black lether.... A payre of fustyans. **1507** Gras 1.703: Shermans sheres the payer. **1532** Finchale ccccxlix: j payr ballans and j balk of yron. **1544** Beck 2.83: X paer of Spaneyshe gloues. **1556** Ibid 84: A peire of gloves. **1567** Barfield Appendix XLVI: Itm payd for a peare of shues. **1569** Remembrance 109: This daye ther do remayn in the Wolle Hows ii bems and i per of skalls, xvii brasson waytts, ii ledyn waytts of xiii neyll apes. **1577** Beck 2.85: ij peir of swete gloves. **1578** Ibid 115: ij pare of Oxford gloves. **1615** Collect. Stat. 465: And a dicker of gloues consisteth of ten paire of gloues. **1679** York Mer. 297: For a pare of paun scales and small exchequer weights from a pound to a dram, 0£. 09 s. 00 d. **1883** Simmonds sv: Pair, a couple or brace...a pair of stockings, gloves, &c.

paire. PAIR

pais, paiss. PACE

pak, pake. PACK

pakett. PACKET

pakk, pakke. PACK

paladr [*]. A m-a for land in Anglesey (c1800-1900) containing 20 1/4 sq yd (16.929 sq m) (Second Rep. 25 and Donisthorpe 213).

palm—1-4 L palma; 1-7 L palmus; 4-6 pame (OED), paume (OED), pawme (OED); 5 paame (OED); 5-6 paulme (OED); 5-7 palme; 7-9 palm [ME paume fr MF paume fr L palma]. A m-l, originally a unit of body measurement referring to a hand's breadth exclusive of the thumb, which was equal to 1/3 SPAN or 1/6 CUBIT. Based on the ft of 12 inches, it was made equal to 3 inches (7.62 cm).—c**1100** Hall 5: Palma extensa est xii digitorum...compressa est iiii digita. c**1300** Ibid 7: Et tres pollices faciunt palmam. c**1325** Rameseia I.76: Palmus autem quatuor digitos habet. **1395** York Mem. 1.142: Et tres pollices faciunt palmam. c**1400** Hall 6: Quattuor palmi faciunt pedem. c**1610** Lingelbach 108: Reduce the palmes of euerie peece of Velvitt into flemische ells, by addinge four palmes and no more to euerie hundred palmes. **1628** Hunt C: A Palme, or Handbreadth. **1665** Assize 6: The foot to contain four palms, and every palm containeth four fingers breadth. **1688** Bernardi 193: Palmus—3 unciæ aut pollices. **1708** Chamberlayne 209: Foot...4 Palm. **1716** Harris 2. sv measure: Palm...3 Inch. **1717** Dict. Rus. sv hand-breadth: A Measure of three Inches. **1820** Second Rep. 25: Palm, sometimes denotes 3 inches. **1832** Edinburgh XII.569: 3 Inches = 1 Palm = 0.0762 [m]. **1888** Fr. Clarke 36: A palm = 3 inches. **1956** Economist 8: Palm...3 inches.

palma, palme, palmus, pame. PALM

panier. PANNIER

pannier—4-7 panyer; 4-9 panier; 5 panyere; 6-7 pannyer; 6-9 pannier [ME panier fr MF panier, pannier fr L panarium, pannarium, bread-basket, fr panis, bread]. A m-c, a large basket, of no consistent size for carrying provisions, fish, or other commodities.—**1387** Higden V.195: A panyer ful of gravel. c**1440** Promp. Parv. 381: Panyere...Panyer, or basket. c**1470** Gregory 161: M...panyers with fyggys and raysonys. c**1550** Welsh 281: 2 paniers glass. **1603** Hostmen 36: Two smale maunds or pannyers full, holdinge two or three pecks apeece. **1664** Gouldman sv basket: A basket, maund, or pannier. Sporta. **1883** Simmonds sv: Pannier, a hamper or basket.

pannyer, panyer. PANNIER

paquet, paquette. PACKET

par, pare. PAIR

pared [*]. A m-l of 3 yd (2.743 m) for cloth in Montgomeryshire (c1800-1900) (Second Rep. 25 and Donisthorpe 213).

parre. PAIR

partica, particata. PERCH

pas, pasce, pase, pass, passe, passus. PACE

paulme, paume, pawme. PALM

payer, payir, payr, payre. PAIR

peace. PIECE

pearch, pearche. PERCH

peare. PAIR

pease, peax, peayce. PIECE

pec, pecca. PECK

peccaid [perh fr E PECK]. A m-c for grain in southern and eastern Wales (c1800-1900) containing 5 to 6 gal (c2.20 to c2.64 dkl) and sometimes synonymous with the HOBED and HOOP (Second Rep. 25 and Donisthorpe 213).

peccum, peccus. PECK

pece, pecia. PIECE

peck—4 L peccum; 4-5 pec; 4-6 pek (Nottingham), pekke; 4-9 peck; 5 pekk; 5-7 pecke, peke; 6 L pecca; 7 L peccus, pect [ME pek fr OF pek, of obscure origin]. A m-c for grain and other dry products, generally containing 2 gal (8.810 l) or 537.6 cu inches and equal to 1/4 Winchester bu. The Irish pk contained 435.2 cu inches (7.133 l) and equaled 2 Irish gal of 217.6 cu inches (Edinburgh XII.571). Since the establishment of the Imperial system in 1824, the pk has contained 554.840 cu inches (9.092 l) or 1/4 Imperial bu of 2219.360 cu inches.—**1315** Ireland xxxv: Quilibet crannocus [avenarum] continebit quindecim pecks cumulatos boni et mundi bladi. **1319** Ibid xxxv: Quiquidem crannocus [avenarum] continebit sexdecim pecks cumulatos boni, sicci et mundi bladi. **1351** Rot. Parl. 2.240: Soient les Mesures, c´est assaver bussell, di. bussell, et pec. **1384** Rot. Parl. 4.185: Videlicet Busselli, dimidii Busselli, & Peck. **1390** Henry Derby 6: Et j pecco auenarum. **1392** Ibid 73: Super officio salsarie per manus eiusdem pro j. pecco farine frumenti per ipsum empto ibidem, iiij scot. pr. c**1400** Brit. Mus. 30.52v: xvj pyntes...j pekk. **1418** trans in Memorials 666: It was ordered, that oysters and mussels should be sold at 4 d. the bushel, 2 d. the half bushel, one penny the pec. **1430** Rot. Parl. 5.432: Videlicet, Busselli, Dimidii Busselli, et Peck. c**1440** Promp. Parv. 391: Pekke, mesure. Batus. **1474** Cov. Leet 396: & halfe Weyght a pekke. **1540** St. Mary´s 61: Ad lx. peccas frumenti et avenarum; ibid 211: Per iii peckes frumenti et quatuor peckes avenarum.

c**1590** Hall 20: 4 peckes makith a bushell of Winchester measure, accordinge to the owld standadt: 2 galons makith a pecke. **1603** Hostmen 36-37: Holdinge two or three pecks apeece. c**1610** Lingelbach 122: Upon a sack or peke. **1615** Collect. Stat. 468: That the said water measure within the shipboard shall only containe fiue pecks after the said standard rased and stricken. **1616** Hopton 162: Whereof are made Pints, Quarts, Pottles, Gallons, Peckes, Halfe-bushels, Bushels. **1621** Stat. Irel. 46: If one pecke of the said graynes exceed the price of ten pence. **1628** Acts Scotland 5.188: Exacting of ane pect to the boll. **1635** Dalton 144: Eight quarts maketh the peck...4 peckes maketh the Bushell. **1665** Sheppard 15: And 4 Pecks make the Bushell. **1682** Hall 30: 1 Last conteynes...80 Bushels, 320 Peckes, 640 Gallons. **1688** Bernardi 69: Bussellus Anglicanus...continens in se Peccos 4. Galones siccos 8. **1708** Chamberlayne 212: 2...Gallons makes a Peck...4 Pecks a Bushel. **1717** Dict. Rus. sv: Peck, an English dry Measure containing two Gallons; the fourth part of a Bushel. **1779** Swinton 31: Peck...537.6 [cu inches]. **1820** Second Rep. 25: Peck, 1/4 bushel = 2 gallons = 4 quarterns...of flour and salt, generally reckoned 14 lbs. **1880** Courtney 162: 8 quarts 1 peck. **1907** Hatch 23: 2 gallons = 1 peck (pk.) = 554.840 cu inches; ibid 35: 1 peck = 9.091926 litres. **1956** Economist 4: Peck: (a) United Kingdom, Imperial system = 554.84 cubic inches. **1969** And. & Bigg 11: 1 pk (peck) = 9.09218 dm^3.

The Scots pk for wheat, peas, beans, rye, and white salt contained 549.333 cu inches (9.004 l) or 5.3125 Scots pt and equal to 0.255454

Winchester bu, and for oats, barley, and malt, 801.381 cu inches (13.135 l) or 7.75 Scots pt and equal to 0.372662 Winchester bu (Swinton 32, J. Sheppard 91, and Cooke 103). Both of these standard Scots pk were equal to 4 LIPPIES or FORPITS, 1/4 FIRLOT, or 1/16 BOLL. Locally (c1600-1800), however, there were many exceptions (Swinton 53-130). North—Nairnshire: wheat, peas, beans, rye, ryegrass-seed, oatmeal, and barleymeal, 670.055 cu inches (1.098 dkl); barley and oats, 893.408 cu inches (1.464 dkl). Sutherlandshire: peas, rye, and beans, 646.275 cu inches (1.059 dkl); oats, barley, and malt, 886.679 cu inches (1.453 dkl). Northwest—Inverness: wheat, peas, beans, rye, ryegrass-seed, and meal, 628.742 cu inches (1.030 dkl); oats, barley, and malt, 879.806 cu inches (1.442 dkl). Ross and Cromarty: wheat, rye, peas, beans, and lime, 620.424 cu inches (1.017 dkl); oats, barley, and malt, 827.232 cu inches (1.356 dkl). Northeast—Aberdeenshire: wheat, rye, peas, beans, meal, and seeds, 672.126 cu inches (1.101 dkl); oats, barley, and malt, 878.934 cu inches (1.440 dkl). Banffshire: wheat, beans, peas, rye, and white salt, 579.062 cu inches (0.924 dkl); oats, barley, and malt, 842.272 cu inches (1.380 dkl). Caithness: oats and barley, 851.467 cu inches (1.395 dkl). Moray (Elgin): wheat, rye, peas, and beans, 586.501 cu inches (0.961 dkl); barley and oats, 843.508 cu inches (1.382 dkl). Central—Perthshire: wheat, peas, rye, and beans, 565.74 cu inches (0.927 dkl); oats, barley, and malt, 834.75 cu inches (1.368 dkl). Stirlingshire: wheat, peas, beans, and rye, 594.573 cu inches (0.974 dkl); oats, barley, and malt, 859.546 cu inches (1.409 dkl).

West central—Dumbartonshire: wheat, peas, beans, and meal, 640.687 cu inches (1.050 dkl); oats, barley, and malt, 854.25 cu inches (1.400 dkl). West—Argyllshire: wheat, rye, beans, and peas, 638.600 cu inches (1.047 dkl); oats, barley, and malt, 859.546 cu inches (1.409 dkl). East—Angus: wheat, peas, and beans, 568.722 cu inches (0.932 dkl); oats, barley, and malt, 830.463 cu inches (1.361 dkl)—both pecks average of Montrose, Forfar, Brechin, Dundee, and Arbroath pecks. Fifeshire: wheat, peas, and beans, 568.722 cu inches (0.932 dkl); oats, barley, and malt, 827.232 cu inches (1.356 dkl). Kincardineshire: wheat, rye, and peas, 620.424 cu inches (1.016 dkl); oats and barley, 853.083 cu inches (1.398 dkl). Kinrossshire: wheat, peas, and beans, 563.875 cu inches (0.924 dkl); oats, barley, and malt, 825.616 cu inches (1.353 dkl). South—Lanarkshire, Glasgow and Lower Ward: wheat, 578.547 cu inches (0.948 dkl); peas and beans, 817.8 cu inches (1.340 dkl); oats and barley, 834.85 cu inches (1.368 dkl). Peeblesshire: wheat, peas, beans, and rye, 588.6 cu inches (0.965 dkl); oats, barley, and malt, 837.12 cu inches (1.372 dkl). Southwest—Ayrshire: wheat, rye, peas, and beans, 614.4 cu inches (1.007 dkl) in Kyle and Carrick, 628.256 cu inches (1.030 dkl) in Cunningham; oats, barley, and malt, 905.44 cu inches (1.484 dkl) and 1008.009 cu inches (1.652 dkl) in Kyle and Carrick. Buteshire and Arran: wheat, peas, and beans, 719.519 cu inches (1.180 dkl); oats, barley, and malt, 1079.279 cu inches (1.769 dkl). Renfrewshire: wheat, the Linlithgow standard; beans, peas, and vetches, 601.357 cu inches (0.985 dkl); oats and barley, 851.467 cu inches (1.395 dkl). Southeast—Berwickshire: all grain, 840.157 cu inches (1.377 dkl). East Lothian: wheat, peas, and beans, 565.490 cu inches (0.926 dkl); oats, barley, and malt, 825.616 cu inches (1.353 dkl). Midlothian: wheat, peas, and beans, 559.028 cu inches (0.916 dkl); oats, barley, and malt, 814.306 cu inches (1.335 dkl). Roxburghshire: wheat, peas, and beans, 568.722 cu inches (0.932 dkl); oats, barley, and malt, 853.083 cu inches (1.398 dkl). Selkirkshire: wheat, rye, beans, and peas, 1140.675 cu inches (1.869 dkl).

pecke, pect. PECK

peec, peece. PIECE

peerch. PERCH

pees, peese. PIECE

peget [perh akin to ME pegge, LG pegel, a stake, MLG pegel, a watermark, a gauge rod, a measure of wine, OE pægel, a wine measure]. A m-c in Wales at Anglesey and Carnarvon (c1800-1900) for corn, 2 HOBEDS equal to 8 Winchester bu (2.819 hl), and lime, 4 Winchester bu (1.409 hl) (Second Rep. 26 and Donisthorpe 213).

peice. PIECE

peir, peire. PAIR

peis, peise. PIECE

pek, peke, pekk, pekke. PECK

penneyweight, pennieweight, penningus, penny. PENNYWEIGHT

pennyweight—5-7 peny; 6 penyeweight; 6-8 penyweight; 6-9 pennyweight; 7 penneyweight, pennieweight, L penningus; 7-9 penny [ME peny fr OE penig, penning, + WEIGHT]. A wt in both the t and tow systems. It was originally the wt of a silver penny which equaled 1/240 of a tow lb, and, as a unit of currency, was called either a denarius or a sterling. The t dwt contained 24 gr or barleycorns (1.555 g) and was equal to 1/20 t oz of 480 gr (31.103 g), while the tow dwt contained 32 wheat gr, or 22 1/2 t gr (1.458 g), and was equal to 1/20 tow oz of 450 gr (29.160 g). The t dwt for weighing pearls, however, contained 30 gr (1.944 g) and was equal to 1/20 oz of 600 t gr (38.880 g). Comparatively the t dwt of 24 gr was 0.877 avdp dr, 0.807 Scots t DROP, and 0.002521 Scots tron lb. In the early modern period, the dwt was occasionally abbreviated pwt or p. wt.—c**1200** Caernarvon 242: Per discrecionem tocius Regni Anglie fuit mensura Domini Regis compoia videlicet quod denarius Anglican qui vocat Sterlingus rotundus & sine tonsura ponderabit xxx...& duo g[ra]na frumenti in medio spici. **1290** Fleta 119: Sterlingus...debet ponderare xxxij grana frumenti mediocra. c**1300** Hall 8: Denarius...ponderabit xxxii grana frumenti rubei, in medio spice assumpta. **1474** Cov. Leet 396: xxxii graynes of whete take out of the mydens of the Ere makith a sterling other-wyse called a peny. **1496** Keith 1.23: That xxxii graynes of Wheate, taken out of the middel of the yeare, weieth a starling, otherwise called a penny, and xx starling maketh an once. **1540** Recorde 133: As 24 Barley-corns dry, and taken out of the middest of the Ear, do make a penny weight, 20 of those peny weights make an ounce. **1588** Hall 46: One penny waight, which is 24 graines. c**1590** Ibid 22: The peny sterling, round, without clypinge, shall way 32 grayns of wheat, dry, out of the midst of the eare.... The coyners in the Towre allowith but 24 grayns to a peny sterlinge waight. **1595** Powell C: Fourescore foure ounces halfe an ounce, and two penye weight Troye. **1606** Hall 38: Ffor 24 graines or barleycornes, drie, out of the middest of the eare, doe make a 1 d. [wt]. **1615** Collect. Stat. 467: And euerie ounce containe twentie sterlings, and euerie sterling be of the weight of two and thirtie cornes of wheat that grew in the middest of the eare of the wheat. **1616** Hopton 159: And this Troy weight containes in euery pound 12 ounces, euery ounce 20 peny weight, euery peny weight 24 graines. **1628** Hunt C: 10 Penneyweight. **1635** Dalton 144: 32 Wheat cornes taken in the midst of the eare, weigheth 1. d. sterling. **1640** Penkethman 1: By Troy Weight, O signifies Ounces; P. Pennyweights. **1651** Violet 122: Weighing 17 pound weight, one Ounce and five pennie weight. **1656** Rawlyns 43: 20 penny weight one ounce. **1660** Bridges 24: 24 Grains...1 Penny-weight. **1665** Assize 1: And two and thirty grains of Wheat make the whole sterling peny. **1677** Roberts 296: The Troy-pound consists of 12 ounces, the ounce of 20 peny weights, the Peny weight of 24 Grains. **1682** Hall 29: Troy weight conteynes: every pound, 12 ounces; every ounce, 20 penny weight; every penny weight, 24 graines. **1688** Bernardi 134-35: Uncia Anglica de Troy. 8 Drachmæ. 8 X 3 = 24 Scripuli. 8 X 3 X 20 [denarii] = 480 grana argenti triticive; ibid 165: Denarius Elizabethæ regumque sequentium, Penningus novus Anglorum...1/240 libræ. **1696** Cocker 111: 24 grains make one penny-weight. 20 penny-weight make an ounce. **1710** Harris 1. sv weights: The Original of all our English Weights, was a Corn of Wheat...32 of these made one Penny-Weight, or were the Weight of the Penny-Sterling: Twenty of the Pence or Penny-Weight, were to make an Ounce. **1717** Dict. Rus. sv: Pennyweight; this consists of 24 Grains...of these 20 make an Ounce Troy. **1728** Chambers 1.360: In Troy Weight, 24...Grains make a Penny-weight Sterling; 20 Penny-weight make an Ounce. **1780** Bald 391: A Penyweight is equal to... .877 [avdp dram]. **1787** Liber xi: The Statute intituled Assisa Panis et Cervisiæ, made in the 51st year of King Henry III, ordained, That an English Penny, called a Sterling, round, and without clipping, should weigh 32 wheat grains in the midst of the ear; and that 20 of those pennyweights should make an ounce. **1790** Jefferson 1.986: According to the subdivision for gold and silver, the ounce is divided into twenty pennyweights, and the pennyweight into twenty-four grains. **1793** Leake 18: A Penny-weight, or the twentieth Part of an Ounce. **1820** Second Rep. 26: Penny-Weight...Formerly, 1/240 of a money or tower pound, weighing 22 1/2 grains...at present 24 grains, 1/240 of troy pound. **1834** Pasley 109: Pearls are weighed by Troy weight, excepting that the Pennyweight contains 30 grains instead of 24.... 30 Pearl Grains...1 Pennyweight. 20 Pennyweights, or 600 Pearl Grains...1 Ounce. **1854** Bowring 93: The tower or easterling pound weighed three-quarters of an ounce troy less than the troy pound.... Its penny, or two hundred and fortieth part, weighed therefore 22 1/2 grains troy; and that was the weight of the thirty-two kernels of wheat from the middle of the ear; ibid 94: In England the term "sterling," originally "easterling," and in France the synonymous term "esterlin," were used to denote the twentieth part of the ounce, also called "penny" in England, and "denier," from denarius, in France. **1880** Courtney 157: The term pennyweight is derived from the weight of the old silver penny. **1903** Warren 100: Grains 24 = 1 pennyweight. **1907** Hatch 35: 1 Pennyweight = 1.555174 grammes. **1956** Economist 7: 24 grains = 1 pennyweight...3 scruples = 1 drachm.

peny, penyeweight, penyweight. PENNYWEIGHT

per. PAIR

percata terre. PERCH OF LAND

perca. PERCH

perch—1-7 L pertica; 2 L particata; 3-7 perche; 5-6 L percha; 5-9 perch; 6 L partica; 6-7 pearche; 6-8 pearch; 7 peerch (OED); ? L perca (Maitland) [ME perche fr OF perche fr L pertica, pole, long staff, measuring rod]. A m-l for land, generally containing 16 1/2 ft or 5 1/2 yd (5.029 m), but perches of 9, 9 1/3, 10, 11, 11 1/2, 12, 15, 16, 18, 18 1/4, 18 1/2, 18 3/4, 19, 19 1/2, 20, 21, 22, 22 1/2, 24, 24 1/2, 25, 25 1/2, 26, and 28 ft (2.743 to 8.534 m) were also used. Perches of 16 1/2 ft and smaller were usually agricultural land measures, while those larger than 16 1/2 ft were used by woodsmen in the forest regions and by town craftsmen engaged in draining, fencing, hedging, and walling operations. The Irish standard perch was 7 yd or 21 ft (6.401 m) and was equal to

1/320 Irish mi of 2240 yd.—c**1100** Hall 4: Duo vero passus decem pedam perticam faciunt.... Pertica ad manus xv pedes habet. c**1150** Acts Scotland 1.387: Particata terre in baronia debet mensurari per sex vlnas que faciunt .xviij. pedes mediocres hoc est neque de maioribus neque de minoribus. particata terre in burgo continet viginti pedes mediocres. **1214** Cur. Reg. 14.283: Sed ad perticam xxvj. pedum. **1221** Eyre 488: Et Willelmo eiusdem loci episcopo et omnibus successoribus suis quietantiam de sex centum et xiiii...acris de essartis mensuratis per perticam continentem in longitudine viginti quinque pedum et dimidium per manupedem. **1229** Close I.186-87: Mandatum est Hugoni de Nevill´ et sociis suis, justiciariis itinerantibus ad placita foreste, quod secundum pertica continere solet vel xxiv vel xxv pedes manupedum temporibus H. regis avi regis, R. regis avunculi, et J. regis patris domini regis, sic placet domino regi et consilio suo quod pertica magis usitata et continente in longitudine xxiiij vel xxv pedes manupedum in essartis mensurandis. c**1272** Hall 7: Et tres pedes faciunt ulnam; et quinque ulne et dimidia faciunt perticam. **1277** Mon. Fran. 285: In longitudine xix. perticarum, pertica sedecim pedes et dimidium continente. c**1289** Bray 10: Et continet quaelibet pertica xvi pedes de pedibus rectis. **1304** Mon. Fran. 294: Et dicunt quod prædicta placea continet in se octo perticatas in longitudine et quinque perticatas et dimidiam in latitudine, per perticam viginti quinque pedum. c**1325** Rameseia I.76: Pertica passus duos, id est pedes decem. c**1400** Henley 8: E la perche le rey est de xvi pez et demi. c**1461** Hall 14: Also v yerdes dim. make a perche, in London, to mete land by. **1474** Cov. Leet 397: And out of the seid yard growith a Rodde to mesure land by the wich Rod conteyneth in lengthe V yardes & halfe. c**1500** Brit. Mus. 6.7: The Lande pearche...xvj fote di; ibid 159: & xl partice in longitudine & iiij...in latitudine faciunt unam acram terræ. c**1500** Hall 8: V virge dimidia faciunt perticam. **1502** Arnold 173: In dyuers odur placis...they mete ground by pollis gaddis and roddis some be of xviij foote some of xx fote and som xvi fote in lengith. **1537** Benese 4: The woodlande perche is communely .xviii. foote in length.... The perche of woodlande is longer than is the perche of fyldelande. **1540** Recorde 207: 5 yardes and a halfe make a Perche. **1589** Bellot 4-5: And all by the pearch of sixteene foote and a halfe. c**1590** Hall 27: 16 foott 1/2 in lenght is a poole or a perche. **1599** Richmond Appendix 2.12: Carucata, Bovata, Virgata, Percha, Acra, Roda. **1603** Henllys 133: For in some place the pole is but ix foot, and in some place xij foote. **1607** Cowell sv acre: The perche differeth, being in some places, and most ordinarily, but 16. foot dimid. But in the Counties of Stafford 24 foote; ibid sv perche: In the Forest of Sheerewood it is 25. foot; ibid sv rodde: Rodde (Pertica) is otherwise called a pearche, and is a measure of 16. foote and an halfe long, and in Stafford Shire 20. foote, to measure land with. **1610** Norden 138: 18 foote and a halfe to a perch. **1616** Hopton 165: 5 yards and a halfe, a pearch. **1635** Dalton 150: Five yards and an halfe (which is 16 foot and an halfe) maketh a pole, rood, or pearch. **1647** Digges 1: Five Yards, 1/2. a Pearch.

1653 Leybourn 1.254: The Pole or Perch of 16 foot and a halfe, but in many places of this Nation (through long custome) there hath been received other quantities, called Customarie, as namely, of 18, 20, 24, and 28 foot to the Pole or Perch. **1665** Assize 6: In many countries [= districts] this Pole or Perch doth vary, as in some places it is 18 foot, and in some other places 21 foot. **1669** Worlidge 330: A Perch, or Lug is sixteen foot and a half Land-measure, but is usually eighteen foot to measure Coppice-woods withal. **1682** Hall 28: A Pearch, or a Rod, or a Pole (by statut) must be 5 yards and an half; or 16 feete and an half. But in some places of England they measure w[ith] a pearch of 12 foote called Tenant right or Court measure. In other places they measure w[ith] a pearch of 18, 20, or 24 foote, called Woodland Measure. **1688** Bernardi 197: Pes est...1/16,5 Perticæ Anglicæ. **1695** Kennett Glossary sv pertica: A Perch, which in the reign of King John was the measure of twenty foot, and was the same as Virga. **1696** Phillips sv pole: In measuring, it is the same with Pearch or Rod, or as some call it Lugg: By Stat. 35 Eliz. this Measure is a length of 16 Foot and a half, but in some Countries [= districts] it consists of 18 Foot and is called Woodland-Measure; in some Places of 21 Foot termed Church-Measure; and in others of 24 Foot under the Name of Forest-Measure. **1708** Chamberlayne 207: 16 Foot and a half make a Perch, Pole or Rod, but there are other Customary Perches or Poles, viz. Eighteen Feet for Fens and Woodland, Twenty one for Forest, Lancashire and Irish Measure, and 18 3/4 Scotch. **1717** Dict. Rus. sv: Perch or

Pearch, a Rod or Pole, with which Land is measur´d.... 18 Foot is the measuring of Coppice-woods.... In Herefordshire, a Perch of Walling is 16 Foot and an half, a Perch of Ditching 21 Foot. **1725** Bradley sv mile: Every Furlong forty Lugs or Poles...every Pole sixteen Foot and a Half. **1789** Hawney 213: But in some Places the Custom is to allow 18 Feet to the Rod...and in some Places...21 Feet. **1819** Cyclopædia sv weights: In building, hedging, and ditching, the perch or pole of eighteen feet is the usual measure [in Berkshire]. **1820** Second Rep. 26: Perch, Pole or Rod...Berkshire: sometimes 18 feet for rough work...Herefordshire: of fencing, 7 yds in length; of walling, 5 1/2...Hertfordshire: sometimes 20 ft...Lancashire: 5 1/2, 6, 6 1/2, 7, 7 1/2, or 8 yards...Leicestershire: of hedging, 8 yards...Oxfordshire: of draining, 6 yds...Westmoreland: near Lancashire, 7 yds...Scotland: 18 1/4 feet. **1829** Palethorpe sv: PERCH, in land-measuring.... In Staffordshire, it is 24 feet; and in the forest of Sherwood 25 feet...and in Herefordshire, a perch of ditching is 21 feet. The perch of walling is 16 1/2 feet, and a perch of denshiered ground is 12 feet. **1860** Britannica 805: Ireland, perch 7 yards. **1880** Britten 174: Perch.... (Worc.), 8 yards. (Guernsey), 7 yards squared for land measure.... (Jersey), 7 1/3 yards = 22 feet.... Of labourers´ work, in some parts of Wales, 6, 7, or 8 yards. **1883** Simmonds sv: Perch, a linear-measure of 5 1/2 yards. See GAD; LUG; POLE; ROD

perche. PERCH

perch of land—2-7 L percata, or particata, or perticata terre (terræ) [perch of land, trans of L percata, or particata, or perticata terre (terrae)]. A m-a for land, of no standard dimensions but usually the square of the linear PERCH common in any region.—**1086** Domesday Book 22: Et arrare in yeme quartem (sic) partem j perticate terre. **1176** Clerkenwell: 10: Et tres percatas terre iuxta vallem; ibid 29: Et tres percatas terre vitra vallem. **1208** Feet 2.131: Et tres percatas terre pro dimidia acra que jacent in Sewardescrot. **1291** trans in Cal. Char. 2.400: A charter, whereby Adam son of Hugh de Glentham gave to the said abbot and canons two acres of arable land and two perches of meadow in Glentham on the east side of that town. **1405** trans in Cal. Close 18:457: 2 acres 6 perches of land and 1 rood of meadow held of the prior of Bylsyngton. **1664** Spelman 453: Continet ergò Particata terræ in integrâ superficie, 40 perticas, l. quartam partem unius acræ; quæ ut supra ostendimus octagies perticam comprehendit.

pere. PAIR; STONE

periot. PERIT

perit—6 peryott (OED); 6-9 periot; 7-9 perit; 8 perrot; 9 peroite [*]. A moneyer's unit of wt equal to 1/20 DROIT or 1/9600 t gr (0.00000675 g). It belonged to a series of imaginary wt used to compute exact coin wt by alternate subdivisions of 20 and 24.—**1665** Sheppard 15: 20 Perits make a Droit. **1707** Justice 4: One Droite into 20 Perits.... One Perit into 24 Blanks. **1725** Bradley sv weights: The Moneyers subdivide the grain thus: 24 Blanks make 1 Perrot; 20 Perrots 1 Dwit; 24 Dwits 1 Mite; 20 Mites 1 grain. **1756** Rolt sv weights: The blank; whereof 24 make a periot. **1816** Kelly 84: The Doit into 20 Periots, and the Periot into 24 Blanks. **1840** Ruding 1.411: Memorandum, Twelve ounces make a pound weight troy, twenty pennyweights an ounce, twenty-four grains a pennyweight, twenty mites a grain, twenty-four droits a mite, twenty perits a droit, twenty-four blanks a perit. **1868** Eng. Cyclo. 822: A droite 20 peroites, and a peroite 24 blanks. **1896** Klimpert 395: 1 Grän hat 20 Mites à 24 Doits à 20 Periots à 24 Blanks.

peroite, perrot, peryott. PERIT

pertica. PERCH

pes. FOOT; PIECE

pese, pess, pesse. PIECE

petra. STONE

peyce. PIECE

peyer, peyr, peyre. PAIR

peyss, pice. PIECE

picher, picheria. PITCHER

piece—3-5 L pecia, pees (OED); 3-7 pece; 4 pise (OED); 4-5 peis (OED), pice; 4-9 piece; 5 peese (OED), pes (OED), pese (Fountains), peyce (OED), pyece (OED); 5-6 pess (OED), pesse (OED); 5-8 peace (OED); 6 pease (OED), peax (OED), peayce, peise, peyss (OED), pysse (OED); 6-7 peece; 6-8 peice; 7 peec (Young II) [ME pece fr OF pece fr (assumed) VL pettia (MedL pecia), of Celt origin]. A m-c, m-l, m-q, and a wt for many products.

The piece occasionally was used for agricultural and metallurgical products: cheese, of uncertain wt; fruit, 4 QUARTERNS (50.802 kg) equal to 1/3 SORT; iron, 1/6 dozen (?) of uncertain wt; lead, generally 176 lb (79.832 kg); rosin, of uncertain wt; steel, 1/30 GARB of uncertain wt; tin, 1 1/2 to 2 1/2 Cwt (76.203 to 127.005 kg); and wax, of uncertain wt.—c**1253** Hall 11-12: La duzeynne de fer est de vi pecis. c**1272** Report 1.414: Garba asseris constat ex triginta peciis.... Duodena ferri ex sex peciis. **1290** Fleta 120: Garba vero aceri fit ex xxx. peciis. **1297** Elton 64: Et in j pecia ferri pro dynelegges faciendis ij d. ob. **1439** Southampton 2.21: Pro 80 peciis casiorum; ibid 24: 1 pecia de rosyn; ibid 52: Pro 169 peciis stanni pond. 41 M.C.; ibid 67: X peciis frute; ibid 87: 1 pecia plumbi pond. 5 C.; ibid 95: ii peciis fructui; ibid 108: 1 pipa continente 100 pecias de formag; ibid 153: 1 pecia fructui. **1443** Brokage II.13: iiii peciis rasemorum...et pro 1 pecia fygus; ibid 19: ii peciis de cera; ibid 259: Cum xxx peciis rasemorum ponderantibus xx [X] C. **1518** St. Peter´s 304: Item, a peise of ledd. **1534** Finchale ccccxl: xij peayce [of lead]...77 stone. **1615** Collect. Stat. 465: The dozen of yron consisteth of 6 peeces. **1820** Second Rep. 27: Piece...Derbyshire: of lead, at the cupolas, or smelting houses, 176 1/4 lb.

The piece was used most often for cloth goods, although in usage the word itself was frequently pre-empted by simply "cloth" (or F drap, chef, cheef, cheff, chiffe, sheet, or caput) or by the name of the particular fabric. Its length (measured by the yd or ELL) and breadth (usually measured by the QUARTER which equaled 1/4 yd) varied with the quality of the fabric, its construction, its monetary value, and its place of origin or manufacture. Hence, even though the standard piece of cloth was 24 yd (c21.95 m) in length and 7 quarters (c1.60 m) in breadth, there were many exceptions: bagging (a coarse cloth) for hops, Worcestershire, 36 yd (c32.92 m) by 31 inches (7.874 dm); broadcloth (a woolen cloth), the standard piece except in Kent, Reading, and Sussex, 28 to 30 yd (c25.60 to c27.43 m) in length; broad Yorkshire, 24 to 25 yd (c21.95 to c22.86 m) by 4 quarters (c0.91 m); buckram (a stiff cotton fabric), 15 yd (c13.72 m) in length; Cheshire cotton, 22 yd (c20.12 m) by 3 quarters (c0.69 m); colored cloth, 26 or 28 yd (c23.77 or c25.60 m) by 5, 6, or 6 1/2 quarters (c1.14, c1.37, or c1.49 m) except in Essex, Norfolk, and Suffolk, 28 to 30 yd (c25.60 to c27.43 m) by 7 quarters (c1.60 m); Coventry white, 29 to 31 yd (c26.52 to c28.35 m) by 7 quarters (c1.60 m); dornick (a heavy damask of silk, wool, or silk and wool), 28 yd (c25.60 m) in length; Dorsetshire flannel, 35 yd (c32.00 m) by 1 yd (c0.91 m); frieze (a coarse woolen cloth with a heavy nap on one side), 35 to 40 yd (c32.00 to c36.58 m) by 3 quarters (c0.69 m); fustian (a stout, twilled cotton fabric with a short nap), generally 13 ells (14.859 m) in length; Hampshire calico (a cotton cloth), 28 yd (c25.60 m) by 1 quarter (c0.23 m); kersey (a coarse woolen cloth made in white, red, blue, and other colors), 16 to 18 yd (c14.63 to c16.46 m) by 4 quarters (c0.91 m) except in Devonshire, 12 to 14 yd (c10.97 to c12.80 m) by 4 quarters (c0.91 m); Lancashire cotton, 22 yd (c20.12 m) by 3 quarters (c0.69 m); Lancashire washer, 15 to 18 yd (c13.72 to c16.46 m) in length; lawn (a fine, sheer, plain-woven linen or cotton cloth), 18 yd (c16.46 m) in length; Manchester cotton, 22 yd (c20.12 m) by 3 quarters (c0.69 m); Montgomeryshire flannel, 100 to 132 yd (c91.44 to c120.70 m) by 7/8 yd (c0.80 m); muslin (a fine cotton fabric), 14 ells (16.002 m) in length; narrow Yorkshire, 17 to 18 yd (c15.54 to c16.46 m) in length; penistone or forest white (a coarse woolen cloth similar to kersey, but always white in color), 12 or 13 yd (c10.97 or c11.89 m) by 3 1/2 to 6 1/2 quarters (c0.80 to c1.49 m); ray (a striped cloth), 28 yd (c25.60 m) by 5 or 6 quarters (c1.14 to c1.37 m); Rochdale flannel, generally 48 yd (c43.89 m) in length; sailcloth, 33 yd (c30.17 m) by 1 quarter (c0.23 m); short buckram, 5 1/4 yd (c4.80 m) in length; short Worcester, 14 to 15 yd (c12.80 to c13.72 m) in length; Shropshire flannel, 100 yd (c91.44 m) in length; sindon (a fine cloth usually made of linen), generally 10 ells (11.430 m) in length; straits, 12 or 14 yd (c10.97 or c12.80 m) by 1 yd (c0.91 m); Suffolk say (a fine, twilled cloth made of wool or wool and silk), 27 to 42 yd (c24.69 to c38.40 m) in length; tartarine (an expensive silk cloth), 10 yd (c9.14 m) in length; Taunton (a type of broadcloth), 12 to 14 yd (c10.97 to c12.80 m) by 7 quarters (c1.60 m); web (a coarse cloth), 90 to 120 yd (c82.30 to c109.73 m) by 3/4 to 7/8 yd (c0.69 to c0.80 m); Wiltshire red, 26 to 28 yd (c23.77 to c25.60 m) in length; Wiltshire white, 26 to 28 yd (c23.77 to c25.60 m) in length; and Worcester white, 29 to 31 yd (c26.52 to c28.35 m) by 7 quarters (c1.60 m).—**1253** Hall 12: Le chef de fustayne est de xiii aunes.... Le chef de cendal est de x aunes. **c1272** Report

1.414: Cheef de fustiano constat ex tresdecim ulnis.... Caput Sindonis ex decem ulnis. **1290** Fleta 120: Pecia autem fustiani consistit ex xiij. vlnis.... Pecia sindonis de cursu xiiij. vlnis. **1303** Gras 1.280: Pro pice acero et panno de worstede. **1308** Ibid 361: Adduxit iiii pecias blanketti. **1350** Rot. Parl. 2.231: Qe la longure de chescun Drap de Rai serra mesure par une corde de sept aunes, quatre soitz mesure par la liste; & la leure...sis quarters de lee mesure par l´aune.... Et des Draps de colour, la longure soit mesure par le dos par une corde des sis alnes & demi, quatre soitz mesure, & la leeure sis quarters & demy mesure par l´aune. **1373** Ibid 318: Les Rayes soleient tener XXVIII aunez en longure, & V quarters de lieure. **1393** Rot. Parl. 3.320: Item suppliont les Communes des countees d´Essex´, Suff´, & Norff´, que Vous plese ordeiner, q´ils eient licence q´ils purront faire lour Draps en manere come ils ont usez de faire devaunt cest temps; issint que chescun piece soit del longure d´une duszeine, & de laeure d´une verge de quatre quarters. c**1400** Gras 1.215: De qualibet pecia integra de fustian. **1406** Rot. Parl. 3.598: Qe le Drap de ray serroit en longure de XXVIII auns, & en leaure VI quarters. **1407** Ibid 618: Qe les Draps de Ray soient en longure XXVIII aulnes mesurez par la list.... Draps de Ray...en longure de XXVIII aulnes, & en laieure V quarters. **1410** Ibid 645: & l´ou le Dussein de Drap [Devonshire kersey] duist teigner XIIII verges. **1443** Rot. Parl. 4.451: Of every Clothe and ych pece of Cloth after the rate.... Clothes called Streytes, holdyng XIIII yerdes in lenght, and yeerde brode unwette; or elles XII yerdes wette.... Clothe of colour should conteigne in lenght XXVIII yerdes, mette by the crest, and in brede VI quarters di. **1439** Rot. Parl. 5.30: For there as they were wonte to mete Clothe by yerde and ynche, now they woll mete by yerde and handfull, the whiche groweth to encrece of the byere, II yerdes of euery Clothe of XXIIII yerdes. c**1461** Hall 15: Fryse schold hold xl. yerdes and more; ibid 16: So that every hole cloth or euery dossynne be hole in lengthe, xxiiij yerdes; ibid 18: A pece fust[yan] cont[aineth] xxx yerdes.... Item all maner of bokerammes hold xl yerdes, save schort bokram ys butt V yerds I quarter.... Dornyk and Bord´ Alysaundyr hold xxviij yerdes...I pece lawne or ump[er]ill, xvj plyte or xviij yerds. **1463** Rot. Parl. 5.501: First, that every hole Wollen Cloth called brode Cloth...after almanere rakkyng, streynyng or teyntyng...be parfitly and thoroughly wette, and...conteigne in lengh XXIIII yerdes, and to every yerde an ynche, conteynyng the brede of a mannes thombe.... Streytes...be parfitly and thoroughly wettė, and...conteigne in lengh XII yerdes...and in brede a yerde.... Every Cloth of Kersey, to...hold and conteigne in lengh XVIII yerdes...and in brede a yerde and the nayle. **1524** Gras 1.196: Pro una pecia de say. **1587** Stat. 83: The length of euerie cloth of raie, by a line of seuen yards, foure times measured by the lyst, and the breadth of euerie raie cloth sixe quarters of measure by the yarde.... Of coloured clothes the length shall be measured by the backe by a line of sixe yardes and an halfe, foure times measured, and the breadth sixe quarters and an halfe; ibid 121: Colored cloth of the length of xxvi. yeardes...and of the breadth of vi. quarters and an halfe; ibid 299: All manner of clothes called streites concerning in length XIIII. yardes...or otherwise xii. yardes watered. c**1590** Hall 25: Every brod cloth...mesured by the crest of the clothe, in lenght 24 yardes...and in bredith 2 yardes, or 7 quarters.... Straites shall contayne in lenght 14 yardes and in bredith one yard.... Collerid clothes made in England, are mesurid by the backe; the lenght is 26 yardes, and the bredith 6 quarters.... The wholle coolerid clothe...shall contayne in lenght 28 yardes; ibid 26: Every Brod Cloth, with the list, shall contayne 7 quarters of a yard...in bredith.... Every brod clothes, Kentishe, Sussex, Reading...shall contayne, in lenght, at the watter throughe weett, bettwixxe 28 and 30 yardes; in bredith 7 quarters.... All collerid clothes of Essex and Northfolke, elleswheare...in lenght, beinge wett, ought to contayne bettwixe 28 and 30 yardes; in bredith 7 quarters; milled and dryed shall way 80 l. eych peece at the lest.... Every kersey with the list shall contayne in bredith one yard.... Devonshires kersis, calid dossens, ought to contayne 12 or 14 yardes in lenght...and in bredith one yard and a nayle at the lest.... Wster clothe [Worcester white] ellswheare being wett shall contayne, in lenght, betwixt 29 and 31 yardes, and in bredith 7 quarters.... Whit clothes, called short Wster, made in the cytty, beinge weett, shall conteyne in lenght bettwixt 14 and 15 yardes; ibid 28: A chiffe of Fustyane consisteth 14 ells, that is 17 yardes and 1/2.... The chiffe of Syndon consisteth 10 ells, that is 12 yardes and 1/2. **1597** Halyburton cxiii: Canves callit tiftit canues ye pece thairof. c**1610** Lingelbach 111: Draper and damaske by the peece. **1612** Halyburton 288: Barberis apronis the peice not contening abone ten elnis; ibid 290: Blankets called Pareis mantles cullored the peice. **1613** Tap 1.64: Carsies. Length yards. Carsies called ordinary the Peece. betweene 16 and 17 wet. The sorting Carsie. betweene 17 and 18 wet. The Devonshire Carsie. betweene 12 and 13 wet; ibid 65: Euery narrow Cloath of the same places or any other of like sort. betweene 24 and 25 wet.... The Penistone, or Forrest white, betweene 12 and 13 wet. The white plaine Straight made in Deuon, or Corn. 12; ibid 66: Euery white and red made in Wiltsh. Glouc. and Sommersetshire...between 26 and 28. Cloth of Ray. 28.... Euery Broadcloath made in Taunton...between 12 & 13 wet; ibid 67: The Lancash. Frise and Rugge. betweene 35 & 37. Manchester Frise and Rugge. 36. **1615** Collect. Stat. 465: A chef of Fustian consisteth of 14. elles.... A chef of Sindon containeth ten elles. **1660** Bridges 29: Of Fustian, 1 Cheff is 14 Ells. Of Fine Linnen, Syndon, or Silks, the Cheff is 10 Ells. **1661** Acts Scotland 7.252: Ffustians ilk three peices. **1665** Sheppard 45: All Whites and Reds in Wilts...must be in length between 26 and 28 yards, and 7 quarters in breadth; ibid 46: The length of Dowseins, must be between 12 and 13 yards.... Manchester, Lancashire, and Cheshire Cottons, must be 22 yards long, and 3 quarters broad; ibid 47: Frizes in Wales and elsewhere...are to be 36 yards at most in length, and 3 quarters in breadth.... Pennystones and Forrest Whites must be between 12 and 13 yards long, and 6 quarters and a half broad; ibid 48: The ordinary Kersey between 16 and 17 yards; ibid 49: Frizes and Rugs thicked and dryed, are to weigh 44 pound a piece, and to be in length between 35 and 37 yards; ibid 54: But in Yorkshire...the narrow to be in length between 17 and 18 yards. **1708** Chamberlayne 208: Taunton and Bridgewater, 7 Quarters, 12 and 13 Yards.... Devonshire Kersies and Dozens, 4 Quarters, 12 and 13 Yards.... Chequer Kersies, Grays, strip´d and plain, 4 Quarters, 17 and 18 Yards.... Penninstons or Forrests, 3 Quarters and 1/2, 12 and 13 Yards.... Washers of Lancashire, 17 and 18 Yards. **1717** Dict. Rus. sv cloth-measure: Taunton, Dunstable, Bridge-water, 7 quarters, 12 and 13 yards.... Devonshire-Kersies and Dozens, 4 quarters, 12 and 13 yards.... Washers of Lancashire, 17 and 18 yards. **1820** Second Rep. 27: Piece...of sailcloth, 33 yards, 1/4 wide.... Dorsetshire: of flannel, 35 yards, yardwide.... Hampshire: of calico, 28 yds, 1/4 wide.... Shropshire: of flannel, 100 yds.... Suffolk, Sudbury: of says, 27, 30 and 42 yds.... Worcestershire: of bagging, for hops, 36 yards, about 31 inches wide.... Wales: of flannels. Rochdale, about 48 yards or less...Montgomeryshire: 100 to 120 or 132 yards or more, 7/8 wide.... Webs, a coarse cloth, 90 to 120 yards, 3/4 to 7/8 wide. In some places a web means two such pieces, making 190 yds.

piere. STONE

pig [ME pigge; so called from the resemblance of the arrangement of the molds in the pig bed to suckling pigs]. A wt for lead or iron of no standard dimensions, although it generally was larger than 1 Cwt (50.802

kg).—**1756** Rolt sv: PIG of lead. The eighth part of a fodder; amounting to 250 lb. weight. **1794** Martin 24: The Derbyshire lead is generally sold by the fodder; 16 pieces, or half pigs, are called a fodder. **1820** Second Rep. 27: Pig...of lead, 21 1/2 stone = 301 lb.... Derbyshire: at the smelting house, 352 1/2 lb.... Northumberland: 1 1/2 cwt = 168 lb. **1829** Palethorpe sv: PIG, a mass of lead, being 1/8th part of a fother of 19 1/2 cwt. or 2 cwt. 1 gr. 21 lb. **1834** Pasley 1.14: 1 Pig of Lead, Northumberland (1 1/2 cwt.)...168 [lb]; ibid 115: 1 Pig of Lead...301 [lb].... 1 Pig of Lead at the smelting house, Derbyshire...352 1/2 [lb]. **1840** Waterston 147: Ballast, pig...lbs. 56. **1849** Dinsdale 96: Pig of lead...A piece of lead of an oblong shape, from 8 to 12 stone in weight. **1882** Jackson 233: The pig of lead = 300 pounds. **1883** Simmonds sv: Pig...an ingot of iron or lead, weighing 3/4 to 1 1/2 cwt. **1895** Donisthorpe 215: PIG: of lead, 21 1/2 stone = 301 pounds.

piling [perh pile + -ing]. A m-q, a bundle, for wheat straw in Staffordshire (c1800-1900) consisting of 3 SHEAVES (Britten 154).

pin [ME pinne fr OE pinn, peg]. A brewery m-c for beer containing 4 1/2 gal (2.046 dkl) everywhere in the United Kingdom except in Ireland, 4 gal (1.818 dkl) (Economist 54).

pinct, pincta. PINT

pint—4-6 pynt, pynte; 5 pintte (OED), pyynte; 5-7 L pinta, pinte; 6 L pincta, point (OED), poynt (OED), poyntt (OED); 6-9 pint; 7 pinct [ME pynte, pinte fr MF pinte fr MedL pincta; see WNID3]. A m-c: for dry products, 33.6 cu inches (0.551 l) or 1/8 gal or 1/64 Winchester bu; for liquids, 4 GILLS equal to 1/2 qt, 1/4 POTTLE, or 1/8 gal and standardized at 28.875 cu inches (0.473 l) for wine and 35.25 cu inches (0.578 l) for ale and beer. Since the establishment of the Imperial system in 1824, the pint both for liquid and dry products has been reckoned at 34.677 cu inches (0.568 l) or 1/8 gal of 277.420 cu inches. In the Imperial ap system the pt of 34.677 cu inches is reckoned as 20 ap fluid oz of 1.733875 cu inches each or the volume of 8750 gr (567.0 g) of distilled water at 62° F. The Scots pt, also known as the JUG or STOUP, equaled 2 CHOPPINS, or 4 MUTCHKINS, or 103.404 cu inches (c1.70 l), although the following were exceptions: Aberdeenshire, 108.89 cu inches (1.785 l); Ayrshire, 110.624 cu inches (1.813 l); Banffshire, 105.284 cu inches (1.726 l); Dumbartonshire, 100.5 cu inches (1.647 l); Dumfriesshire, 114.0 cu inches (1.868 l); Elgin, 105.438 cu inches (1.728 l); Inverness, 115.161 cu inches (1.887 l); Nairnshire, 111.676 cu inches (1.830 l); Peeblesshire, 104.64 cu inches (1.715 l); Perthshire, 104.344 cu inches (1.710 l); and Sutherlandshire, 115.161 cu inches (1.887 l) (Swinton 53, 57, 62, 70, 72, 77, 86, 101, 108, 110, 124). The Irish pt contained 27.2 cu inches (0.446 l) and was used both for liquid and dry products (Edinburgh XII.571-72).—c**1400** Hall 36: Una libra facit I pynt. c**1400** Brit. Mus. 30.52v: iiij pyntes...j potell. **1416** York Mem. 1.213: Et pynte et dymy pynt. **1425** Acts Scotland 2.12: The firlote sal contene twa galonis ande a pynte Ande Ilk pynt sal contene be wecht of cleir watter of tay xij vnce that is for to say ij pundes & ix vnce troyis Swa weyis. c**1440** Promp. Parv. 401: Pyynte, mesure. Pinta. **1474** Cov. Leet 396: ii pyntes maketh a quart; & ij quartes maketh a Pottell; & ii Pottels makith a Gallon; and viij Gallons makith a Buysshell. **1517** Hall 48: viii pyntes to the galon´. **1525** Jacobus 3: BUTELARIA. Empt´ j [X] c [+] iiij lagine ij quarte j pincta ceruisie. **1526** Ibid 82: j pincta olij. c**1590** Hall 21: 2 pyntes makith a quart.... Euery pynt waieth one pounde troye. **1603** Henllys 138: Ffor liquid or wette measures...wee use heere the usuall pinte, by which wee proceede to make all other measures of greater accompte. **1618** Acts Scotland 4.586: They fand that the same conteined Twentie ane pincts and ane mutchkin of just Sterline Jug. **1624** Huntar 4: The Scottish pinte or standerd Iug of Sterling, is found to conteine 3 pound 7 ounce Weight of the water of Leith, everie pinte is devided in 2 choppins and 4 muchkins.... The pinte doth weigh 55 ounces. **1635** Dalton 144: 8 pintes maketh the gallon...64 pints maketh the Bushell. **1665** Assize 4: From the said gallons made by the said eight pints. **1678** Du Cange sv galo: Mensura liquidorum apud Anglos, quarum unaquæque octo continet pintas Anglicanas. **1688** Bernardi 150: Pinta denique arida, 1/8 Galonis, seu congii frumentarii, et 1/8X8 = 1/64 Brusselli. **1707** Acts Scotland 11.407: The Scots pint or eight part of the Scots Gallon. **1708** Chamberlayne 210: The ordinary smallest Receptive Measure is called a Pint...2 Pints make a Quart. **1779** Swinton 29: Scotch Measures of CAPACITY for LIQUORS, raised from the pint, which weighs, of river water, 3 lb. 7 oz. Scotch Troye, or 3 lb.

11 oz. 13.16 dr. Avoird.... Pint...103.404 [cu inches]. **1791** Keith 2.3: In Scotland the Pint by the Stirling Jugg is 103.404 [cu inches]. **1816** Kelly 87: WINE MEASURE. 1 Pint...28,875 [cu inches].... ALE AND BEER MEASURE. 1 Pint...35,25 [cu inches]...0,5776 [l]; ibid 88: DRY MEASURE. 4 Gills...1 Pint...33,6 [cu inches]...0,55053 [l]; ibid 93: The Stirling pint jug is the unit of both the liquid and dry measures of Scotland. It contains 103,404 English cubic inches. **1820** Second Rep. 27: Pint, 1/8 gallon = 1/2 a quart = 28 8/10 cu. inches, wine measure; 35 1/4 customary ale measure; 33 6/10 Winchester measure.... Scotland: 2 choppins, about 105 cu. inches, 3 ale pints E. **1829** Palethorpe sv: The former wine pint, that is, that in use prior to the 25th of May, 1825, contained 28.875 cubic inches; the pint for ale and beer measure, 35.25 cubic inches, and the pint for dry goods, 33.6 nearly. **1860** Britannica 805: Scotland...pint, 103.4 [cu inches]. **1907** Hatch 23: 4 gills = 1 pint (pt.) = 34.6775 cubic inches; ibid 24: 1 pint (O.) = 20 fluid ounces = 34.6775 cu inches.... 1 pint (O.) is the volume of 8,750 grains of distilled water at 62° F; ibid 35: 1 pint = .56825 litre. **1920** Stevens 2: 1 Gallon, C. = 8 Pints, = 160 Fluidounces. 1 Pint, O. = 20 Fluidounces. **1956** Economist 4: Pint: (a) United Kingdom = 34.6775 cubic inches. **1969** And. & Bigg 11: 1 pt (pint) = 0.568261 dm^3 = 0.568 litre.

pinta, pinte, pintte. PINT

pipa. PIPE

pipe—4-5 L pipa; 5-7 pype; 5-9 pipe [ME pipe fr MF pipe, a cask for wine, a pipe, fr (assumed) VL pippa, alter of pipa]. A m-c for dry and liquid products, generally synonymous with the BUTT: cider, Guernsey and Jersey, approximately 120 gal (c4.54 hl); currants, 15 to 22 Cwt (762.030 to 1117.644 kg); peas, generally 12 bu (c4.23 hl); salmon, 84 gal (c3.18 hl) equal to 1/6 salmon LAST; salt, generally 16 bu (c5.64 hl); whale oil, "Large Butt," 332 BI gal (c15.09 hl), "Small Butt," 223 BI gal (c10.14 hl), "Long Pipe," 118 BI gal (c5.36 hl), and "Common Pipe," 98 BI gal (c4.45 hl) (Pasley 50); wine, oil, and honey, generally 126 gal (c4.77 hl), occasionally 120 and 125 gal (c4.54 and c4.73 hl), and equal to 1/2 TUN. Since the establishment of the Imperial system the pipe of ale has been reckoned at 108 gal (4.910 hl) everywhere in the United Kingdom except Ireland, 104 gal (4.728 hl).—c**1300** Topham 9: Una pipa vini. **1324** Gras 1.381: Uno dolio et una pipa vini. **1390** Henry Derby 24: Roberto Gobon pro j tonella et j pipa de Rynen, in toto vj s. **1406** Rot. Parl. 3.596: Oille en groos par tonell ou par pipe. **1423** Rot. Parl. 4.256: The Pipe xx [X] vi [+] VI galons. **1439** Southampton 2.2: Pro 1 pipa des pes´ continente 12 bosshell; ibid 4: 2 pipis des pes´ continentibus 3 quarteria; ibid 32: 4 pipis salis continentibus 8 quarteria. **1440** Palladius 57: Lete close hem in a barel or a pipe. **1443** Brokage II.58: Cum i pipa bastarde. c**1461** Hall 15: The pipe cont[aineth] i [X] c [+] xxv galounes.... The pype vi [X] xx gallounes. **1517** Ibid 49: The pype, vi [X] xx et vi galons´. **1587** Stat. 267: The pipe C.xxvi galons. c**1590** Hall 21: The pipe contenith a butt which is 1/2 of a tunne, 126 gallons. c**1610** Lingelbach 112:

Oyle Civile by the pype. **1615** Collect. Stat. 467: Euerie Pipe sixe score and six gallons. **1616** Bullokar sv pipe: A measure of halfe a Tunne; that is, 126. Gallons. **1635** Dalton 148: Wine, Oyle, and Honey: their measure is all one, sc. the...Pipe, 126. gallons. **1664** Gouldman sv: A pipe or half a tun. Hemi-dolium. **1665** Assize 4: Every Pipe cxxvi gallons. **1704** Mer. Adven. 243: Ffor every pipe of wine the said bounds. **1708** Chamberlayne 210: Of these Gallons...a Pipe or Butt holds 126. **1717** Dict. Rus. sv butt: Butt or Pipe of Wine, contains two Hogsheads, or One hundred twenty six Gallons; and a Butt of Currans from Fifteen to Twenty-two Hundred weight. **1725** Bradley sv butt: Butt, or Pipe, a Liquid Measure, whereof two Hogsheads make a Butt or Pipe, as two Pipes or Butts make one Tun. **1816** Kelly 87: WINE MEASURE...2 Hogsheads...1 Pipe or Butt...476,9018 [1]. **1820** Second Rep. 27: Pipe...Guernsey and Jersey: of cider, 240 pots, about 120 gallons. **1850** Alexander 88: Pipe old measure; for wine...126.—gallons. **1895** Donisthorpe 83: It seems probable that the pipe, butt, and cask were different names for the same thing.

pipot [prob a dim of PIPE]. A m-c for liquids during the fifteenth century equal to 1/2 PIPE (Southampton 1.82).

pise. PIECE

pitchaer, pitchard. PITCHER

pitcher—3-5 picher; 4 L picheria; 4-6 pycher (OED), pychere (OED); 5 pychare (OED); 5-6 pychar (OED); 6 pitchaer (OED), pitchard (OED), pytcher (OED); 6-9 pitcher (OED) [ME picher fr OF pichier fr MedL bicarius, goblet, beaker]. A m-c for liquids generally containing 1 gal (c3.78 l).—**1390** Henry Derby 21: Clerico buterie super vino, per manus Payn pro ij sextariis di. picheria vini albi per ipsum emptis pro ollis unius dolii. **1392** Ibid 160: Et per manus eiusdem pro xx sextariis ij picher vini vasconie. **1393** Ibid 256: Item pro xij sextariis j picheria di. vini Vasconie ad ij s. viij d.

plack [*]. A m-a for land in Leicestershire (c1850) containing 5 sq yd (4.180 sq m) (Sternberg 81).

plewland, ploughland, ploweland, plowelande, plowelonde. PLOWLAND

plowland—5 plewland, plowelonde, plowlond, plowlonde (OED); 6 plowelande (OED), plowlande (OED); 6-9 ploughland, plowland; 7 ploweland [ME plowlonde fr plow, plough, plow, + lond, londe, land, land]. Equivalent to HIDE.—c**1400** Acts Scotland 1.387: The plew land that ordanit to contene .viij. oxingang/the oxgang sall contene .xiij. akeris. c**1435** Amundesham 1.453: Wyth two plowe londe in hys demayns. c**1440** Promp. Parv. 405: Plowlond. Carrucata. **1603** Henllys 135: viii oxelandes make a ploweland being...64 acr. x plowlands make a knightes ffee being...640 acr. **1635** Dalton 71: But a plow-land, or Carve of land, is called in Latine, Carucata terræ, that is, quantum aratrum arare potest in æstivo tempore. **1638** Bolton 274: An hide of land (or plowland or carue of land which are all one) are not of any certaine content. **1642** R. Powell 86: A plow-land in severall parishes. **1755** Willis 358: Some hold an Hide to contain 4 yard Land, and some an 100, or 120 Acres; some account it to be all one with a Carucate or Plough-Land. **1777** Nicol. and Burn 613: Ploughland, as much as can be cultivated in a year by one plough. **1820** Second Rep. 27: Ploughland...Wales: 8 oxlands = 64 customary acres. **1888** Round 3.192: Except in their virtual and independent agreement that the ploughland, the essential unit, was 120 acres; ibid 193: We cannot deny that the real hide was in fact the equivalent of the ploughland; ibid 211: The yardland of 30 acres, and...the ploughland composed of four such yardlands.

plowlande, plowlond, plowlonde. PLOWLAND

poak, poake. POKE

poale. POLE

poccet. POCKET

pock. POKE

pockate. POCKET

pocke. POKE

pocket—4 pokete; 4-6 poket; 5-6 pokett; 5-8 pockett; 6 pockate, pockette (OED), pokit (OED); 6-9 pocket; 7 poccet (OED) [ME poket fr AF and ONF pokete, dim of ONF poke, poque, bag, pouch]. A m-c, a coarse bag or sack, for several products: hops, generally 1 1/2 to 2 Cwt (76.203 to 101.604 kg), but Kent and Surrey, 1 1/4 Cwt (63.502 kg); nails, in varying quantities, depending on the size and type of nail; umber, of uncertain wt; and wool, generally 1/2 pack or 120 lb (54.431 kg).—**1350** trans in Memorials 262: Also, in the Chapel there, in a pokete, 2500 of wyndounail. **1476** Stonor II.5: The sarpler, the pooke, and the ij pokets woll. **1507** Gras 1.700: Mather called crope or umbero the pocke...and the pockett. **1524** Ibid 195: Pro six pokettis hopps. **1545** Rates 1.20: Hoppes the pockate. **1565** Rich 129: Every sarpler...poke or pockett of good marche woulles. **c1610** Lingelbach 113: Hoppes by the sacke or pockett. **1701** Hatton 3.232: Pocket of Wool...Part of a Pack, about half. **1820** Second Rep. 27: Pocket of wool; 1/2 a pack = 120 lbs...of hops: Kent, 1 1/4 cwt, Surrey, 1 1/4 cwt, measuring about 5 1/4 ft in circumference, 7 1/2 long; 4 lb being allowed for the weight of the canvas. **1834** Pasley 114: 1 Pocket of Hops, in Kent and Surrey...140 [lb]. **1840** Waterston 148: Hops, pocket, cwt. 1 1/2 to cwt. 2. **1882** Jackson 227: Pocket of wool...120 Lbs. av.

pockett, pockette. POCKET

poddle [prob a corruption of POTTLE]. A m-c in Cornwall (c1800-1900) equal to a qt (c1.10 l) (Britten 154).

poik. POKE

point—3-6 pointe (OED), poynte (OED); 3-9 point; 7 L punctum [ME point, pointe fr OF point, a prick, dot, fr L punctum, a dot, fr pungere, to prick]. A m-l equal to 1/72 inch (0.035 cm) or 1/6 LINE. It was adopted by type-setters for designating the sizes of type by the number of seventy-seconds of an inch height of the type face.—**1678** Du Cange sv alna: Pes Regius est 12. pollicum; pollex 12. linearum; linea 12. [sic] punctorum. **1931** Naft 14: 1 Point (Printers')...1/72 inch...0.352 millimeters. **1956** Economist 8: Point...1/72 inch; ibid 69: Sizes of types: The unit of depth is the point (72 points = 1 inch).

point. PINT

pointe. POINT

pok, poka. POKE

poke—3-9 poke; 4-5 L poka; 4-6 pok; 5 poyke (OED); 5-6 poik (OED), pokke (OED); 5-7 pocke, pooke; 6 polk (OED); 6-7 poake (OED); 7 poak (OED); 7-9 pock; 9 pooak (OED), pook (OED), pouk (OED), powk (OED), puock (OED), puck (OED), pwoak (OED), pwok (OED), pwoke (OED) [ME poke fr ONF poke, poque, bag, pouch]. A m-c for a variety of products. It was a large bag or sack whose size varied according to the quality and wt of the product enclosed. In particular it was used to transport raw wool.—**1228** Gras 1.157: 1 poke de alum. **1276** Ibid 225: xl sackes et 1 poke de laine marchans. **1304** Ibid 172: Et i poka lane. **1396** Ibid 443: ii pokys farine. **1420** Ibid 469: Pro 1 fat i poka cum iiii pannis. c**1440** Promp. Parv. 21: Bagge, or poke...pocke. c**1461** Hall 16: Also Woll ys sold by numbre and schipped to, as by sacks, sarplers, and pokys. ii sacks make a sarpler, and x sarplers make a laste [sic], and the poke ys att no serteyne, butt aftre as ytt weys. c**1475** Gras 1.193: Of a poke mader. **1476** Stonor II.5: The sarpler, the pooke, and the ij pokets woll. **1480** Cely 31: I have wyll understand and that ze have solde vj sarplerys & pok of my medell woll. **1507** Gras 1.699: Hoppys the pocke; ibid 700: Mather called crope or umbero the pocke. **1524** Ibid 194: Pro septem pokes hopps. **1538** Mer. Adven. 63: Of a pok of woll. **1545** Rates 1.20: Hoppes the pooke. **1553** Mer. Adven. 66:

And that thar shall no man lay no more deke above hys poke of the gretest bod iij stone at the most, and every poke of lesse quantyte to taike lesser dek accordinge to the greatnes of hys poke. **1562** York Mer. 169: Six small pokes of Brassel to a tonne, and the greater as they be rayted forthe of the ship. **1590** Rates 2.20: Hops the poke containing iiii C. **1612** Halyburton 323: Onioun seed the pock. **1704** Mer. Adven. 243: Ffor a poke or bail of mather. **1895** Donisthorpe 215: POKE: of wool, 20 cwt.

poket, pokete, pokett, pokit. POCKET

pokke. POKE

pole—4 pool (OED); 4-6 poole; 4-9 pole; 5-6, 8 poll; 6 polle, poule (OED); 6-7 powle (OED); 7 poale [ME pole fr OE pāl, pole, stake, fr L palus, stake]. Equivalent to PERCH and occasionally abbreviated po.—**1502** Arnold 173: In dyuers odur placis...they mete ground by pollis gaddis and roddis some be of xviij foote some of xx fote and som xvi fote in lengith. **1534** Fitzherbert 21: xvi. fote and a half to the perche or pole. **1589** Bellot 5: The pole of sixteene foote and a halfe; ibid 6: The acres are not all alike, for in some countreys [= districts] they doe measure by the pole of eighteene foote, and...by the pole of twentie foote, and...by the pole of foure and twentie foote. c**1590** Hall 27: 16 foott 1/2 in lenght is a poole or a perche. **1595** Powell C2: In many countries [= districts] this polle or pearch doth vary, as in some places it is eighteene foote, and in some other places xxi. foote. **1603** Henllys 133: For in some place the pole is but ix foot, and in some place xij foote. **1616** Rathborne 131: An Acre measured by the Pole of these feet, 12 18 20 24 24 1/2 28.... Wood-land grounds, whose quantities are required to be of the Acre of 18. foot Pole. **1635** Dalton 150: Five yards and an halfe (which is 16 foot and an halfe) maketh a pole, rood, or pearch. **1638** Bolton 274: Most places in Ireland 21. foot goeth to the pole. **1639** Bedwell B2: And every Lugge or Poale, to containe 26 [sic] foot and an halfe. **1665** Assize 6: In many countries [= districts] this Pole or Perch doth vary, as in some places it is 18 foot, and in some other places 21 foot. **1682** H. Coggeshall 2.71: An Acre is 160 Square Poles; 100,000 Square Links of this Chain. Therefore 625 Square Links make 1 Pole. **1682** Hall 28: A Pearch, or a Rod, or a Pole (by statut) must be 5 yards and an half; or 16 feete and an half. **1696** Phillips sv pole: In measuring, it is the same with Pearch or Rod, or as some call it Lugg: By Stat. 35 Eliz. this Measure is a length of 16 Foot and a half, but in some Countries [= districts] it consists of 18 Foot and is called Woodland-Measure; in some Places of 21 Foot termed Church-Measure; and in others of 24 Foot under the Name of Forest-Measure. **1701** Hatton 3.10: 5 Yards and 1/2.—1 Poll or Perch; ibid 14: In some Countries [= districts] they have 7, 7 1/2, and 8 Yards to the Poll. **1708** Chamberlayne 207: 16 Foot and a half make a Perch, Pole or Rod, but there are other Customary Perches or Poles, viz. Eighteen Feet for Fens and Woodland, Twenty one for Forest, Lancashire and Irish Measure, and 18 3/4 Scotch. **1717** Dict. Rus. sv: Perch or Pearch, a Rod or Pole, with which Land is measur´d. **1725** Bradley sv mile: Every Furlong forty Lugs or Poles...every Pole sixteen Foot and a Half. **1755** Postlethwayt II.190: For fens and woodlands they reckon eighteen feet to the pole, and for forests twenty-one. **1816** Kelly 86: The Woodland Pole of 18 Feet, the Plantation Pole of 21 Feet, the Cheshire Pole of 24 Feet, and the Sherwood Forest Pole of 25 Feet. **1820** Second Rep. 26: Perch, Pole or Rod. **1829** Palethorpe sv acre: The length of the pole differs from 16 1/2 feet to 28 feet in several parts of the kingdom. **1850** Alexander 88: Pole, or Perch...5,50 yards. **1882** Jackson 282: Pole = 5 1/2 yards; (also poles of 6, 7, and 8 yards, and of 25 feet). **1907** Hatch 35: 1 pole = 5.0291956 metres.

polk. POKE

poll, polle. POLE

pollex. INCH

poncheon, ponchion, ponchyn, poncion. PUNCHEON

pond, ponde. POUND

pondus. WEY

pontion, pontioune. PUNCHEON

pooak, pook, pooke. POKE

pool, poole. POLE

poot. POT

pot—2-8 pott; 3-9 pot; 4-5 poot (OED); 4-7 potte; 5 poyt (OED), putte (OED) [ME pot, pott fr OE pott; akin to MDu pot, MLG pot, put]. A m-c for ale and other liquids, generally equal to a qt (cl.15 1), except in

Guernsey and Jersey, 2 qt or 1/2 gal (c2.06 l), and butter, 20 lb or 6 lb (2.722 kg) for the pot and 14 lb (6.350 kg) for the butter. In Worcestershire a pot of apples, potatoes, etc. contained 5 pk (c4.40 dkl). A pot of apples in Wolverhampton weighed 75 lb (34.019 kg), and of beans and peas in Warwickshire, 40 lb (18.144 kg).—**1439** Southampton 2.82: l pott synziberys veridis. **1512** Clode 97: Itm, of the gifte of Robert Wilford, one Ale pott. **1545** Rates 1.28: Oyle called baume oyle the potte. **1673** Stat. Charles 159: And the Pot of Butter ought to weigh Twenty pounds, viz. fourteen pounds of good and Merchantable Butter Neat, and the Pot Six pounds. **1701** Hatton 3.232: Pott...In Guernsey and Jersey half the Gallon, or 126 Cubical or solid Inches. **1820** Second Rep. 28: Pot of ale, generally a quart...of butter, 14 lbs. **1880** Britten 174: Pot, of potatoes, apples, &c. (Worc.), 5 pecks. **1895** Donisthorpe 215: POT...Guernsey and Jersey, about 2 quarts. **1896** Wagstaff 35: A pot of apples in Wolverhampton = 75 lbs. A pot of beans and peas in Warwickshire = 40 lbs.

potel, potell, potella, potelle, potellum, potellus, potle. POTTLE

pott, potte. POT

pottel, pottell. POTTLE

pottle—3-4 L potellus; 4 L potella; 4-5 potel; 4-7 potell; 5 potelle (OED); 5-7 pottel, pottell; 6-7 potle; 6-9 pottle; ? potellum (Prior) [ME potel fr OF potel, dim of pot]. A m-c containing 2 qt (c1.89 l), used principally for liquids; when used for dry products, the pottle sometimes was called a QUARTERN, or 1/4 pk. The Irish pottle contained 108.8 cu inches (1.783 l) or 2 Irish qt of 54.4 cu inches (Edinburgh XII.571-720). Occasionally it was abbreviated pot.—**1287** Select Cases 2.19: Robertus Scot de London, potellus falsus et quarta falsa et alia quarta falsa, et quia vendidit pro xvj. d. **1291** Ibid 40: Johannes Lysegong cum vino reneys, potellus bonus, quartus bonus et signatus. **c1340** Oxford 267: Agn´ la Tappestere habet I potellum & I quartam fals´. **1376** Gross II.104: Et de vno potello vini. **1379** Rot. Parl. 3.64: C´est assavoir, Galon, Potel, & Quart. **1390** Henry Derby 6: Clerico Buterie super vino per manus Johannis Taverner pro ij potellis et j quarta vini Vasconie, xxij d. ob. **1395** York Mem. 2.10: Et unum strikill ligni, lagena, potella et quarta eris pro vino. **c1400** Hall 36: 4 libre faciunt I potell´. **1467** Cov. Leet 334: Gallon & potell & quarte. **1474** Ibid 396: ii quartes maketh a Pottell. **c1500** Brit. Mus. 24.17v: In bushell pecke gallon potle and quarte. **1523** Coopers 68: Paid for a galon and a potell muscadell. **c1590** Hall 20: 2 quartes makith a pottell. **1600** Hill 67: 2. Quartes...1. Pottell. **1603** Henllys 138: Ffor liquid or wette measures...wee use heere...ii quarts to a pottle. **1607** Clode 307: 21 gallons and a potle.... For 3 potles of redd wine. **1615** Collect. Stat. 464: That is to wit, bushells, halfe a quarter bushells, gallons, pottels and quarts. **1616** Hopton 160: Ale and Beere...are measured by Pints, Quarts, Pottles, Gallons...and these and such like bee Concaue measures. **1628** Hunt C: 4 Pints in a Pottell. **1635** Dalton 144: Two quarts, maketh the pottle.

1665 Sheppard 7: Two Quarts make a Pottle; Two Pottles make a Gallon. **1708** Chamberlayne 210: 2 Quarts make a Pottle...2 Pottles make a Gallon. **1717** Dict. Rus. sv: Pottle, (in English liquid and dry Measure) is two Quarts. **1834** Pasley 42: 2 Quarts...1 Pottle or Quartern. 2 Pottles or Quarterns...1 Gallon. **1877** Leigh 159: Pottle...A measure of two quarts. **1956** Economist 8: Pottle...2 quarts.

pouk. POKE

poule. POLE

pound—1-4, 6 pund; 3 L liber, F liver; 3-7 L libra; 3-9 pound; 4-5 F livre, punde; 4-6 pond (OED), ponde (OED), pownd; 4-7 pounde; 5 F livere; 5-6 L libre, pownde [ME *pound* fr OE *pund* fr L *pondo*, pound, originally "in weight;" akin to L *pondus*, a weight; parallel to *lb* fr MedL *libra*, pound, fr L *libra*, Roman pound of twelve ounces]. A wt in the ap, avdp, merc, tow, English t, and Scots t and tron systems. Its abbreviations, l, lb, li, and lib, have all derived from the L *libra*.

The ap lb contained 5760 gr (373.242 g), or 288 s of 20 gr each (1.296 g) or 96 dr of 60 gr each (3.888 g) or 12 t oz of 480 gr each (31.103 g). The number of gr (barleycorns) in the ap lb was the same as in the t lb.—*c*1600 Hall 36: Scrupuli...is 20 barley cornes.... 3 scruples contain a drachme.... 8 drachmes, an ounce.... Libræ...is a pound. **1688** Bernardi 137: Vel more Pharmacopolarum: Libra de Troy, 12 Unciæ ℥ 96 = 12 X 8 drachmæ ʒ : Scripili ℈. 288 = 96 X 3: grana monetaria rursus 5760 = 288 X 20. **1699** Hatton 1.20: *The Apothecaries*

Weights.... 20 Grains make A Scruple...℈. 3 Scruples make A Drachm...ʒ. 8 Drachms make An Ounce...℥. 12 Ounces make A Pound...℔. **1708** Chamberlayne 205: 12 ounces 1 Pound ℔. **1728** Chambers 1.360: The Apothecaries also use the Troy Pound, Ounce, and Grain. **1829** Palethorpe sv: 12 oz. of 8 drms. of 3 scruples of 20 grs. each, make 1 ℔. apothecaries´ weight. **1903** Warren 100: Apothecaries´ Weight.... Grains 5,760, or 12 ounces = 1 pound.

The avdp lb contained 7000 gr (453.592 g), or 256 dr of 27.344 gr each (1.772 g) or 16 oz of 437 1/2 gr each (28.350 g), and was used for all products not subject to ap or t wt. There was much confusion in medieval and early modern texts concerning the exact wt of this lb. First, the t oz of 480 gr (31.103 g) was sometimes erroneously ascribed to the avdp scale, and thus the avdp lb was miscalculated as 16 oz of 480 gr apiece or a total of 7680 gr (497.648 g). Second, the ap scale of 20 gr to the s, 3 s to the dr, and 8 dr to the oz (again, 480 gr to the ap oz) was incorrectly used in the conversion, and the avdp lb was once more taken as 16 oz of 480 gr each, totaling 7680 gr. Occasionally the avdp dr was mistaken for the ap dr of 60 gr (3.888 g), and the avdp oz was therefore reckoned as 16 dr of 60 gr each, a total of 960 gr (62.208 g), causing the avdp lb to be 15,360 gr (995.328 g) or 16 oz of 960 gr each. In all three cases, the avdp lb was computed to be heavier than the standard of 7000 gr. Comparatively the avdp lb was 1.215625 English t lb, 0.91938 Scots t lb of 7616 t gr, and 0.735504 Scots tron lb.—*c***1400** Hall 36: Una libra faciunt 1 pynt, 7,680 grana. *c***1461** *Ibid*

13: The lb. of thys weyght conteynyth xvi unces of Troy weyght; ibid 19: Pownd conteynythe xvi unces. **1474** Cov. Leet 396: The seid xxxij graynes of whete take out of the myddes of the Ere makith a sterling peny & xx sterling makith a Ounce of haburdepeyse; and xvj Ouncez makith a li. **1496** Seventh Rep. 29: The same tyme ordeined that xvi uncs of Troie maketh the Haberty poie [pound]. c**1510** Pauli 17: Notmegges for 6 or 7 [d] the pownde. **1517** Hall 47: And owte of that make the habar de payse pownde as ye shall´ hereafter knowe; ibid 48: So makyth the whete afore namyd the Habar de Payse once.... And xvi of that onces the trewe habar de poix lib. c**1525** Ibid 40: Item xvi onces Habar de Payce ys. a lib.... Item xviii onces di. of Troy weygthe makys xvi onces Habar de payse. c**1570** Fox 35: iiii. pownde of wex; ibid 36: iiii pounde of wex. **1588** Hall 46: The waight now used, every pownd conteineing 16 ounces.... The whole ounce is 16 drams. **1597** Halyburton cxiv: Leimondis the pund thairof.... Licoras the pund thairof. **1603** Henllys 138: And all spice, Iron, Rosen, pitche and other drugges uttered by the mercers are sold by the haberdepoies pound. **1606** Hall 38: This waight of Haberdepois is allowed alsoe by Statute being 16 oz. to the pound waight with the which is wayed all phisick drugges, grocery wares, rozen, wax, pitch, tarr, tallowe, sope, hempe, fflaxe, all metalles and mineralles. **1665** Assize 2: There is also another weight named Avoirdupois weight, whereunto there is 16 ounces for the pound. **1682** Hall 29: Aver-du-pois conteynes: every pound, 16 ounces; every ounce, 8 drgmes [sic]; every dragme, 3 scruples; every scruple, 20 graines. **1688** Bernardi 137-38: Libra equidem Avoirdupois qua solent populares mei graviores mercium æstimare quam pretiosiores, 1/112 Hundredi sui sive centenarii crassi, 16 unciæ, 128 = 16 X 8 drachmæ; ibid 138: Habet et libra Avoirdupois scripulos suos 384 = 128 X 3, gravans nobis 1,2169, sed ratione Wybardica 17/14 = 1,2413 libræ de Troy. **1708** Chamberlayne 206: But the Avoirdupois Pound is more than the Troy Pound, for 14 Pound Avoirdupois are = to 17 Pound Troy-Weight. **1710** Harris 1. sv weight: And the other is called Averdupois, containing 16 Ounces in the Pound. **1717** Dict. Rus. sv dram: Dram or Drachm, the just Weight of sixty Grains of Wheat; in Avoir-du-pois Weight, the sixteenth part of an ounce; ibid sv hundred-weight: But ordinarily a Pound is the least Quantity taken notice of in Aver-du-pois Gross Weight; ibid sv pound: A sort of Weight containing 16 Ounces Avoir-du-pois. **1742** Account 1.553: The single Averdupois Bell Pound, against the flat Averdupois Pound Weight, was found...to be heavier by Two Troy Grains and a half. **1778** Diderot XXVI.420: L'avoir-du-pois est de seize onces. **1790** Jefferson 1.986: So that the pound troy contains 5760 grains, of which 7000 are requisite to make the pound avoirdupois. **1829** Palethorpe sv: An avoirdupois pound is equal to 1 lb. 2 oz. 11 dwts. 16 grs. troy. **1855** Hooper 5: The avoirdupois pound weight...is to the troy as 175 to 144; and is therefore = 7000 grains troy. But its ounce, which is the 1/16 part of it, is = 437.5 such grains. **1893** Mendenhall 145: 1 pound avoirdupois = 1/2.2046 kilo. **1907** Hatch 15: 1 Pound = 453.5924277 Grammes.

In Scotland (c1600-1800) the avdp lb was used extensively in the various shires for the following products (Swinton 53-130, Kelly 96-112, and Cyclopædia sv weights). North—Nairnshire: flour, foreign goods, and groceries. Sutherlandshire: English goods and groceries. Northwest—Inverness: flour and groceries. Ross and Cromarty: flour and meat. Northeast—Aberdeenshire: English goods, groceries, and salted butter; also, if sold wholesale, butter, cheese, flesh, hog's lard, tallow, and wool. Banffshire: English goods and groceries. Moray (Elgin): English goods, flour, groceries, and salt. Central—Perthshire: all goods brought by English or foreign merchants. Stirlingshire: bread, dressed wool, and groceries. West central--Dumbartonshire: English goods and groceries. West—Argyllshire: flour, groceries, iron, and salt. East—Angus: English goods and groceries. Fifeshire: English goods and groceries. Kincardineshire: English goods and salt. Kinrossshire: groceries. South—Lanarkshire: English goods and groceries. Peeblesshire: groceries. Southwest—Ayrshire: fine barley, flour, groceries, iron, and salmon. Kirkcudbrightshire: English goods, groceries, and meat. Renfrewshire: English goods and groceries. Wigtownshire: groceries and meat. Southeast—Berwickshire: at Berwick and Eymouth, all goods except sweet butter and fish; at Dunse, groceries and shop goods; at Coldstream, English goods, groceries, and meat. East Lothian: English goods and groceries. Midlothian: coal, groceries, and most merchant's goods. Roxburghshire: English goods and groceries. Selkirkshire:

English goods and groceries.

The merc lb (libra mercatoria; also called merchants' pound and commercial pound) contained 6750 t gr (437.400 g), or 15 merc oz of 450 t gr each (29.160 g), and was equal to 5/4 tow lb of 5400 t gr (349.920 g). However, the merc lb was actually defined in terms of wheat gr, being 9600 wheat gr or 15 oz of 640 wheat gr each. It was used in England for all goods except electuaries, money, and spices until sometime in the fourteenth century, when it was replaced by the avdp lb.—c**1253** Hall 11: En letuaris e confeciuns la liver est de xii uncis; en tutes autre chosis la li. est de xv uncis. c**1272** Report 1.414: Libra vero omnium aliarum rerum consistit ex viginti quinque solidis, uncia vero in electuariis consistit ex viginti denariis, et libra continet xii uncias. In aliis vero rebus libra continet quindecim uncias. **1290** Fleta 119: Item denarius sterlingus, sicut dictum est, ponderat xxxij. grana frumenti, et pondus xx. d. facit vnciam, et quindecim vncie faciunt libram mercatoriam. c**1300** Brit. Mus. 13.29: Item in electuariis...libra continet .xij. vncias. In aliis rebus libra continet .xv. vncias. c**1300** Hall 8: Uncia debet ponderare viginti denarios. Quindecim uncie faciunt libram Londonie. **1495** Brit. Mus. 28.156: Quindecim vncie faciunt libram. **1607** Cowell sv sarpler: Also that 15. ounces of the quantitie aforesaid doe make a merchants pounde. **1665** Sheppard 13: And all our Weights and Measures have their first Composition from the penny Sterling, which ought to weigh Two and thirty wheat corns of a middle sort: Twenty of which pence make an ounce, and

12 such ounces a pound of 20 shillings, but Fifteen ounces make the Merchants pound. **1793** Leake 30: The Merchants Pound, which Fleta says, was fifteen Ounces. **1805** Macpherson I.471: The pound of twelve ounces is used only for money, spices, and electuaries, and the pound of fifteen ounces for all other things. **1882** Jackson 282: Merchants´ pound = 15 oz. = 25 shillings = 9600 grains. **1888** Pell 583: Libra Mercatoria of Fleta, 437.335 grammes...(6750 Troy). **1896** Colles 515: The commercial pound of fifteen ounces.

The tow lb, also called the sterling, easterling, coinage, Saxon, goldsmith´s, or moneyer´s lb, contained 5400 t gr (349.920 g), or 12 tow oz of 450 t gr each (29.160 g), and was equal to 4/5 merc lb of 6750 t gr (437.400 g). The tow lb, however, was actually defined in terms of wheat gr, being 7680 wheat gr or 12 oz of 640 wheat gr each. It was used in England generally for electuaries, money, and spices until 1527, when Henry VIII declared it illegal and it was replaced by the t lb.—c**1253** Hall 11: E fet asauer ke lib. de deners e de especis confectiouns, si cum d´eletuari, si est de le peys de xx sol.... En letuaris e confeciuns la liver est de xii uncis. **1290** Fleta 119: Et vnde xx. d. faciunt vnciam et xij. vncie faciunt libram xx. s. in pondere et numero. c**1300** Hall 8: Denarius qui vocatur sterlingus rotundus sit, et sine tonsura; et ponderabit xxxii grana frumenti rubei, in medio spice assumpta. Et xx denarii faciunt i unciam. Et xii uncie faciunt i libram, videlicet xx s. sterlingorum. c**1400** Ibid 6-7: C´est assavoir, que le denier d´Engleterre round et sanz tonsure poisera xxxii greins de froument en my le spic. Et xx d. font la unce; et xii unces font la livre. **1474** Cov. Leet 396: That xxxij graynes of whete take out of the mydens of the Ere makith a sterling otherwyse called a peny; and xx sterling maketh an Ounce; and xij Ounce maketh a Pounde for siluer, golde, bred & Mesure. **1545** Rates 1.52: A pounde of Tower Wayght wayeth of the Troy. xi. ounces .i. quarter. **1587** Stat. 540: The pounde Towre shall be no more used, but all manner of golde and sylver shall be wayed by the pounde Troye which maketh XII oz. Troye and which excedith the pounde Towre by III quarters of the ounce. **1606** Hall 39: Tower waight for Troy waight, which was the Prince´s prerogative, gayned thereby 3 quarters of an ounce in the exchange of each pound waight. **1615** Collect. Stat. 464: That an English penie, which is called the sterling, round without clipping, shall weigh two & thirty grains of dry wheat in the middest of the eare and twenty d. make an ounce, and 12. ouncis make a lb. **1755** Postlethwayt II.188: The pound used at the mint before that time, called the tower or the moneyers pound, was equal to 5400 Troy grains. **1819** Cyclopædia sv weight: The old Saxon pound was 540/576 of the present troy pound; and...54/70 of the present avoirdupois pound. **1820** Second Rep. 28: Of silver coins, a pound sterling; the money pound, or Tower pound of the Anglo Saxons, used for some centuries after the Conquest, contained 12 ounces of 450 grains each = 5400 grains. **1840** Ruding 1.7: Pound Tower equal to 5400 grains, or 11 1/4 oz. Troy. **1870** Third Rep. 33: The most ancient system of weights in this Kingdom was that of the Moneyer´s pound, or the money-pound of the Anglo-Saxons, which was continued in use for some centuries after the Conquest, being then known as the tower pound, or sometimes the goldsmith's pound. **1896** Colles 516: And the sterling pound of 12 ounces, being 1/16 lighter than the troy pound, contained 5,400 troy grains; whence the commercial pound of 15 of the same ounces contained 6,750 troy grains. **1905** Chadwick 32: It is indeed often assumed that the old English pound was identical with the later Cologne or Tower pound (5400 gr.).

The English t lb contained 5760 t gr (373.242 g), or 240 dwt of 24 t gr each (1.555 g) or 12 t oz of 480 t gr each (31.103 g). It was introduced into England sometime during the fourteenth century and was used principally for electuaries, gold, precious stones, and silver. It became the standard for these and other items when Henry VIII abolished the tow lb in 1527.—**1414** Rot. Parl. 4.52: & establer le pris de chescun livere, quartron, ou ounce.... & q'ils preignent pur la libre de Troy orree XLVI s. VIII d. a pluis. **1443** Ibid 256: Silver is bought and soold unkoyned atte pris of XXXII s. the pound of troie. **c1461** Hall 13: And by thys weyght is bought and sold Gold, Sylver, Perlys, and odyr precius stonys, and iuwells and beeds schold be sold by this weyght. **c1525** Ibid 40: xii ownces Troye weyghte ys juste a lib. Troy weyghte, for gold & syluer. **1588** Ibid 45: Troy Weight is most used by the ounce for Gold and Silver.... The pownd waight is 12 oz. **c1590** Ibid 20: The gouldsmyth countith but 12 ounces to his pownde waight; ibid 22: 12 ounces makyth the pound troy. **1606** Ibid 37: The

Troy waight cont[aineth] 12 oz. to the pound waight; ibid 38: Ffor 24 graines or barleycornes, drie, out of the middest of the eare, do make a 1 d.... Soe 480 graines, 1 oz. or 20 d.... Soe the pounde waight is 12 oz...or 5760 graines. **1616** Hopton 159: This Troy weight containes in euerie pound 12 ounces, euery ounce 20 peny weight, euery peny weight 24 graines. **1635** Dalton 143: Troy weight is by law; and thereby are weighed gold, silver, pearle, pretious stones, silke, electuaries, bread, wheat, and all manner of graine, or corne.... This hath to the pound xii. ounces. **1664** Spelman 366: Libra numaria nomen à pondo cepit, quòd libram quam vocant Troianam, hoc est, uncias 12. olim pendebat. **1675** Vaughan 243: In England the pound Troy is 12 oz, each oz. 20 peny weight, each peny weight 24 grains, in all 5760. **1682** Hall 29: Troy weight conteynes: every pound, 12 ounces; every ounce, 20 penny weight; every penny weight, 24 graines. **1688** Bernardi 137: Libra Anglica de Troy. 12 Unciæ...12 X 20 = 240 p.w. sive penningi veteres Regum Edvardorum...qui sunt penningi novi ac decurrentes 720 = 240 X 3; aut 5760 = 240 X 25 grana tritici. **1708** Chamberlayne 205: In Troy-Weight...there are 480 Grains in the Ounce and 5760 Grains in the Pound.... By Troy-Weight we Weigh Bread, Corn, Gold, Silver, Jewels and Liquors. **1716** Harris 1. sv weight: Troy Weight, is that by which Gold, Silver, Jewels, Amber, Electuaries, Bread, Corn, Liquors, &c. are weighed; and from this Weight all Measures of wet and dry Commodities are taken. **1728** Chambers 1.360: In Troy Weight, 24 of these Grains make a Penny-weight Sterling; 20 Penny-weight make an Ounce; and 12

Ounces a Pound. **1742** Account 2.187: The English Troy Pound of Twelve Ounces or 5760 Grains. **1755** Willis 361: The antient Way of paying Money into the Exchequer was either Pondere or Numero, by Weight or Tale; hence ad Pensam signifies such Payment by Weight wherein the Payer was obliged to make the Pound Sterling to be full 12 Ounces Troy. **1820** Second Rep. 28: Pound Troy: 12 ounces of 480 grains each = 5760 grains. **1855** Hooper 3-4: Troy-weight has a pound of 12 ounces: and when used for money, each ounce is of 20 penny-weights; and a pennyweight, of 24 grains. **1903** Warren xiv: This Tower pound of 5,400 grains Troy was our standard pound at the Mint until the reign of Henry VIII, when it was replaced by the Troy pound of 5,760 grains Troy.

The Scots t lb had two different weights and two different scales: for gold and silver, 5760 t gr (373.242 g), or 192 DROPS of 30 t gr each (1.944 g) or 12 t oz of 480 t gr each (31.103 g); for meal, meat, hemp, unwrought pewter, flax, lead, iron, Baltic and Dutch goods, 7616 t gr (493.517 g), or 256 drops of 29.75 t gr each (1.928 g) or 16 t oz of 476 t gr each (30.845 g). The latter Scots t lb sometimes was called Dutch (Amsterdam) or French (Paris) weight. The Scots tron lb, used for home productions (see regional breakdown below), contained 9520 t gr (616.896 g), or 320 drops of 29.75 t gr each (1.928 g) or 20 t oz of 476 t gr each (30.845 g). Comparatively the English t lb was 0.822622 avdp lb, 0.756302 Scots t lb of 7616 t gr, and 0.605042 Scots tron lb; the Scots t lb of 7616 t gr was 1.322222 English t lb, 1.087689 avdp lb, and 0.8 Scots tron lb; and the Scots tron lb was 1.652777 English t lb, 1.359611 avdp lb, and 1.25 Scots t lb of 7616 t gr.—**1425** Acts Scotland 2.12: ITEM thai ordanit ande statute the stane to wey Irne woll ande vthir merchandice to contene xvj pundes troyis Ilk troyis punde to contene xvj vnce. **1563** Ibid 541: Thay find to mak ane vniuerfall wecht of the stane of the wecht .xvj. pund trois. **1587** Ibid 3.521: Ilk trois pund contening sextene wnce. **1618** Ibid 4.587: And that Weght called of old the Trone weght to bee allvtterlie abolisched and discharged/and neuer hereafter to be received nor vsed. **1761** Thomson viii: The Scots pound is equal to 7616 Troy grains. **1779** Swinton 36: In Scotland, Gold and Silver are weighed by the...[English troy] ounce and pound; but the ounce is divided into 16 drops, and the drop into 30 grains; ibid 38: The Act 1617 makes the French Troye weight the standard for Scotland; and declares, That the standard Stirling pint-jug contains 55 French Troye ounces of river water. This pint is found to contain 103.404 cubic inches, and the cubic inch of such water to weigh 253.18 English Troy grains; by consequence the Scotch Troye pound [for meal, meat, hemp, and iron] weighs...7616 English Troy grains; ibid 38-39: Trone. For home-productions, according to the custom of Edinburgh.... This weight, though abolished by act in 1617, is still in constant use, and is different almost in every country [= district]. According to the custom of Edinburgh, the present Trone pound contains 20 Scots Troye ounces [of 476 t gr each]. **1813** Cooke 96: Trone. 20 Ounces...1 Pound. **1816** Kelly 92: Scotch Troy Weight, also called Amsterdam and French Weight.... The pound, 16 of which compose a stone, contains 7,616 troy grains.... Trone Weight. This weight was abolished by act in 1618. Its name is still retained for selling butter, cheese, tallow, wool, lint, hemp, hay, and some other home commodities. **1820** J. Sheppard 90: Scotch Troy Weight...16 drops make 1 ounce...16 ounces...1 pound. **1832** Edinburgh XVIII.500: Another kind of Troy weight was used in Scotland, called Dutch weight, and sometimes Amsterdam, and French weight, for weighing iron, hemp, flax, Baltic and Dutch goods, meal, butcher´s meat, unwrought pewter and lead.... One stone or 16 lbs. of this weight as used in Glasgow, is = 17.442482 imperial avoirdupois lbs.

In the Scots shires (c1600-1800), however, the tron lb and the Scots t lb of 7616 t gr were used for many products which were not included in the above national or standard list (Swinton 53-130, Kelly 96-112, and Cyclopædia sv weights). North—Nairnshire: Scots t, hemp, hides, and twine for nets; tron, butter, cheese, tallow, and wool. Sutherlandshire: tron, butter, cheese, tallow, and wool. Northwest—Inverness: iron, butter, cheese, and wool. Ross and Cromarty: tron, butter, cheese, fish, flax, and tallow. Northeast—Aberdeenshire: Scots t, butter, cheese, coal, hog´s lard, tallow, and wool (only if sold by retail, if wholesale then avdp lb applied to these products); tron, feathers and hay. Banffshire: Scots t, coal and green hides; tron, butter, cheese, hay, tallow, and wool. Caithness: tron, butter, cheese, feathers, tallow, and wool. Moray: tron, butter, cheese, hay, and wool. Central—Perthshire: Scots t, coal; tron, butter, cheese, and rough tallow. Stirlingshire: Scots t, coal, hay, and salmon; tron, butter, cheese, feathers, rough hides, tallow, and wool. West central—Dumbartonshire: iron, butter, cheese, fish, and meat. West—Argyllshire: tron, butter, cheese, fish, flesh, hay, tallow, and wool. East—Angus: Scots t, coal; tron, butter, cheese, flax, and wool. Fifeshire: tron, butter, cheese, hides, and wool. Kincardineshire: Scots t, English and Scots coal and hay; tron, butter, cheese, tallow, and wool. Kinrossshire: tron, butter, cheese, hay, rough hides, and wool. South—Lanarkshire: Scots t, barley; tron, cheese, butter, meat, and wool. Peeblesshire: Scots t, barley; tron, butter, cheese, coal, hay, hides, tallow, and wool. Southwest—Ayrshire: tron, butter, cheese, hay, meat, and wool. Buteshire and Arran: tron, beef, butter, cheese, hay, hemp, mutton, raw hides, straw, tallow, and wool. Kirkcudbrightshire: tron, all local production. Renfrewshire: tron, butter, cheese, fish, meat, and tallow. Wigtownshire: tron, butter, cheese, and wool. Southeast—Berwickshire: Scots t, flour; tron, butter, cheese, hay, tallow, and raw hides. East Lothian: tron, all local production. Midlothian: Scots t, feathers; tron, butter, cheese, hay, hemp, tallow, and wool. Roxburghshire: Scots t, barley and flour; tron, butter, cheese, hay, raw hides, tallow, and wool. Selkirkshire: Scots t, barley and fish; tron, butter, cheese, hay, raw hides, tallow, and wool.

Besides the ap, avdp, merc, tow, English t, and Scots t and tron lb, local markets in England, Wales, and Scotland (c1800-1900) employed special "market" lb for certain products; the following were always reckoned on the avdp scale, except where indicated: butter—12 oz (340.200 g) in Westmorland, 16 oz (453.592 g) in Westmorland and at Campbeltown in Argyllshire, 16 to 20 oz (453.592 to 567.000 g) in the East Riding of Yorkshire, 16 or 24 oz (453.592 or 680.400 g) in Wigtownshire, 16 to 24 oz (453.592 to 680.400 g) in the North Riding of Yorkshire, 17 oz (481.950 g) in Buckinghamshire, Derbyshire, Shropshire, and South Wales, 18 oz (510.300 g) in Cheshire, Cornwall, Devonshire, Dorsetshire, Gloucestershire, Herefordshire, Lancashire, Louth in Lincolnshire, Shropshire, Wolverhampton in Staffordshire, Westmorland, South Wales, and Berwick in Berwickshire, 18 to 21 oz (510.300 to 595.350 g) in North Wales, 20 oz (567.000 g) in Westmorland, and in the West Riding of Yorkshire, 21 oz (595.350 g) in Stanhope and Westmorland, 22 oz (623.700 g) in Durham, Angus, and Perthshire, 23 oz (652.050 g) in Dumbartonshire and Peeblesshire, 24 oz (680.400 g) in Stockton, South Wales, Inverary, Ayrshire, and Banffshire, 26 oz (737.100 g) in Glamis, 27 oz (765.450 g) in Kirriemuir; cheese—22 oz (623.700 g) in Perthshire, 22 1/2 oz (637.875 g) in Berwickshire, 23 oz (652.050 g) in Peeblesshire, 24 oz (680.400 g) in Banffshire; groceries—16 oz (453.592 g) in Ayrshire; hay—22 oz (623.700 g) in East Lothian, 23 oz (652.050 g) in Peeblesshire, 24 oz (680.400 g) in Ayrshire and Banffshire; hides—22 oz (623.700 g) in East Lothian; meal—17 1/2 oz (496.125 g) in Banffshire; meat—17 1/2 oz (sometimes in t oz) (496.125 g) in Banffshire and Stirlingshire, 22 oz (623.700 g) in Stirlingshire, 24 oz (680.400 g) in Ayrshire; tallow—22 oz (623.700 g) in East Lothian; and wool—23 oz

(652.050 g) in Peeblesshire, 24 oz (680.400 g) in Banffshire and Berwickshire (Pasley 111-12, Cyclopædia sv weights, and Britten 174-75).

pounde. POUND

poundlar. PUNDLAR

pounsioun. PUNCHEON

powk. POKE

powle. POLE

pownd, pownde. POUND

poyke. POKE

poynt. PINT

poynte. POINT

poyntt. PINT

poyt. POT

prickle [*; see second citation]. A m-c for fruit in northern England. It was a wicker or willow basket of uncertain size.—c**1634** Hall 53: It is deliuered out of the Ketches and Boates to those poore persons in Baskettes called Prickles, conteyneing 4 Peckes and sometimes not soe much. **1829** Brockett 236: Prickle, a basket or measure of wicker work among fruiterers. Formerly made of briers. Hence, perhaps, the name.

punchen. PUNCHEON

puncheon—5 ponchyn, poncion (OED), punshyn, pwncion (OED); 5-6 punchin; 6 poncheon, ponchion (OED), pontion, pontioune (OED), pounsioun, puncheoun (OED), punchione, punchon (OED), puncioune (OED), punschioun (OED), punshion, punsion (OED), punsioun, puntion (OED); 6-7 punshon; 6-8 punchion; 7 punchen, punsheon, punsheoun; 8-9 puncheon [ME poncion fr MF ponchon, poinçon, of unknown origin]. A m-c, a large wooden vessel resembling a cask, used for several commodities: beer, 72 gal (c3.33 hl); dried fruit, perhaps 10 to 12 Cwt (508.020 to 609.624 kg); soap, of uncertain size; and brandy, rum, gin, molasses, and wine, 84 gal (c3.18 hl), synonymous with the TERTIAN and double the wine TIERCE of 42 gal (c1.59 hl). The Irish wine puncheon (c1800) contained 84 Irish gal (c3.00 hl) and was double the Irish wine tierce of 42 gal (c1.50 hl) (Edinburgh XII.572). It sometimes was abbreviated pun.—**1443** Brokage II.37: Cum ii punshyns vini.... Cum iiii ponchyns saponis; ibid 64: 1 ponchyn saponis albi. **1525** Jacobus 67: Item empt´ j pounsioun acetj; ibid 74: j punsioun acetj. **1547** Cal. Pat. 23.397: In buttes, pypes, hoggesheddes, pontions or barrelles. c**1550** Welsh 8: 20 poncheons raisions; ibid 232: 2 punchins prunes. c**1590** Hall 21: The tertiane or punchione of a tunne, which is 1/3 part of a tunne, contenith 84 gallons. **1612** Halyburton 289: Beiff the punsheoun; ibid 308: Girds of Irone for punsheones or pypes the hundreth weght. **1646** H. Baker 276: The punchen to hold...84 Gallonds. **1661** Acts Scotland 7.259: Disburthen the saids herrings...and...dry and load the same in barrells & punshons; ibid 260: All their barrells or punshions may be marked. **1682** Hall 29: 1 Tunne conteynes...3 Punchions. **1701** Hatton 3.232: Puncheon.... Of Wine = 84 Gallons; of Pruons 10 to 12 hundred weight. **1707** Justice 2: Of these Gallons...84 a Punchion. **1708** Chamberlayne 210: A Puncheon 84 Gallons. **1710** Harris 1. sv measures: The common

Wine Gallon sealed at Guild-Hall in London...is supposed to contain 231 Cubick Inches; and from thence...the Punchion 19404. **1717** Dict. Rus. sv wine-measure: A Punchion 84 Gallons. **1816** Kelly 87: WINE MEASURE...2 Tierces...1 Puncheon...317,9345 [1]. **1830** Second Rep. 29: Puncheon of beer, in London, 72 gallons...of wine, 84 gallons. **1829** Palethorpe sv: PUNCHEON, a cask used for brandy, rum, gin, molasses, &c. which contained 84 gals. of the late wine measure. **1850** Alexander 90: Puncheon; old measure...84.—gallons.

puncheoun, punchin, punchion, punchione, punchon, puncioune. PUNCHEON

punctum. POINT

pund, punde. POUND

pundlar—7 poundlar (OED); 7-9 pundlar, pundler (OED) [altered form of ON pundari, steelyard, fr pund, POUND]. A type of steelyard (see first citation) in the Orkney and Shetland Islands, principally used for weighing barley and malt.—**1779** Swinton 104: The Pundlar is a beam of wood about six feet long, and about three inches in diameter at one end, tapering gradually to the other. A hook is fixed to the greater end for suspending the goods. About six inches from that end, a tongue and shears, like those on the beam of a balance, are fixed; and, at the upper end of the shears, there is a large iron ring, through which, when the instrument is used, there is put a cross-beam for suspending the machine; and this cross-beam is generally supported by two men on their shoulders. The Pundlar is marked with notches at proper distances, corresponding to, and exhibiting the weight, from three Setteens upwards, to twelve; and the weight of the commodity is ascertained by a stone of the weight of a Setteen hung upon the Pundlar by an iron ring, which may be shifted from notch to notch, till the tongue between the shears, as in a steelyard, discovers the instrument to be in equilibrio. **1820** J. Sheppard 134: In the Orkneys and Shetland grain of all descriptions is sold by weight; and the malt pundlar is generally made use of for that purpose. See AUNCEL; BISMAR; SETTEEN

pundler. PUNDLAR

punnet—8-9 punnet, punnit (OED) [of obscure origin; perh fr pun, dial var of pound, + et]. A m-c, a small, round, shallow chip basket, for vegetables and fruit varying in weight from 3/4 to 1 lb (340.194 to 453.592 g).—**1896** Wagstaff 36: A ´punnet´ of strawberries in Greenock varies from 3/4 lb. to 1 lb. A punnet of vegetables in Middlesex = 1 lb.

punnit. PUNNET

punschioun, punsheon, punsheoun, punshion, punshon, punshyn, punsion, punsioun, puntion. PUNCHEON

puock, puok. POKE

putte. POT

pwn [W pwn, a pack, a burden]. A m-q for straw in North Wales (c1800-1900) weighing 160 lb (72.574 kg) (Britten 175).

pwncion. PUNCHEON

pwoak, pwok, pwoke. POKE

pwys [*]. A wt of 2 lb (0.907 kg) for wool in South Wales (c1800-1900) equal to 1/13 MAEN (Second Rep. 29 and Donisthorpe 216).

pychar, pychare, pycher, pychere. PITCHER

pyece. PIECE

pynt, pynte. PINT

pype. PIPE

pysse. PIECE

pytcher. PITCHER

pyynte. PINT

Q

quaer, quaier, quair, quaire, quar. QUIRE

quarantain, quarantana, quarantena, quarantene. QUARENTINE

quare. QUIRE

quarenteina, quarenteine, quarentena, quarentene, quarenteyne, quarentina. QUARENTINE

quarentine—1-7 L quarentena; 2 L quarenteina; 3-4 quarenteyne; 7-9 quarentine; 8 quarantain; 9 quarantene; ? L quarantana (Prior), L quarantena (Prior), quarenteine (Bello), quarentene (Prior), L quarentina (Maitland) [MedL quarentena fr OF quarantaine, period of 40 days, set of forty]. A m-l containing 40 linear perches (see FURLONG), and a m-a containing 40 sq perches (see ROOD).—c**1100** Bello 11: Leuga autem Anglica duodecim quarenteinis conficitur. Quarenteina vero quadraginta perticis. **1664** Spelman 474: Quarentena, næ.... Stadium, Angl. a furlonge. Agri spacium quod secundum strigarum seu arationis longitudinem, ad perticas extenditur quadraginta (Gall. quarante) atque inde nomen: Nam quod ex quadraginta aliquibus rebus consistit, Galli quarantaine appellant.... Chart. Withlasii Regis Merciorum apud Ingulf.—Quatuor caracatas terræ arabilis, continentes in longitudine 8 quarentenas, & 8 quarentenas in latitudine.... Liber MS. Crabhusiæ fol. 8.—Le Messuage de Crabhus...en lungure Ouwoc la terre de la Rive tendaunt vers occident, desque a la fosse de le marcys, conteynt treys quarenteynes, & trente & oyt perchez. Checun quarenteyne par sey conteynt quaraunte perchez. **1678** Du Cange sv quarentena: QUARENTENA, Modus agri apud Anglos, constans 40. perticis. **1695** Kennett Glossary sv quarentena: A Quarentine, a Fourty long, or Furlong.... A measure of fourty Perches.... In the Doomsday Survey, it was the usual mensuration of woodland. So in Burcester there was—Silva unius quarentenæ longitudine & unius latitude. **1756** Rolt sv quarantain: A measure, or extent of land, containing 40 perches. **1888** Pell 564: The division of the quarantene into acres. **1888** Round 3.220: The measure which we find in Domesday in conjunction with the lineal acre is the ´quarentena´. Now the quarentena is the ´furlong´, that is, the side of the areal acre. **1897** Maitland 432: 1 league = 12 furlongs or quarentines or acre-lengths = 480 perches.

quarr. QUIRE

quart—3 L quartus; 3-6 L quarta; 4-7 quarte; 4-9 quart; 5 qvarte (OED), qwhart (OED) [ME quart fr MF quarte fr OF quarte, fem n, fr quart, fourth, fr L quartus, fourth]. A m-c: for dry products, 2 pt (c1.10 l) and equal to 1/4 gal, 1/8 pk, and 1/32 bu; for liquids, 2 pt (ale and beer = c1.16 l, wine = c0.95 l) and equal to 1/2 POTTLE and 1/4 gal. Since the establishment of the Imperial system in 1824, the qt both for liquid and dry products has been reckoned as 69.355 cu inches (1.136 l) and equal to 1/4 gal of 277.420 cu inches. The Scots qt, for liquid and dry products, contained 206.808 cu inches (c3.40 l) or 2 Scots pt or 4 CHOPPINS or 8 MUTCHKINS or 32 GILLS. The Irish qt, for liquid and dry products, contained 54.4 cu inches (c0.89 l) or 2 Irish pt of 27.2 cu inches.—**1287** Select Cases 2.19: Willelmus le Barbur, potellus falsus et quarta bona, et quia vendidit pro xvj. d.... Robertus Raven de Ely, lagena bona, potellus bonus et quartus bonus. c**1340** Oxford 267: Johanna de Leghe habet I quartam fals´. **1351** Rot. Parl. 2.240: Soient les Mesures, c´est assaver bussell, demi bussell, & pec, galon, potel, & quarte. **1390** Henry Derby 6: Pro ij potellis et j quarta vini Vasconie, xxij d. ob. c**1400** Hall 36: 2 libre faciunt unum quart´, 15,360 grana. c**1420** Evesham 308: Unum caponem cum una quarta vini. **1474** Cov. Leet 396: ij pyntes maketh a quart; & ii quartes maketh a Pottell; & ij Pottels makith a Gallon; & viij Gallons makith a Buysshell. **1525** Jacobus 28: lxvij lagine j quarta ceruisie. c**1590** Hall 20: 4 quartes makith a gallon; 2 quartes makith a pottell; ibid 21: 2 pyntes makith a quart. **1603** Henllys 138: Two pintes to a quart, ij quarts to a pottle, ij pottles to a gallon. **1615** Collect. Stat. 464: First, sixe lawfull men shall bee sworne truely to gather all the measures of the towne, that is to wit, bushells, halfe a quarter bushells, gallons, pottels and quarts, as well of Tauernes as of other places. **1635** Dalton 144: Eight quarts maketh the peck...32 quarts maketh the Bushel. **1657** Jenkins 17: A full Ale-quart of the best Ale or Beer for a peny. **1665** Sheppard 7: Of dry things...Two pounds or pints make a Quart. **1708** Chamberlayne 210: 2 Pints make a Quart...2 Quarts make a Pottle...2 Pottles make a Gallon. **1816** Kelly 87: WINE MEASURE...2 Pints...1 Quart...57,75 [cu inches]...0,9463 [l].... ALE AND BEER MEASURE...2 Pints...1 Quart...70,5 [cu inches]...1,1552 [l]; ibid 88: DRY MEASURE...2 Pints...1 Quart...67,2 [cu inches]...1,10107 [l]. **1820** Second Rep. 29: Quart, two pints, whether of wine measure or ale measure.... Scotland: two Scotch pints. **1832** Edinburgh XII.571: 2 [Irish] Pints = 1 Quart = 54.4 [cu inches]. **1860** Britannica 805: Scotland...quart, 206.8 [cu inches]. **1907** Hatch 23: 2 pints = 1 quart (qt.) = 69.355 cubic inches; ibid 35: 1 quart = 1.13649 litres. **1956** Economist 4: Quart: (a) United Kingdom = 69.355 cubic inches. **1969** And. & Bigg 11: 1 qt (quart) = 1.13652 dm^3 = 1.137 litres.

quarta, quartarium, quarte. QUART

quarter—3 L quartarium; 3-7 L quarterium; 3-9 quarter; 4-7 L quarteria; 5 quartere, quartre, quartur, qwartur; 6 quartyr [ME quarter fr OF quartier fr L quartarius, a fourth part, fr quartus, fourth]. A m-c for grain equivalent to the SEAM, a m-l for cloth containing 9 inches (22.86 cm) or 1/4 YARD, and a wt equivalent to the QUARTERN of 28 lb (12.700 kg) or 1/4 Cwt of 112 lb. It was sometimes abbreviated q., qr., or qtr.—**1200** Cur. Reg. 8.218: De ordeo xj. sceppas et j. quarterium et xxxj. summas avene, unde tercia pars crevit super tenementum quod recuperavit. **1228** Gras 1.156: Quodlibet quarterium bladi. **1256** Burton 376: Quarterium frumenti venditur pro iii. s. vel xl. d. et hordeum pro xx. d. vel ii. s. et avena pro xvi. d. c**1272** Hall 7: Et viij lagene faciunt busshelum Londonie, quod est viij pars quarterii. **1283** St. Paul's 160: Per mensuram regis xvj. quarteria. **1298** Neilson 18: xxvii ring. quæ fecerunt i quartarium et dimidium. **1298** Falkirk 1-2: In precio Dlviij quarteriorum iij bussellorum frumenti CCCxxxviij quarteriorum pisarum et CCiiijxxj quarteriorum avene; ibid 2: Idem computat in Dlviij quarter. iii bus. frumenti; ibid 5: Et in defectu mensure j quarterium avene. c**1300** Hall 8: Et viii buselli bladi faciunt i quarterium. **1316** Neilson 41: Inde in missione apud Rameseiam xviii ring. quæ fecerunt unum magnum quartarium. c**1325** Rameseia III.158: Memorandum, quod octodecim communes ringæ faciunt unam magnam quarteriam. c**1350** London 39: Le quarter de furment pur xxxviij. s. **1357** Select Cases 3.182: Johannes Houpere, j quarterium...Johannes Webbe, dimidium quarterium...Walterus Aylward, iii quarteria. **1390** Rot. Parl. 3.281: Oept Busselx pur le Quarter rasez & nient comblez. **1392** Gras 1.527: Pro ccc quarteriis frumenti. c**1400** Hall 7: Et viii galons de froument sont le bussell´ de Loundres, qest le oeptisme partie du quarter. **1413** Rot. Parl. 4.14: C´est assavoir, viii Busselx pur la Quartre, & qe chescun Bussell contiendra oept Galons.... La Quarter de Furment noef Busselx par une Mesure use deins la dit Cite [London] appelle le Faat. c**1425** Gross II.377: Vnum quarterium frumenti.... i. quarterium brasei. **1433** Rot. Parl. 4.450: Oept busshels rasez pur le quarter. **1439** Rot. Parl. 5.31: That where as in a Parlement late at Westmynster holden, it was ordeigned, that no Whete shulde passe out of this land, yf the price of a Quarter of Whete passed or exceded the somme of VI s. VIII d., nor of Barly undur the same fourme. c**1440** Promp. Parv. 419: Quartere, of corne...Quarterium. **1443** Brokage II.9: Pro iiii quarteriis frumenti. c**1450** Common 174: A qwartur of whete...a quartur of maulte.... Item for iiij quarturs meselyn. c**1461** Hall 12: And viii gallons of wyne make a boschell of whete...wiche is the viii parte of a quarter whete. **1474** Cov. Leet 396: And viij Gallons makith a Buysshell, and neyther hepe nor Cantell...and viij Buysshelles makith a Quarter. c**1475** Gras 1.193: Of the quarter of iche corn. **1517** Hall 49: So that iiii busshelles wey ii c...weyghte of habar de poix powndes; the quartyr weyghte iiii c...weyghte. c**1530** St. Peter´s 310: Octodecim quarteria whete...viginti quatuor quarteria barly. **1540** St. Mary´s 59: i quarterium frumenti. **1549** Gras 1.708: Pro xix [X] xx quarteriis bracii. **1555** York Mer. 156: Item, a quarter of salt of Yorkes mesure, fyve pence. Item, a quarter of any other grayne of Yorkes measure, fore pence. **1562** Ibid 168: All manner of grayne Yorke measure the quarter, vi d.; All manner of grayne Hull measure the quarter, vij d. c**1590** Hall 21: The quarter or seame is 8 bushells. **1615** Collect. Stat. 465: And that none from henceforth doe buy in the Citie of London...no maner corne nor malt, but after eight bushels the quarter. **1616** Hopton 162: Whereof are made...Coombes, or halfe Quarters, Quarters, or seames. **1624** Huntar 4: The English quarter of corne, conteines hard by 2. bowes of Scottish measure. **1635** Dalton 144: 512 pints [or] 256 quarts [or] 64 gallons...[or] 8 bushels maketh the Quarter. **1641** Best 176: In the high Garner foure quarters of malte. **1646** H. Baker 169: 28 li (which is the quarter of a C.). **1657** Tower 547: That all the Kings Purveyors do take eight bushels of corn only to the quarter striked. **1661** Hodder 12: Note that 4 nails is one quarter of a yard, one yard 4 quarters...one ell English 5 quarters; ibid 15: For 28 pound carry one quarter. **1678** Du Cange sv quarteria: Quarteria et Quarterium, Mensuræ species. **1695** Kennett Glossary sv quarterium: A Quarter, a Seam, or eight bushels of corn. **1696** Cocker 111: 28 pounds make a quarter...4 quarters make an 100 weight, or 112 pound. **1708** Chamberlayne 207: 4 Bushels the comb or Curnock...2 Curnocks make a Quarter, Seam or Raff. **1717** Dict. Rus. sv dry-measure: Two Curnocks make a Quarter, Seam or Raff, and ten Quarters a Last. **1820** Second Rep. 29: Quarter...of salt, 4 cwt...Devonshire: of Welsh coal or culm, 16 heaped bushels...Derbyshire: of lime at the wharfs, 8 level bushels: at the kilns, 8 heaped bushels...Yorkshire: of chopped bark, in some parts, 9 heaped bushels; ibid 32: Seam or Seem, sometimes a quarter of corn or malt. **1829** Palethorpe sv: QUARTER, a measure of capacity, containing 8 bushels. It is nominal, that is, not a real vessel, or measure, but is used to express a certain quantity of other measures. **1832** Badcock 7: In the middle of the 15th century, importation was not permitted till the price attained 6 s. 8 d. per quarter (equal to about 85 s. in present currency). **1880** Britten 175: Quarter.... (Guernsey and Jersey), of potatoes, 240 lb. **1882** Jackson 413: Quarter = 8 bushels. **1907** Hatch 20: 28 pounds = 1 quarter (qr.) = 448 ounces; ibid 34: 1 quarter = 12.70059 kilograms. **1931** Naft 22: 1 British Quarter...2 coombs...2.9094 hectoliters. **1951** Trade 28: Quarter = 28 pounds. **1966** O´Keefe 670: 1 quarter...2.909 hl; ibid 673: 1 quarter = 12.70 kg.

quartere, quarteren. QUARTERN

quarteria, quarterium. QUARTER

quartern—3-7 quartron (OED); 4 quartroun (OED), quartrun (OED), quaterone (OED); 4-5 quarteroun (Southampton 2), quarton (Nottingham); 5 quarteren (OED), quarterone (OED); 5-8 quarteron; 6 quateren (OED); 6-7 quarterne; 6-9 quartern; 7 L quartronus [ME quarteroun, quartron fr OF quarteron, the fourth part of a pound, or of a hundred; see QUART, QUARTER]. A wt of 28 lb (12.700 kg) for fruit equal to 1/4 PIECE or 1/12 SORT. It was used occasionally as the equivalent of the quarter lb (0.113 kg), or quarter stone (1.587 kg), or quarter pt (see GILL), or quarter pk (see POTTLE), or quarter Cwt (see HUNDRED).—**1439** Southampton 2.65: 16 quarterons fructui. **1474** Cov. Leet 396: xx [X] v for the C, the wich kepes weyght & mesure l li. the halfe C, xxv li the quartern. **1566** Recorde K iii: There be greater waights which are called an Hundred, halfe a hundred, & a quarterne, and also halfe a quarterne. **1600** Hill 66: 14. Poundes...maketh 1. halfe quartern of an C.... 28. Poundes...maketh 1. quar. called also a Tod. **1606** Hall 38: A quarterne is...28 pounde. **1678** Du Cange sv quartronus: Quarta pars libræ, Gall. Quarteron. **1707** Justice 58: The 100, which is divided into Quarterons. **1756** Rolt sv quartern: Or Quarteron. A diminutive of quart, signifying a quarter of a pint, as a quart does a quarter of a gallon. **1829** Palethorpe sv: QUARTERN, a liquid and dry measure; the former containing the 1/4th part of a pint, and the latter the 1/4th part of a peck. **1883** Simmonds sv: Quartern, a name given in London to the gill, the fourth of a pint; also the fourth part of a peck. **1934**

Int. Traders´ 82: Quartern (dry)...United Kingdom...1/4 peck.... Quartern (liquid)...United Kingdom...1/4 pint.... Quartern (stone)...United Kingdom...1/4 stone.

quarterne. QUARTERN

quarternium. QUIRE

quarteron, quarterone, quarteroun, quarton. QUARTERN

quartre. QUARTER

quartron, quartronus, quartroun, quartrun. QUARTERN

quartur. QUARTER

quartus. QUART

quartyr. QUARTER

quateren, quaterone. QUARTERN

quayer, quayere, quayr, quayre, quear, queare, queer, queere, quere, quier. QUIRE

quintal—3-4 L quintallus; 5-6 kyntal (OED), kyntall; 5-9 quintal; 6 kyntayl (OED), quintale (OED); 6-7 kentall (OED), quintall; 6-8 kintall; 6-9 kintal; 7 kental (OED), quintell (OED); 9 kentle (OED) [ME quintal, hundredweight, fr MF quintal fr MedL quintale fr Ar qinṭār]. Equivalent to HUNDRED (weight) and abbreviated q. or ql.—c**1195** Benedict II.204: Et pondus quintalli est pondus centum librarum auri. c**1205** Hoveden III.165: Quintallus est pondus c. librarum. c**1303** Gras 1.161: Quintallus cere...quintallus vermilun. **1545** Rates 1.55: A kyntall of pepper. c**1550** Welsh 131: 5 quintals brass. **1577** D. Gray 38: Quintalles, containyng 100 li. weight. **1607** B. J. 18: Of the Kintall or hundredweight. **1616** Bullokar sv kintall: A certaine weight of about an hundred. **1646** H. Baker 207: Quintal...the 100 li. weight. **1665** Sheppard 17: A Kintall or Quintall, is a certain measure or weight of Wood, Iron, or such like thing or Merchandize, to the value of an hundred, or something over or under, according to the divers uses of sundry Nations. **1701** Hatton 3.228: Kintall...of Fish 100 l. Weight. **1707** Justice 58: The Quintal, making 100, 104, 105, 110, and sometimes 112 Pound, or more, according to the Custom of each Place. **1716** Harris 2. sv weight: Weights of great Content; as Hundreds, Kintals, Centeners, Talents, Thousands, Weighs, Skippounds, Charges, Lispounds, Rooves.... Cantars, Centeners, or Kintals, sometimes wrote Quintals, accounted by Merchants as Hundreds. **1717** Dict. Rus. sv: Quintal or Kintal, an Hundred Pound-weight, at six-score per Cent, of Iron, Lead, or other Metal. **1780** Paucton 825: Hundred ou quintal = 112 liv. **1820** Second Rep. 30: Quintal...properly 100 lbs; sometimes written kintal. **1840** Waterston 147: Cod-fish, quintal...lbs. 112. **1866** Thor. Rogers 1.170: The quintal of iron is probably the same as the hundred-weight. **1895** Donisthorpe 216: QUINTAL...of cheese, in some counties, 120 pounds. **1956** Economist 4: Quintal: (a) Hundredweight...United Kingdom = 112 lb. See CENT

quintale, quintall, quintallus, quintell. QUINTAL

quire—3 cwaer (OED), quaer (OED); 4 L quarternium; 4-5 quayer; 4-6 quayre (OED); 5 quaier (OED), quayere (OED), qvayr (OED), qwayer (OED), qwayre (OED); 5-6 quair (OED), quar (OED), quare (OED), quarr (Finchale), qvare

(OED); 5-7 quaire (OED); 6 quayr (Dur. House), quear (OED), queare, quere (OED), quyr; 6-7 queere (OED), quyre; 6-7, 9 queer; 7 quier; 7-9 quire [ME quaer, quair fr OF quaier, caern (F cahier), a book of loose sheets, fr (assumed) VL quarternum, sheets of paper (usually 4) packed together, fr L quaterni, by fours, fr quater, four times]. A m-q for paper, consisting of either 24 or 25 sheets and equal to 1/20 REAM. It was originally a set of 4 sheets of parchment or paper folded so as to form 8 leaves, and this was the unit most commonly used for medieval mss. Sometimes it was abbreviated qr.—**1392** Henry Derby 159: CLERICO speciarie per manus Wilbram pro ij magnis quarterniis papiri pro officio thesaurarii per ipsum emptis ibidem, ij s. ij d.... Et pro uno quarternio papiri, vj d. c**1440** Promp. Parv. 418: Quayer. Quarternus. **1545** Rates 1.30: Paynted papers the queare. c**1590** Hall 25: Euery reame hathe 20 quyrs of paper; euery quyre hathe 25 sheettes. **1607** Clode 307: For 2 quier of paper. **1616** Hopton 164: A Quire is 25 sheetes. **1635** Dalton 150: A bale of paper, is ten reame; a reame is twenty quires; a quire is 25 sheetes. **1665** Sheppard 18: A Ream is 20 Quire, a quire is 25 sheets. **1682** Hall 31: A Quyre is 25 Sheetes; a Reame, 20 Quyre; a Bale, 10 Reame. **1708** Chamberlayne 205: Of Paper 24 or 25 Sheets to the Quire; 20 Quire to a Ream. **1820** Second Rep. 30: Quire of paper, 24 sheets. **1829** Brockett 239: Queer...a quire, as of paper. **1849** Dinsdale 100: Queer...A quire of paper. **1956** Economist 8: Quire...1/20 ream.

quirren [*]. A m-c for butter in some parts of Ireland (c1800) reckoned equal to a POTTLE totaling 4 lb (1.814 kg) in weight (Kelly 115).

quyr, quyre, qvare. QUIRE

qvarte. QUART

qvayr. QUIRE

qwartur. QUARTER

qwayer, qwayre. QUIRE

qwhart. QUART

R

raff [*]. Equivalent in early eighteenth century to SEAM.—**1707** Justice 3-4: 4 Bushels a Comb, or Cumock [prob an error for *curnock*], 2 Cumocks a Quarter, Seam, or Raff. **1708** Chamberlayne 212: 4 Bushels the Comb or Curnock...2 Curnocks make a Quarter, Seam or Raff. **1717** Dict. Rus. sv dry-measure: Four Bushels the Comb or Curmock, two Curnocks make a Quarter, Seam or Raff.

ras, rasa. RASER

raser—6 razier (OED), raziere (OED); 6-7 ras; 6-? raser (OED); 7 L rasa [MF *rasier*, *rasiere*; cf MF vb *raser* fr (assumed) VL *rasare*, to scrape often]. A m-c for grain, containing approximately 2 bu (*c*7.00 dkl) or 1/4 SEAM. It was a level measure, as opposed to a CANTEL or COMBLE.—**1678** Du Cange sv rasa 2: Rasa, Mensura frumentaria, in agro Dumbensi Ras: ubi plerumque continet quatuor cupas.

razier, raziere. RASER

realme. REAM

ream—4-6 rem; 5 reeme; 5-6 reme, rym (OED); 5-7 reame; 6-7 realme; rim; 6-9 ream; 7 reym (Halyburton), rheme (OED); 7-8 rheam (OED) [ME *rem*, *reme* fr MF *raime* fr Ar *rizmah*, a bale or bundle]. A m-q for paper, consisting of 20 QUIRES of 24 or 25 sheets each and equal to 1/10 BALE. A "printer´s ream," however, is commonly 21 1/2 quires or 516 sheets, while a "stationer´s ream" is just 504 sheets. Occasionally it is abbreviated *rm*.—**1392** Henry Derby 154: Et pro j rem papiri. **1411** trans in Cal. Close 20.148: One ´reme´ of paper. **1439** Southampton 2.108: 12 remys papiri pro wastyng. *c***1440** Promp. Parv. 429: Reeme, paper. **1507** Gras 1.701: Paper called wyte the reme. **1509** Ibid 573: xx remys papiri. **1524** Ibid 196: Pro xxx reames paper. **1545** Rates 1.30: Paynted papers the realme. c**1590** Hall 25: Euery reame hath 20 quyrs of paper. **1607** Clode 313: For a realme of capp paper. c**1610** Lingelbach 113: White paper by the twelve Realmes. **1612** Halyburton 323: Broun paper the bundle contening tuo rim...cap paper the rim. **1616** Hopton 164: A Bale of Paper is 10 Reame or 200 Quires, a Reame is 20 Quires, or 500 sheetes: a Quire is 25 sheetes. **1635** Dalton 150: A reame is twenty quires. **1665** Sheppard 18: A Bale of Paper is 10 Ream, a Ream is 20 Quire, a quire is 25 sheets. **1708** Chamberlayne 205: Of Paper 24 or 25 Sheets to the Quire; 20 Quire to a Ream; 10 Ream to a Bale. **1820** Second Rep. 30: Ream of paper, 20 quires. **1883** Simmonds sv: Ream, a package of paper containing 20 quires. **1956** Economist 8: Ream, Printers...516 sheets.... Ream "Stationer"...504 sheets.

reame. REAM

reda. ROOD

reel [ME reel fr OE hrēol]. A m-l for thread and yarn (c1700-1900): Clydesdale, 2 1/2 yd (2.286 m); Essex, wool, 1 1/4 and 1 1/2 yd (1.143 and 1.372 m); and Hampshire, flax, 2 yd (1.829 m) (Acts Scotland 9.311, Second Rep. 30, and Britten 175).

reeme. REAM

rees [prob fr E dial vb ree, sift, fr ME reien]. A m-q for herrings, consisting of 15 GLEANS or 375 in number.—**1805** Macpherson I.471: 25 herrings 1 glen, 15 glens 1 rees. **1820** Second Rep. 30: Rees of herrings, 15 gleans = 375. **1895** Donisthorpe 216: REES; of herrings, 15 glenes = 375.

rem, reme. REAM

reode [*]. A m-c for wine containing 2 TUNS or 500 gal (c18.92 hl).—**c1461** Hall 15: Off the mesure of Lycoure. There is a mesure of wyne whyche is called a reode it cont[aineth] ii tunnys, that is v [X] c galons.

reym. REAM

rhandir [W rhandir, share land (Laws Wales)]. A m-a for land in Wales (c1300) containing 4 TYDDYNS or 16 ERWS (c5.78 ha) (Laws Wales 1004).

rhaw [*]. A m-c for peat in Wales (c1800-1900). It was a pile or heap containing 120 or 140 cu yd (91.747 or 107.038 cu m) (Second Rep. 30 and Donisthorpe 216).

rheam, rheme. REAM

ridge [ME rigge fr OE hrycg; akin to rig, the space between the furrows of a plowed field]. A m-l for land in Wales containing 3 LEAPS or 20 1/4 ft (6.176 m). In England it was a m-a similar to BUTT OF LAND, RIG, and SELION.—**1639** Gray 454: Two yardlands of glebe lands contayning in number Three score and one ridges or lands arable. **1664** Spelman 488: Sive ab eo quòd Angli hodie dicimus a rig, or ridge of land. **1665** Sheppard 22: A Selion...otherwise called a Ridge of Land. **1688** Holme ii: 3 Ridges, Butts, Flats, Stitches or small Butts, Pikes. **1820** Second Rep. 30: Ridge of land, Wales, formerly 20 1/4 feet, or 3 leaps.

rig—6-9 rig; 7 rigg [ME (northern dial) rig, back, ridge, fr OE hrycg, ridge]. A m-a for land in northern England and Scotland, synonymous with the RIDGE, SELION, or BUTT OF LAND, being the strip of ground or the pathway between two parallel furrows of the open field.—**1664** Spelman 488: A rig, or ridge of land. **1681** Acts Scotland 8.295: Three riggs of land lyand contigue in the field called the said Ryebank of Rosmarkie.... Two Riggs of land thereof lyand contigue in the field called the Gallowbank.... That other rigg or butt of land of the samen lyand in the ffield called the Gallowbank. **1888** Taylor 180: The breadth of the rig or sellion to be ploughed.

rigg. RIG

rim. REAM

ring—3-? ring; 3-4 L ringa; 4 L rynga [ME ring fr OE hring; prob referred to a band around the rim of the measure]. A m-c at Ramsey and Elton containing 1/2 bu (cl.59 hl) and equal to 1/18 SEAM of 9 bu.—**1297** Neilson 3: Idem reddit compotum de v s. iii d. de iiii ring. ii bu. avenae; ibid 5: Idem computat in vi ring. fabarum emptis ad opus celerarii, xiiii s.: pretium ringæ, xxviii d; ibid 10: In missione apud Rameseiam xviii ring. de novo grano quæ fecerunt i quartarium. **1297** Elton 82: Et de ix s. iiij. de iij ringis ij busellis tolcor´ venditis inter Pascham et Gulam augusti. **1324** Ibid 282: Et sic de incremento j Ringa ij buselli; ibid 283: Item preposito de Weston´ vj Ryngas fabarum et pisorum per talliam. c**1325** Rameseia III.158: Memorandum, quod octodecim communes ringae faciunt unam magnam quarteriam.... Et quatuor communes ringæ, duo busselli, faciunt mittam gruti.

ringa. RING

roale. ROLL

rod—1-6 rodd; 5-7 rodde; 5-9 rod; 7, 9 rood [ME rod fr OE rodd; akin to ON rudda, club]. Equivalent to PERCH and occasionally abbreviated r. or rd.—**1474** Cov. Leet 397: And out of the seid yard growith a Rodde to mesure land by, the wich Rod conteyneth in lengthe V yardes & halfe. **1502** Arnold 173: In dyuers odur placis...they mete ground by pollis gaddis and roddis some be of xviij foote some of xx fote and som xvi fote in lengith. **1607** Cowell sv perche: A Rodde or Pole of 16. foote and a halfe in length. **1608** Stevin D2: Call the Pearch or Rood. **1638** Bolton 274: Five yards and a halfe...make the pole, Rood or peach [sic]. **1651** Jager 65: Eight furlongs, or 320 Rods. **1682** Hall 28: A Pearch, or a Rod, or a Pole (by statut) must be 5 yards and an half; or 16 feete and an half. **1696** Phillips sv pole: In measuring, it is the same with Pearch or Rod, or as some call it Lugg: By Stat. 35 Eliz. this Measure is a length of 16 Foot and a half, but in some Countries [= districts] it consists of 18 Foot and is called Woodland-Measure; in some Places of 21 Foot termed Church-Measure; and in others of 24 Foot under the Name of Forest-Measure. **1708** Chamberlayne 207: 16 Foot and a half make a Perch, Pole or Rod. **1717** Dict. Rus. sv perch: Perch or Pearch, a Rod or Pole, with which Land is measur´d. **1789** Hawney 213: But in some Places the Custom is to allow 18 Feet to the Rod...and in some Places...21 Feet. **1819** Cyclopædia sv weights: In the Lancashire

Report on Agriculture...the rod is of no less than six different lengths in different parts of the county; namely, the statute, or 5 1/2 yards, 6, 6 1/2, 7, 7 1/2, and 8 yards. **1820** Second Rep. 26: Perch, Pole or Rod. **1830** Crüger 157: 1 Pole, Rod, Rood, Lug oder Perch 5 1/2 Yard. **1832** Edinburgh XII.569: 5 1/2 Yards = 1 Pole or Rood = 5.0291 [m]. **1880** Britten 175: Rood...often provincially used for rod. (Ches.), of hedging, 8 yards.... (Cumb.), 7 yards. (Derb.), of bark, seems to be a pile 7 yards in length; of draining or fencing, 7 or 8 yards.... (Durh.), of wall-building, 7 yards. (Norf.), 21 feet. (Nhumb.), 7 yards. (Sal. and Staff.), of hedging, 8 yards.... (Yks.), in the moorlands, of fencing, 7 yards. (Wales), of ditching, draining, and hedging, 8 yards. (Berwicksh.), of labourers´ work, 6 or 7 yards.... (Dumfriessh.), of draining, 19 feet.... (Wigtonsh.), lineal, 20 feet. (Fifesh.), of fencing, 6 yards. (W. Lothian), of draining, 6 yards. **1888** Fr. Clarke 36: 5 1/2 yards make 1 rod, perch, or pole. **1897** Maitland 374: Then as to rods...in Hertfordshire, 20 feet; in Wiltshire, 15 or 16 1/2 or 18; ibid 375: There is much evidence that rods of 20 and 21 feet were often used in Yorkshire and Derbyshire. Rods of 18, 19 1/2, 21, 22 1/2 and 24 feet were known in Lancashire. A writer of the thirteenth century speaks as if rods of 16, 18, 20, 22 and 24 feet were in common use. **1909** Curtler 3: The rod...from 12 to 24 feet. **1956** Economist 7: 5 1/2 yards = 1 rod, pole or perch. **1966** O´Keefe 667: 1 rod...5.0292 m.

rod, roda. ROOD

rodd, rodde. ROD; ROOD

rode, roed, roide. ROOD

rol, role. ROLL

roll—3-7 rolle; 4-9 roll; 5-7 rol (OED), rowle; 6 row (OED), rowlle; 6-7 roole (OED), roule, roull; 6-8 rowl (OED); 6-9 role; 7 roale; 7-8 roul (OED) [ME rolle fr OF rolle fr VL rotulus, a roll, fr L rotulus, dim of L rota, a wheel]. A m-q for parchment, consisting of 60 skins, and a wt for butter, 24 avdp oz (680.388 g). Occasionally it was used in place of the PIECE as a measure for cloth.—**1507** Gras 1.699: Harffordes the rowle; ibid 700: Mynster´ clothe the rowlle. **1509** Ibid 578: x rolles cours canvas. **1545** Rates 1.3: Bokeram the rowle; ibid 20: Hannouers the roule conteyning .vi C. elles. c**1590** Hall 25: The parchement rowle is 5 dossen, conteninge 60 skynns. **1612** Halyburton 292: Buckram of the eist cuntrey the roull or half peice. **1616** Hopton 164: A Rowle of parchment is 5 dozen, or 60 skins. **1628** Hunt C: [5] Dozen in a Role of Parchment.... [60] Skinns is a Roale of Parchment. **1635** Dalton 150: A roule of parchment is five dozen, or sixtie skins. **1656** Rawlyns 70: A roll of Parchment contains...Dozens 5. **1665** Sheppard 18: A Roll of Parchment is 5 dozen or 60 skins. **1708** Chamberlayne 205: Of Parchment, Twelve Skins make a Dozen; and five Dozen a Roll. **1717** Dict. Rus. sv: Roll of Parchment...is the quantity of 60 Skins. **1820** Second Rep. 30: Role of parchment, 72 [sic] sheets. **1883** Simmonds sv: Roll...5 dozen skins. **1956** Economist 53: Butter...1 roll = 24 ounces.

rolle. ROLL

rondelet, rondelett, rondellettum, rondlet, ronelet, ronelete, ronlet, ronlett. RUNDLET

rood—1-6, 8 rod; 1-7 L roda; 3-6 rode; 4-7 roode; 5 roed (OED), rowd (OED), rude, rwd (OED); 5-6 L reda (Finchale), rud (OED); 6 rodde, roide (OED), roud (OED); 6-7 rodd, ruid; 6-9 rood [ME rod, roode fr OE rōd, a cross, measure of land, rod, pole]. A m-l containing 660 ft (2.012 hm) and equal to 1/8 mi of 5280 ft (see FURLONG), and a m-a containing 40 sq PERCHES (0.101 ha) and equal to 1/4 statute acre of 160 sq perches (see FARTHINGDALE); this latter rood was commonly called a "quarter acre." The Scots rood contained 40 sq FALLS or 1440 sq ELLS and was equal to 13,838.4 English sq ft (1285.587 sq m) or 0.3177 English acres (Huntar 7 and Swinton 26). In Ireland the rood contained 40 sq perches or 1960 sq yd or 17,640 sq ft (1638.756 sq m). In the superficial measurement of stone, brick, or slate work, 36 sq yd (30.100 sq m) made a rood (Britannica 808). Occasionally it is abbreviated r. or ro.—c**1065** St. Edmunds 32: Ærdman II acras et I rodam. **1198** Feet 3.65: Et pro hoc fine et concordia et quieta clamantia...predictus Radulfus dedit predicto Willelmo iiij acras terre et iij rodas. **1200** Ibid 107: Et iij rodas juxta Culuerdespit. **1201** Feet 2.12: Et dimidiam acram et decem rodes in Mikelholm´ et dimidiam acram in Quakefen. **1202** Feet 3.196: Et dimidiam rodam prati juxta domum ipsius Simonis uersus orientem. c**1260** Clark 100: Ivo Belamy pro una roda. **1278** Gray 459: In Estfeld quater viginti acre et tres rode. c**1289** Bray 8: Julia et

Matilda Burgeis x acras i rodam; ibid 9: Nicholas Pewere unum mesuagium et xxxviij acras et dimidiam et unam rodam. **1300** Elton 92: Reginaldas Kayston emit tres acras et tres rodas terre de diuersis hominibus. **1312** Ibid 189: Vna virgata terre existentis iiij acrarum et j rode. c**1400** Acts Scotland 1.387: The aker sall contene four rude/the rude .xl. fallis The fall sall hald .vj. ellis. c**1400** Henley 68: E devet sauer ke lacre ke est mesuree par la verge de xviii peez fet i acre & vne rode. **1409** Gray 361: Una roda terre vocata Shamelondesbutte. c**1440** Promp. Parv. 435: Rode, of londe. Roda. c**1461** Hall 14: The roode of grownde. c**1475** Nicholson 77: Fourty perchys in lengyth makyth a Rode of Lande; put iiij thereto in brede, and that makyth an Acre. **1505** Davenport lxxx: Item, I beqweth to Margery my belchelde whan she is of lawfull age ii. acres and a rod londe lyynge in Watkers Feld. **1537** Benese 4: The quarter of an acre (other wayes called a roode) conteyneth in it xl. perches. **1540** Recorde 208: A Rod of land which some call a roode, some a yard lande, and some a farthingdale. **1566** Ibid Kv: A Rodde of lande. **1577** Val. Leigh 91: Tenne daye workes or fourtie Pearches maketh—a rode or quart of an Acre. **1624** Huntar 7: 5. Yardes and a halfe maketh a pearch. 40. Pearches are a Rood. 4. Roodes are an Acre. **1647** Digges 1: So an Acre by Statute ought to containe 160. Pearches...a Roode, commonly called a quarter, 40. Pearches. **1653** Leybourn 1.248: Every Rood of Land 40 square Perches. **1664** Spelman 453: Alias roda dicta, quod vide, Anglicè a Rood, Scotice ane Ruid of Land; ibid 489: Roda terræ.... Vox agrimensorum, quartam acræ partem designans; Rodd enim Anglis est pertica. **1665** Sheppard 19: Particata terræ...is a Rood of Land.... A Rood of Land...is a certain quantity of Land, the fourth part of an acre. **1678** Du Cange sv roda: Anglis, Quarta pars acræ, quæ et Farding deale, seu Farundel dicitur, juxta Cowellum, ex Anglico Rodd, Pertica. Continet autem acra, secundum stadii longitudinem 40. rodas, seu perticas; in latitudine tantum quatuor. Perinde etiam Roda terræ 40. perticas in longitudine, unam vero solummodo in latitudine. **1699** Hatton 1.22: 1 Pearch in breadth, 40 in length, do make a Rodd of Land, which some call a Rood. **1701** Ibid 3.11: 40 Poll long, & 1 broad...1 Rod of Land or Qr. of an Acre. 4 Square Rods...1 Acre. **1717** Dict. Rus. sv: Rood, a Measure being the fourth part of an Acre, and containing 40 Square Pearches or Poles. **1798** Cullyer viii: Forty of which Perches make one Rood. **1816** Kelly 95: 36 Square Ells...1 Square Fall. 40 Square Falls...1 Square Rood. 4 Roods...1 Acre. **1820** Second Rep. 30: Rood of land, properly 1/4 acre = 40 perches = 1.210 sq. yards; but the term is often provincially used for rod, or a measure approaching to it. **1888** Fr. Clarke 36: 4 roods make 1 acre, or 43,560 square feet. **1907** Hatch 37: 1 rood (40 perches) = 10.117 ares. **1951** Trade 28: Rood = 1,210 square yards.

rood. ROD

roode. ROOD

rook [a var of RUCK]. A m-q in Yorkshire (c1800-1900) for 4 bean sheaves set up to dry in a field. It was also known as a ruckle of beans

(Britten 156).

roole. ROLL

rope [ME rope, rap fr OE rāp]. A m-q for onions and garlic. The tops of 15 heads, or 1/15 C of 225, were braided together, giving the appearance of a rope. In Somersetshire it was a m-l of 20 ft (6.096 m) for wall-building.—c**1590** Hall 28: Nottes of the 100 of Onyons and Garleke. The Hundred consisteth of 15 ropes and euery rope 15 heades. **1660** Bridges 30: A Rope is 15 Heads, and every hundred 15 Ropes. **1665** Sheppard 58: A hundred of Garleck consisteth of 15 Ropes, and every rope containeth 15 heads. **1816** Kelly 86: A Rope in some kinds of measurement is reckoned 20 Feet. **1855** Jessop 14: The rope = 20 feet. **1880** Britten 175: Rope.... (Som.), of wall-building, 20 feet in length. **1883** Simmonds sv: A row of things tied together, as a rope of onions. **1934** Int. Traders´ 83: Rope...United Kingdom...20 feet.

roud. ROOD

roul [perh a special use of ROLL]. A m-q for eels (c1800-1900), numbering 1500 (Second Rep. 31 and Donisthorpe 217).

roul, roule, roull. ROLL

roundelettus, roundellettus, roundlet. RUNDLET

row. ROLL

rowd. ROOD

rowl, rowle, rowlle. ROLL

ruck [ME ruke, roke, of Scand origin; akin to Nor dial rūka, heap, ON hraukr, rick]. A m-q for bark in some parts of Derbyshire (c1800) consisting of 5 1/4 cu yd (4.014 cu m), stacked (Second Rep. 31).

rud, rude, ruid. ROOD

rundelet, rundellus. RUNDLET

rundlet—3-4 L roundellettus (Liber), L rundellus (Liber); 4-6 rondelet; 5 rondelett, L rondellettum (Southampton 2), ronlett (OED); 5-6 ronelet (Nottingham); 6 ronelete (OED), rundelet; 6-7 rondlet, roundlet, runlett; 6-8 rundlett; 6-9 rundlet; 7 ronlet (OED), L roundelettus; 7-9 runlet [ME rondelet fr MF rondelet, dim of rondel fr OF rondel, rondelle, a little tun, fr ronde, round]. A m-c for wine generally containing 18 or 18 1/2 gal (c6.81 or c7.00 dkl) and generally equal to 1/14 TUN. When used for products other than wine, it was synonymous with the KILDERKIN. The Irish rundlet (c1800) contained 3916.8 cu inches (6.420 dkl) or 18 Irish gal of 217.6 cu inches (Edinburgh XII.572).—**1420** Gras 1.499: Pro viii rondeletts saponis albi. c**1550** 294: 3 rondlets muskedine. **1566** Recorde Kiiij: Of wine and oyle the Rondelet holdeth 18 1/2 Gallons. c**1590** Hall 21: The Rundelet, which is 1/14 part of a tunne, contenith 18 galons 1/2. **1607** Cowell sv roundlet: Roundlet, is a certaine measure of wine, oyle, &c. containing 18. gallons and a halfe. **1607** Clode 307: For the 2 Runletts. **1615** Collect. Stat. 467: Euerie Rundlet to contain eighteen gallons and an halfe. **1619** Young II.152: 2 rundletts of clarett. **1635** Dalton 144: Two Firkins maketh the Kilderkin...halfe Barrell...[or] Rondlet. **1665** Sheppard 59: The Barrell 31 Gallons and a half, and the Rundlet 18 Gallons and a half. **1678** Du Cange sv roundelettus: Mensura liquidorum...continet decem et octo galones cum dimidio. **1682** Hall 29: 1 Tunne conteynes...14 Rundlets, 252 Gallons. **1696** Jeake 72: 1 Rundlet or Rondlet = 18 1/2 Gallons. **1701** Hatton 3.233: Rundlett...An uncertain Quantity of Liquids from 3 to 20 Gallon. **1708** Chamberlayne 210: Of these Gallons, a Runlet of Wine holds 18. **1717** Dict. Rus. sv: Rundlet or Runlet...of Wine is to hold 18 Gallons. **1790** Jefferson 1.983: Two firkins, or bushels, make a measure called a rundlet or Kilderkin. **1820** Second Rep. 31: Rundlet or Runlet of wine, 18 gallons. **1851** H. Taylor 58: Besides these, are various denominations of casks, chiefly employed for wine and spirits, as kegs, ankers, runlets, tierces, pipes, and tuns.

rundlett, runlet, runlett. RUNDLET

ruskey, ruskie. RUSKY

rusky—8-9 ruskey, ruskie (OED), rusky (OED) [Gael rusgan]. A m-c of no standard dimensions for corn (c1800-1900) in Scotland. It was a basket made of twigs and straw (Britten 157).

rwd. ROOD

rym. REAM

S

saac, sac, sacc, sacca, sacchus, saccke, saccum, saccus, sache.

SACK

sack—1 sæcc (OED); 1, 4 saac; 1, 5-6 sacc; 3-4 L sacca, L saccum, sec (OED), seck (OED); 3-5 secke (OED); 3-5, 8 sac; 3-6 sakke; 3-7 L saccus, sacke; 4-5 sak, sekke (OED); 4-6 sek; 4-9 sack; 5 cek (OED), saccke (OED), sache (OED), L saculus, sakk; 5-6 sake; 7 L sacchus [ME sac, sak, sack, bag, sackcloth, fr OE sacc, sæcc fr L saccus, sack, bag, fr Gr sakkos, sack, bag, of Sem origin]. A m-c for dry products: apples, Kent, 3 1/2 bu (cl.23 hl), Worcestershire, 4 bu (cl.41 hl); ashes, Hertfordshire, 5 bu (cl.76 hl); charcoal, 4 bu (cl.41 hl), except in Essex, 8 pk (c7.05 dkl); cloves, mace, or nutmegs, 300 lb (136.077 kg); coal, various, from 1 Cwt to 1 TON or more; flour, generally 5 bu weighing 2 1/2 Cwt or 280 lb (127.005 kg); grain, generally 4 heaped bu (cl.80 hl); hemp, 3 Cwt (152.406 kg); meal, 5 bu totaling 2 1/2 Cwt or 280 lb (127.005 kg); potatoes, 168 to 280 lb (76.203 to 127.005 kg); salt, 5 bu (cl.76 hl); sheep skins, Scotland, 500; wheat, North Wales, 1 1/2 HOBEDS totaling 260 lb (117.933 kg); and shorn wool, generally 364 lb (165.107 kg), or 2 WEYS or 13 TODS or 52 CLOVES or 26 STONE of 14 lb each equal to 1/12 LAST, but occasionally 350 lb (158.756 kg) or 28 stone of 12 1/2 lb each. When woolfells were exported, 240 skins or fells generally were considered equivalent to a sack of shorn wool. Occasionally it was abbreviated sk.—**1200** Cur. Reg. 8.144: Willelmus filius Roberti optulit se iiij. die versus priorissam de Svine de placito x. saccorum lane et de x. marcis argenti que ei debet ut dicit.

1228 Gras 1.156: 1 sacca lani. **1229** Close 1.260: Quod unam lestam coriorum et dimidiam et tres saccos lane. c**1243** Select Cases 3.lxxxvi: In nauta vero fuerunt iiij [X] xx sacci lane. **1249** Gross II.359: Et de quolibet sacco lane duos denarios. c**1253** Hall 11: Et xii sacs sunt un last. c**1272** Ibid 10: Et due waye faciunt unum saccum. Et duodecim sacci continent le last. **1275** Gras 1.225: xl sackes et l poke de laine. **1290** Fleta 119: Et due waye lane faciunt vnum saccum, et xij. sacca faciunt vnum lestum.... Et tales xij libre xxviij petre faciunt vnum saccum lane. c**1300** Hall 8: Duodecim libre et dimidia faciunt petram Londonie. Saccus lane debet ponderare viginti octo petras. **1311** Neilson 35: In xii ulnis canobi pro sackis, xix d. c**1330** Gross II.229: Dautre part pour un gros sak de leine. c**1340** Pegolotti 254: Lana si vende in Londra e per tutta l´isola d´Inghilterra a sacco, di chiovi 52 pesi per l sacco, e ogni chiovo pesa libbre 7 d´Inghilterra. **1341** Rot. Parl. 2.133: Primes, ce q´est coilly & leve de les xxM. saks de Leyne autrefoitz grantez a notre Seign´ le Roi en maner d´avoir recompensation de ycele de la Noesisme de l´an secounde, soit recoupe & allowe es Countees ou les Leines sont levees, & les persones paiez, & les Commissions de xxM. sacks repelles. **1343** Ibid 142: La pere serroit de XIIII li. & XXVI petr´ facent un sak. **1350** Ibid 230: Et fount les Custumers de ccc un sac de Leyne. **1351** Ibid 240: Le Sak ne poise que vint & sys pieres, & chescune pere poise xiiii livres. c**1360** Hale 136: Merchant estrange mesne leyne outre le mere payer per un saac que tient 2 peises 6 d. et pur cockett 2. **1389** Rot. Parl. 3.272:

Qatorsze livers al Pere, & vint & sis Peres al Sak. **1439** Southampton 2.28: 4 sakk´ de hoppys; ibid 76: 1 saculo de haberdasshe; ibid 85: Pro 1 saculo amygdolarum; ibid 86: 1 saculo grani pro panno. **1443** Brokage II.174: 1 parvo sacco piperis. c**1460** Capgrave 222: Of a sak wolle. c**1461** Hall 13: Also woll is weyd by this weyght [avoirdupois], butt itt is nott rekynnyd soo, for ytt is bowght odyr by the Nayle, or the Stone, or the Todde, or els the Sakk; ibid 16: A sakk, sarpler, poke, last [of] Woll; ibid 19: Woll is bowght and sold...by the Sakke.... That ys to say...Sacke content´ iii [X] c [+] lxiiij. c**1475** Gras 1.192: Of a sak wulle. **1478** Stonor II.62: xix marcs le sacc. **1478** Ricart 84: Item, that all maner of colyers that bryngeth colys to towne for to sille, smale or grete, that they bryng their sakkes of juste measure...so that every sak be tryed and provid to be and holde a carnok, and the ij. sakkes to hold a quarter. **1507** Gras 1.698: Flexe hyckeled the sake. **1565** Rich 147: And it is further ordeined that yf any pockett to be wayed at the beame...shalle excede the waight of one sacc iiii naile. **1587** Stat. 116: So that the sacke of woll wey no more but xxvi. stones, and euerie stone to wey fourteene pound. c**1590** Hall 31: A Last is 12 Sackes; a Sack, 2 Weyes; a Weye, 6 1/2 Toddes; a Todd, 2 Stone; a Stone, 14 pound; a Cleave, half a stone. **1594** Rates 2.11: Cullen hemp or other hemp the sacke containing iiiC...the C. containing v [X] xx [+] xii li; ibid 19: Hemp the sack containing iii c. weight. **1595** Powell F2: Euery sacke of charre Coales, must conteine and holde foure bushels of good and cleane coales. **1597**

Halyburton cxvi: Ilk sek of scheip skynnes contenand v [X] c. **1615** Collect. Stat. 465: And two weights of wooll make a sacke, and twelue sackes make a last. **1616** Hopton 163: The Sack of Coles is 4 bushels. **1635** Dalton 149: Wooll, 14 pound weight goeth to the stone of wooll, 28 pounds goeth to the Tod, and 26 stone goeth to the sacke. **1657** Tower 41: It is enacted, That a Stone of Wooll shall contain but fourteen pounds; and that twenty-six Stone make a Sack. **1660** Bridges 29: A full Sack of Charcoal should bee 4 Bushels. **1665** Sheppard 17: Of Wooll, 12 sacks are said to make a Last; ibid 64: A Sack of Wooll. (Sacchus Lanæ...) is a quantity of wooll that containeth 26 stone, and a stone 14 pounds. **1678** Du Cange sv saccus: Ponderis lanarii species. Constat autem 28. petris, petra vero 12. libris et dimidia. **1682** Hall 30: Coales must conteyn in every sacke, 4 bushels. **1708** Chamberlayne 207: Wooll is weigh´d by the...Sack, 364 Pounds. **1717** Dict. Rus. sv: Sack...of Sheeps-woll, 26 Stone, every Stone containing 14 Pounds, but in Scotland 24 Stone, and each Stone 16 Pounds. **1750** Reynardson 13: The Sack of Wool...was to weigh but 26 Stone, 14 Pounds to each Stone. **1778** Diderot XXVI.422: Les marchands de laine ont aussi leurs especes de poids particulieres; le sac...le tod...toutes mesures angloises sans termes françois. **1779** Swinton 37: 20 stones = Sack of Flour. **1820** J. Sheppard 84: A sack [of flour], or five bushels, is 280 lb. **1820** Second Rep. 31: Sack...of flour or meal, 280 lbs...of salt, 5 bushels...Essex: of charcoal: 8 pecks...Hertfordshire: of ashes, 5 bushels...Kent: of apples and potatoes, about 3 1/2 bushels...Somersetshire: of potatoes, 240 lb...Surrey: of potatoes, 3 bushels of 60 lb. each...Worcestershire: of apples, 4 bushels...N. Wales: of wheat, 1 1/2 hobaid, to weigh 260 lb. **1834** Pasley 114: 1 Sack of Potatoes in Surrey...180 [lb].... 1 Sack of Meal or Flour, legal (2 1/2 cwt.)...280 [lb].... 1 Sack of Cloves, Mace or Nutmegs, neat weight, legal...300 [lb]. **1850** Alexander 100: Sack for wool...364.—pounds. **1956** Economist 50: Sack: Flour and meal = 280 lb.

sacke, saculus, sæcc. SACK

sæm. SEAM

sak, sake, sakk, sakke. SACK

sarpelar, sarpeler, sarpelere, sarpelerium, sarpeller, sarplair, sarplar, sarplare, sarplarius. SARPLER

sarpler—3 L sarpelerium, sarpeller; 4 sarpuler (OED); 4-9 sarplar, sarpler; 5 sarpelar (OED), sarpeler (OED), sarpelere (OED), sarplair (OED), sarplere, sarpleth (OED), serplar; 5, 7 sarpliar; 5-7 sarplare; 6 sarplier (OED); 7 L sarplarius, serplaith, serplath, serplathe, serpliathe, sirplithe; 8 serpler; ? L sarplera (Prior), serpliath (Prior) [ME sarpler fr MF sarpilliere]. A m-c for wool. It was a large, coarse canvas bag generally equal to 2 SACKS, totaling 728 lb (330.213 kg) or 1/6 of a 4368 lb LAST. However, there were variations ranging from 1/2 sack to more than 2 sacks.—**1208** Bish. Winch. 6: In sarpeleriis ad lanam ponderandam, vj s. xj d. In saccis, iij s. **1275** Gras 1.227: vi saches de laine en vi sarpellers apaie. **1350** trans in

Cal. Close 8.222: Also that no sarplar shall contain more than 1 1/2 sacks. **1397** trans in ibid 16.38: Robert de Howom paid custom at Kyngeston for 4 sacks 14 stone and 1 clove of wool in two sarplers...6 sacks 17 stone in three sarplers...9 sacks 25 stone 1 clove in five sarplers. **1457** Acts Scotland 2.49: Thre serplares of his awne gudes. **c1461** Hall 16: Also Woll ys sold by numbre and schipped to, as by sacks, sarplers, and pokys. ii sacks make a sarpler, and x [sic] sarplers make a laste; ibid 19: That ys to say...Sarpler content´ ii Sackes.... The Sarplere ys made off Sackes. **1478** Stonor II.62: iiij serplar ffynne Cottes wolle ffor xix marcs le sacc. **1565** Rich 149: v [X] c felles smalle tale rekened for a sarpler. **1597** Halyburton cxv: And until mair perfytt knawledge be haid of the iust quantitie of the serplaith, twa tun of fraucht to be comptit to the sek, and twa sek fraucht to the serplaith. **1607** Cowell sv: Sarpler (Sarplera lanæ) is a quantitie of woll. This in Scotland is called Serplathe, and conteineth fourescore stone, for the Lords in the counsell in anno 1527. decreed foure serpliathes of packed wolle to containe 16. score stone of woll. **1624** Huntar 3: The Sirplithe of goodes, which is the common fraughting of Marchandice betwixt this Countrey [Scotland], and the Easterne Countreyes, is esteemed, to wey 80. stone weight, or 1280 pund weight. **1664** Spelman 513: Serplath, & Serplaith...Sarcina apud Scotos petras 80. continens. **1665** Sheppard 64: A Cark of Wooll is said to be a quantity, whereof 30 make a Sarplar.... A Sarplar...is a quantity of Wooll, and seems to be all one with a Weigh of Wooll..... A Sarplar (otherwise called a Pocket) is a half Sack. **1678** Du Cange sv sarplare: Sarplarius, Ponderis lanarii species sacco major, dicitur, quod lanis involvendis sarpilleriis statutæ mensuræ utuntur præcipue apud Anglos. **1701** Hatton 3.233: Sarpliar.... A piece of Canvas to warp [sic] Wares in. **1717** Dict. Rus. sv cark: A certain Quantity of Wooll, the thirtieth part of a Sarplar. **1787** Hale 153: A pockett of wooll contained half a sack [sic], and so did a serpler. **1829** Palethorpe sv sarplar: SARPLAR OF WOOL, a quantity of wool, otherwise called a pocket or half-sack; and contains 11 stone of wool, at 14 lbs. to the stone. **1883** Simmonds sv sarplar: A large bale or package of wool, containing 80 tods, or a ton in weight.

sarplera, sarplere, sarpleth, sarpliar, sarplier, sarpuler. SARPLER

scain, scan, scane. SKEIN

scape, scappe. SKEP

scayne. SKEIN

sceaftmund. SHAFTMENT

scep, scepe, scepp, sceppa, sceppe. SKEP

schaffa. SHEAF

schafftmon, schaftemonde, schaftmon, schaftmond, schaftmonde, schaftmone, schaftmonthe, schaftmount. SHAFTMENT

schaine. SKEIN

schaldre. CHALDER

scheef, schef, schefe, scheff, scheffe. SHEAF

schepe, schepp. SKEP

schide. SHIDE

schiefe. SHEAF

schock, schocke, schokke. SHOCK

schopin. CHOPPIN

schore. SCORE

schudde, schyd, schydd, schyde, scid. SHIDE

scoare, scoir, scoore. SCORE

scope [prob fr SKEP]. Equivalent in size and application (c1400) to CORF (Salzman 1.15).

scor. SCORE

score—3-6 scor (OED); 3-9 score; 4-5 schore (OED); 4-6 skor (OED); 4-7 skore; 5 scoyr (OED), skowre (OED); 5-7 scoure; 6 scoore (OED), scower (OED), skoir (OED); 7 scoare (OED), scoir [ME scor fr ON skor, notch, tally]. A m-q generally numbering 20 of any item, but there were exceptions: barley, beans, and oats, Liverpool, 21 bu (c7.40 hl); coal, Newcastle, 21 CHALDERS (124,656 lb or 56,542.710 kg); grain, Roxburghshire and Selkirkshire, 21 BOLLS (c58.59 hl); lime, Derbyshire, 20 to 22 heaped bu (c9.01 to c9.91 hl); and sheep, Dumbartonshire, 21 in number.—**1440** Palladius 48: Ffeet scores nyne in length. c**1460** Capgrave 13: A hundred IIII. score and VIII; ibid 45: Foure score thousand and fyve thousand. **1562** York Mer. 168: Iron sex skores endes to the tonne. **1563** Acts Scotland 2.540: Ane thousand fyue hundreth thre scoir sax yeiris. **1577** Val. Leigh 91-92: So an acre containeth

Clx. perches, halfe an acre foure scoure Perches. **1616** Hopton 164: Coney, Kid, Lambe, Budge...haue fiue score in the hundred. **1635** Dalton 149: Six score herrings shall goe to the hundred. **1682** Hall 29: A skore is 20 yards. **1704** Mer. Adven. 244: Ffor a score of round letts or great ffish; ibid 245: Ffor sorting and laying up every score of round wood belonging to a ffreeman or fforreigner. **1708** Chamberlayne 213: On Shipboard they allow 21 Chaldron to the Score. **1717** Dict. Rus. sv timber: Other Skins six score to the Hundred. **1819** Cooke 72: Scores of 20 lb. **1820** Second Rep. 31: Score...Derbyshire: of lime, 20 to 22 heaped bushels...Liverpool: of barley, beans and oats, 21 bushels...Newcastle: of Chaldrons of coals, 21...Dumbartonshire: of sheep, sometimes 21...Roxburghshire and Selkirkshire: of bolls of grain, sometimes 21. **1847** Speed 3: Weight...18 score 12 lbs. **1854** Bowring 2: The groupings in scores...two tens...is a common mode of representing numbers. **1888** Jolly 32: The Live and Dead Weight in Imperial Stones...in Smithfield Stones...in Cwts...and in Scores. **1956** Economist 8: Score...20. Score long...21; ibid 58: 1 score [of wool] = 20 lb.

scoure, scower, scoyr. SCORE

scriple, scripule, scripulus. SCRUPLE

scruple—5 scriple, scripule, scrypule; 5-7 L scrupulus; 6-7 scrupul (OED); 6-8 scrupule (OED); 6-9 scruple; 7 L scripulus [ME scriple fr L scripulum, scrupulum, a small weight, fr scrupulus, small stone, pebble, dim of scrupus]. A wt in the ap system containing 20 gr (1.296 g) and equal to 1/3 ap dr of 60 gr (3.888 g) or 1/24 ap oz of 480 gr (31.103 g). Comparatively the s was 0.731 avdp dr, 0.672 Scots t DROP, and 0.002101 Scots tron lb. It was sometimes mistakenly assigned to the avdp system by early modern writers. It sometimes was abbreviated sc. or scr.—**1440** Palladius 59: A scriple...and half a scriple. c**1450** Hall 33: Scrupulus 3 pars dragme. Dragma octava pars uncie; ibid 34: Scrupulus constat ex 20 granis, nec maximis nec minimis, ordei. c**1475** Ibid 35: A scripule ys the thridde part of a dragme & is thus Wryte ℈... A scrypule weyeth a peny. c**1600** Ibid 36: Scrupuli ℈ is 20 barley cornes. **1606** Ibid 38: A scruple is 20 graines. **1616** Hopton 160: You must note that the Auerdupois pound is diuided into Graines, Scruples, Dragmes, and so to Ounces. **1682** Hall 29: Aver-du-pois conteynes...every dragme, 3 scruples; every scruple, 20 graines. **1688** Bernardi 137: Vel more Pharmacopolarum: Libra de Troy, 12 Unciæ ℥, 96 = 12 X 8 drachmæ ʒ: Scripuli ℈. 288 = 96 X 3. **1696** Cocker 108: a scruple. **1708** Chamberlayne 205: The Apothecaries reckon 20 Grains Gr. make a Scruple ℈. **1728** Chambers 1.360: The Ounce into 8 Drachms; the Drachm into 3 Scruples; and the Scruple into 20 Grains. **1778** Diderot XXVI.420: C´est aussi les poids de apoticaires, mais qui se divise autrement; vingt grains sont un scrupule, trois scrupules une dragme et huit dragmes une once. **1790** Jefferson 1.986: The drachm into 3 scruples; The scruple into 20 grains. **1816** Kelly 84: 20 Grains...1 Scruple. **1880** Courtney 158: 20 grains (gr.) 1 scruple, marked sc. or ℈. **1907** Hatch 35: 1 scruple = 1.29598 grammes. **1951**

Trade 22: 20 grains...1 scruple. 3 scruples...1 drachm.

scrupul, scrupule, scrupulus, scrypule. SCRUPLE

seam—1-7 L summa; 3 sæm (OED); 3-4 sem; 3-7 seme; 3-9 seem; 4 L sema, L summagium; 5 ceme, zeme (OED); 6 seayme (OED), seym (OED), sheme (OED); 6-7 seame; 6-9 seam; 7 seeme, L suma (Select Pleas 1); 9 zame (OED), zeam (OED) [ME seem, sem fr OE sēam fr (assumed) VL sauma, packsaddle, fr LL sagma]. A m-c and a wt, identical to the QUARTER, for dry products: chopped bark, Yorkshire, in some parts, 9 heaped bu (c4.05 hl); dung, Devonshire, 3 Cwt (152.406 kg); glass, generally 120 lb (54.431 kg) or 24 STONE of 5 lb each, but occasionally 100 lb (45.359 kg) or 20 stone of 5 lb each; grain, generally 8 striked or leveled bu (c2.82 hl) of 8 gal each and equal to 1/4 CHALDER (after the establishment of the Imperial system the grain seam increased in size slightly (2.909 hl) because of the larger bu), but variations from 7 to 9 bu (c2.47 to c3.17 hl) were not uncommon (by custom, however, the seam of grain contained 8 "heaped" bu equivalent to 9 "striked;" when this was prohibited by law, the corn-dealers popularized a measure of 9 striked bu called a FATT); lime, Derbyshire, 8 striked bu (c2.82 hl) at the wharves and 8 heaped bu (c3.60 hl) at the kilns; potatoes, Guernsey and Jersey, 240 lb (108.862 kg); salt, 4 Cwt (203.208 kg); and Welsh coal, Devonshire, 16 heaped bu (c7.20 hl).—**1086** Sussex 14: Pro forisfactura villanorum ix libræ et iii summas de pisis. **1200** Cur. Reg. 8.218: De ordeo xj. sceppas et j. quarterium et xxj. summas avene, unde tercia pars crevit super tenementum quod recuperavit. **1206** Feet

2.47: De redditu iiij summarum bladi singulis annis. c**1220** Evesham 219: Octo summas frumenti. c**1225** Osmund 1.310: Quod quatuor summas frumenti. **1228** Gras 1.156: Quodlibet quarterium bladi.... Unum seme ferri. c**1253** Hall 12: La sem de veyr est de xxiiii peris, e checune pere est de v li; e si est le sem de vi [X] xx lib. c**1272** Ibid 10: Et ita continet le seem [vitri] sexies viginti libre. **1290** Fleta 120: Item summa vitri constat ex xx. petris, et quelibet petra ex quinque libris, et sic continentur in summa que dicitur le seem quinquies viginti libre. c**1300** Brit. Mus. 13.29: Sema vitri constat ex .xxiiij. petris & quelibet petra ex .v. libris. Et ita continet le sem .sexies xx. libras. **1304** Gras 1.169: Pro xlvii summis ordei. c**1320** Thorpe 11: Et iste quatuor mine cum Gatea que dicitur Gundulfi faciunt tres summas. c**1350** Swinthun 79: De quolibet summagio bladi. c**1440** Promp. Parv. 65: Ceme, or quarter of corne. Quarterium. c**1590** Hall 21: The quarter or seame is 8 bushells. **1603** Henllys 139: Lyme ys sold by the Bushell, and so by the hundred, and not by the seame, or horseloade, as in other places. **1616** Hopton 162: Whereof are made...Coombes, or halfe Quarters, Quarters, or seames. **1656** Howes 3-4: 2 Coombs...8 Bushels...a Quarter or a Seame. **1664** Gouldman sv seme: A seme of corn. **1665** Sheppard 57: The Seem of Glass containeth 24 stone, and euery stone 5 pound, and so the Seeme containeth Six-score pound. **1708** Chamberlayne 207: A Seam of Glass is 24 Stone, 5 Pounds to the Stone, make 120 Pounds.... 4 Bushels the Comb or Curnock...2 Curnocks make a Quarter, Seam or Raff. **1716** Harris 2. sv measures: Seem or Quarter.

1717 Dict. Rus. sv dry-measure: Two Curnocks make a Quarter, Seam or Raff, and ten Quarters a Last. **1805** Macpherson I.471: 5 pounds of glass, 1 stone, 24 stones 1 seem. **1820** Second Rep. 32: Seam or Seem, sometimes a quarter of corn or malt. **1880** Britten 176: Seam (Dev.), of dung, 3 cwts. **1882** Jackson 227: Seam of glass...120 Lbs. av. **1966** O´Keefe 671: 1 quarter or seam = 8 bu. = 2.909 hl.

seame, seayme. SEAM

sec, seck, secke. SACK

seem, seeme. SEAM

seilion, seillon, seilon. SELION

sek, sekke. SACK

seldra. CHALDER

selion—3 L seilion, L seillon, L seilon, L seylion; 3-5 L seylon; 3-? selion; 4 L seyllon; 5, 9 sellion; 6 selyon (OED); 7 selione (OED), sillyon (OED) [ME selion, sellion fr MF seillon, a measure of land, fr OF sillon, ridge, furrow]. A m-a for the strip of land or pathway between two parallel furrows of the open field. Similar to BUTT OF LAND, RIG, and RIDGE.—**1201** Feet 1.12: Scilicet in orientali parte ville unum seillonem inter terram Walteri filii Willelmi et Walteri filii Adelstan´...et duos seillones inter terram Siwathe. **1202** Ibid 15: Quatuor seilones terre sue qui jacent inter culturas predicti Johannis qui uocantur Micheles acras; ibid 73: Duos seilones super Swikes et ij seilones super Kirkefurlang´; ibid 78: Scilicet unum seillonem ad Aldewellesti et unum seilonem ad Hagethornes et duos seillones ad Baligat'. **1208** Ibid 128: Et in escambium cuius tofti et cuius seilionis predicta Basilia dedit et concessit predicto Hugoni...et unum seilionem in campo de Goldcroft. c**1272** Gray 254: Decim seliones terre...duas dimidias seyliones...et unam seylionem...et duas seyliones...et duas dimidias seliones...et unam dimidiam seylonem. **1290** Ault 56: Newton Longville, Bucks.... Item quod nullus pauper infra seliones fabas coliget set ad capita et ad divisas selionium. c**1310** Malmesbury II.202: Ita tamen quod medietas proximi seylionis quæ jacet juxta dictam semitam in parte australi. **1411** Ault 73: Elmley Castle, Worcs.... xxx selliones. **1616** Gray 244: In the same feild two selions. **1665** Sheppard 22: A Selion...otherwise called a Ridge of Land...of no certain quantity, but sometimes containeth half an acre, sometimes more and sometimes less. **1874** Hazlitt 438: Selion.—Half an acre. **1888** Taylor 180: The perch, or virga, was itself doubtless merely the oxgoad, which, laid upon the ground at the headland, would conveniently measure the breadth of the rig or sellion to be ploughed. **1897** Maitland 383: In our Latin documents these ridges appear as selions.

selione, sellion, selyon. SELION

sem, sema, seme. SEAM

seron—6-9 seron; 6, 9 serone; 9 ceroon (OED), seroon [Sp serón, a pannier, hamper, crate]. A m-c for dry products: almonds, generally 2 Cwt (97.976 kg); aniseed, 3 to 4 Cwt (152.406 to 203.208 kg); barilla, 3 Cwt (152.406 kg); castle-soap, 2 1/2 to 3 3/4 Cwt (127.005 to 190.507 kg);

and cochineal, 140 lb (63.503 kg). It was a large bale or bundle that was tightly wrapped in animal´s hide.—**1545** Rates 1.55: A serone of sope. c**1550** Welsh 62: 1 serone white soap; ibid 170: 4 serones divers goods. **1696** Phillips sv: Seron of Almonds, the Quantity of Two Hundred Weight: Of Anis-seeds from 3 to 4 C: Of Castle-Soap from 2 1/2 C to 3 3/4 C. **1701** Hatton 3.233: Seron...Of Barillia 3 C. Almonds 2 C. Anniseeds 3 to 4 C. and Castle-soap 2 1/2 C. to 3 3/4 C. **1717** Dict. Rus. sv: Seron of Almonds, the quantity of two Hundred-Weight. Of Anis seeds, from 3 to 4 C: Of Castle-soap, from 2 1/2 C. to 3 3/4 C. **1840** Waterston 147: Almonds, seron, cwt. 1 1/4 to cwt. 2.... Cochineal, seron...lbs. 140. **1883** Simmonds sv: Seron, Seroon, a kind of skin package...cochineal, indigo, and various drugs are imported in this form. See HUNDRED

serone, seroon. SERON

serplaith, serplar, serplath, serplathe, serpler, serpliath, serpliathe. SARPLER

sesster. SESTER

sester—1 sestre (Select Doc.); 1-7 L sextarium, L sextarius; 1-9 sester; 3 sestier, L sextertium; 4 cestre (Prior), cistern (Prior), sesster (OED), sextarye (Prior); 5 cestron, sesteryn, sexter, sexterne; 6 cester (Prior), sesterne, sestur, systern, systerne; 8 sextar; ? cistra (Prior), sextarie (Prior), L sextercium (Prior), sextur (Prior), L sistarius (Prior), sisterne (Prior) [OE sester; see OED]. A m-c for dry and liquid products: ale and beer, generally 12 gal (c5.54 dkl) but occasionally 13 to 19 gal (c6.01 to c8.78 dkl); grain, generally 1 SEAM or 8 bu (c2.82 hl); lime, 3 to 4 seams (c8.46 to c11.28 hl); wine, oil, and honey, generally 4 gal (c1.51 dkl) but occasionally 5 to 6 gal (c1.89 to c2.27 dkl). The Scots sester of wine was reckoned at 3 gal (c4.08 dkl).—**c1000** Brit. Mus. 4.106: Sextarium mellis. **c1050** Select Doc. 79: Unum sextarium mellis triginta duarum unciarum. **1086** Sussex 98: Silva lxx porcorum et xx porci de gablo et ii sextaria mellis. **c1150** Acts Scotland 1.312: Assisa vini secundum constitutionem regis David.... Item sextarium debet continere tres lagenas. **c1150** Gross I.292: Singulis vero noctibus prout justum est ordinatis ac distributis quisque decanus ad hospitium suum unum sextarium habeat, notarius vero dimidium sextarium habeat. **c1220** Evesham 209: Duodecim sextaria mellis; ibid 218: Duas justas cerevisiæ quarum quælibet continebit duas caritates, quarum caritatum sex faciunt sextarium regis. **1233** Close 1.223: Quod nullus mercator, ducens vina venalia in Angliam vel vina Wasconie...vel aliunde, decetero post has nundinas Sancti Botulfi venire faciat in Angliam aliquod dolium vini, quod minus contineat secundum numerum sextertiorum quam continere consuevit temporibus Henricus regis. **1246** trans in Cal. Char. 1.308: Of the gift of Robert de Maconio, twenty sestiers of corn yearly. **1290** Fleta 120: Doleum vini iij. sextaria vini puri debet continere et quodlibet sextarium quatuor ialones. **1390** Henry Derby 15: Clerico buterie super vino, per manus eiusdem pro j sextario iij potellis di. vini Vasconie, altero per ipsum empto ibidem, sextarium ad ij s. viij d. **1421** Cov. Leet 25:

That no brewster sell no derre a Cestron ale to noo hukster but for xviij d. **1425** Acts Scotland 2.12: The ald boll first maid be king Dauid contenit a sexterne the sexterne contenit xij galonis of the ald met. **1440** Palladius 58: In half a sexter aged wyne do shake; ibid 100: Sex sester old wyne; ibid 169: In sesters XII of aisel that soure harde is. c**1461** Hall 15: The barell cont[aineth] xxxi gallones I quart, there sesteryn cont[aineth] iiij gallouns. **1507** Gras 1.706: He that ys a gawner owght to understonde there ys in a tunne lx systerns and every systerne ys iiii galons be yt wyne or oylle.... Any amme of Andwarpe butt xxxvi gallons for ix sesternes ys an ambether. **1521** Cov. Leet 678: And yf the price of Malte be vndur the price of a noboll then the seyd bruers to sell ther ale for xviij d. a sestur; and that thei sell xiiij galondes to the sestur. **1678** Du Cange sv sextarium: Apud Anglos Sextarius vini continet 4. jalones.... Variæ fuit capacitatis sextarius, pro variis locis, cum in aridis tum in liquidis. **1745** Fleetwood 52: A Sester or Sextarius was what we now call a Quarter, or a Seam, containing 8 bushels; ibid 58: But Sir H. Spelman says, that at Paris, a Modius Vini holds 36 Sextarios, and that a Sextar is 8 Pints. **1820** Second Rep. 32: Sester of wheat; before the Conquest was a horse load. **1872** Robertson 69: The Sester of London was a measure of four gallons, according to Fleta, and 52 sesters of pure wine went to the cask.

sesterne, sesteryn, sestier, sestre, sestur. SESTER

seteen. SETTEEN

setteen—8-9 setteen; 9 seteen [*]. Equivalent to LISPOUND in the Orkney and Shetland Islands.—**1779** Swinton 105: 1 mark to 24 marks, which make a Setteen or Lyspund; ibid 106: Setteen or Lyspund...32.6306 [avdp lb in Orkney Islands]; ibid 107: Setteen or Lyspund...30.4553 [avdp lb in Shetland Islands]. **1820** J. Sheppard 134: 24 marks make 1 seteen or lyspund.... 1 seteen...32.64 [avdp lb in the Orkneys]; ibid 135: 24 marks make 1 seteen which is equal to 30.48 [avdp lb in Shetland Islands]. See MARK

sextar, sextarie, sextarium, sextarius, sextarye, sexter, sextercium, sexterne, sextertium, sextur. SESTER

seylion, seyllon, seylon. SELION

seym. SEAM

shaff, shaffe. SHEAF

shaffment, shafman, shafment, shafmond, shaftemente, shaftman, shaftmen. SHAFTMENT

shaftment—1 sceaftmund (OED); 4 schaftmonde (OED); 5 chaftmonde (OED), schafftmon (OED), schaftemonde (OED), schaftmon (OED), schaftmond, schaftmone (OED), schaftmonthe, schaftmount (OED); 5-6 shafmond (OED); 6 shaftemente (OED), shaftman (OED); 6-9 shaftment (OED); 7 shafman (OED), shaftmen (OED), shaftmet, shaftmont (OED); 7, 9 shafment (OED); 8-9 shathmont (OED); 9 shaffment (OED) [OE sceaftmund fr sceaft, shaft, + mund, hand]. A m-l generally regarded as containing 6 inches (c15.24 cm) and defined as the distance from the tip of the extended thumb across the breadth of the palm.—c**1400** Acts Scotland 1.387: The tong salbe the lynth off a schaftmonthe and .j. ynche. **1474** Cov. Leet 399: And his ffagott of wodde of an ob. schal-be iij schaftmond and a halfe a-bout and a yerde of lenthe. And his ffagott of j d. schal-be vij schaftmond a-bout, kepyng the same lenght. **1677** Roberts 300: In the measure of Broad-Cloth by retail in Drapers Shops, allowing to the Buyer a Shaftnet [sic] upon each yard...which commonly may be about 5...6 inches in length.

shaftmet, shaftmont, shathmont. SHAFTMENT

shayff. SHEAF

sheaf—4-5 scheef (OED), schef (OED), shef (Fab. Rolls); 4-6 sheef (OED), shefe; 5 chyfe (OED), sheeffe (OED), sheiff (Fab. Rolls); 5-6 schefe (OED), scheff, scheffe (OED), sheff (OED); 5-7 sheffe; 6 schaffa, schiefe, shaff (OED), shaffe, shayff, sheaffe (OED); 6-7 sheafe (OED); 6-9 sheaf; 7 sheave (OED), sheive (OED); 8 sheaff (OED) [ME sheef, shef, schef fr OE scēaf; see WNID3]. A m-c, m-q, and wt for several products: glass, of uncertain wt; grain, generally 1/12 to 1/24 THRAVE, but in Ireland, 1/17 CRANNOCK of 2 SEAMS or approximately 1 bu (c3.52 dkl); and steel, 30 GADS or PIECES of uncertain wt, equal to 1/6 or 1/12 BURDEN.—c**1461** Hall 17: And xxx gaddes make a scheff, and xii scheff make a burdon. **1507** Gras 1.703: Stelle the barelle wyche owght to be iiii [X] xx burden and xxx gaddes makythe sheffe. **1508** Fab. Rolls 353: x shaffe Renysh glase. **1532** Finchale ccccxlvii: 8 shayff brymmys-glasse. **1534** Fitzherbert 37: And let hym caste out the .x. shefe in the name of god. **1597** Skene 1. sv schaffa: Ane schiefe of steile. **1787** Liber 363: To hold the produce of 17 sheaves of corn. **1805** Macpherson I.471: 30 pieces of steel 1 sheaf. **1883** Simmonds sv: Sheaf, a bundle of corn bound up in the field. **1896** Pearman 41: The sheaf is a bundle, which may be of any size. But it was limited by the provision that it should be such a bundle as could be tied together by a band formed of the corn.

sheafe, sheaff, sheaffe, sheave. SHEAF

shede. SHIDE

sheef, sheeffe, shef, shefe, sheff, sheffe, sheiff, sheive. SHEAF

sheldra. CHALDER

sheme. SEAM

shid. SHIDE

shide—1-3 scid (OED); 3 sid (OED); 4 chide (OED), szhide (OED); 4-6 schide (OED), schyde (OED); 4-9 shide; 5 chyde (OED), schudde (OED), schyd (OED), schydd (OED); 5-6 shyde, shyyd (OED); 6 shede (OED), shyd (OED); 6-8 shid [ME shide fr OE scīd]. A m-l of 4 ft (1.220 m) for firewood with variations in circumference of 16 to 38 inches (4.064 to 9.652 dm). In the citations the "carfe," a var of carf, is the cut part at the end of a piece of wood, while the "tall shide," "taleshid," and "tale-shide," var of talshide, refer to a cut shide.—**1559** Fab. Rolls 353: In byllot or shydes. **1587** Stat. 171: And that euerie tall shide, conteine in length foure foot besides the carfe. And euerie tall shide named of one, to conteine in greatnesse within a foot of the middest sixteene inches about. And euerie tall shide named of two...three and twentie inches about...three, to conteine...eight and twentie inches...foure, to conteine...three and thirtie inches about...and five to conteine...eight and thirtie inches about. c**1590** Hall 27-28: Euery taleshid conteyn´ in lenght 4 foott, besyde the carfe, in lenght eiche alike. The taleshid namid one ought to be 16 ynches about. The taleshid namid 2...23 ynches about. The taleshid namid 3...28 ynches about. The taleshid namid 4...33 ynches about. The taleshid namyd 5...38 ynches. **1616** Hopton 163: All Shids must be foure foot long beside the carfe, and upon them is 1. 2. 3. 4. or 5. markes or notches, and then they must bee in compasse about the middest 16. 23. 28. 33 or 38 inches, according as it hath number of markes. **1665** Assize 18: Item, every Tale-shide must contain in length four foot. **1682** Hall 30: Shids must be 4 foote long, besides the Carfe. They are noted with 1, 2, 3, 4 or 5, and must accordingly be in compasse, about the midst, 16, 23, 28, 33, [38] inches. **1756** Rolt sv measures: A shid is to be 4 feet long.

shock—4-5 shokk (OED); 4-5, 8-9 schock; 4-9 shock; 5 schokke; 6 schocke (OED), shoke (OED); 6-7 shocke (OED) [ME shock; akin to MDu schoc, schocke, heap, pile, group of sixty, MLG schok, shock, group of sixty, MHG schoc, heap, pile, group of sixty]. A m-q for canes, ropes, iron plates, trays, boxes, etc., numbering 60, and in Derbyshire, for corn, equivalent to 12 SHEAVES.—c**1440** Promp. Parv. 447: Schokke, of corne. **1573** Tusser 130: Corne tithed...to gather go get and cause it on shocks to be by and by set. **1590** Rates 2.4: Bast ropes the shock containing lx. ropes; ibid 5: Boxes the shock containing lx; ibid 7: Cannes of wood the shock containing lx. cannes; ibid 13: Double iron plates called doubles the shock containing sixe bundels...Doubles the bundel containing tenne in euery bundel; ibid 37: Trayes the shock containing lx. **1701** Hatton 3.233: Shock...of Soap-boxes, Canes, Wood-Trays, &c...60. **1783** Beawes 865: Anchor and Locks the Schocks of 60; ibid 866: Deals, of Oak or Fir, above 20 Feet the Schock; ibid 868: Oars, great, the Schock; ibid 869: Wood, Shovels, the 10 Schocks.... Wainscot, Boards, the Schock. **1818** Rördansz 95: A shock is 60; ibid 145: The schock of sixty pieces. **1820** Second Rep. 32: Shock...of canes or boxes, 60.... Derbyshire, seems to mean 12 sheaves of corn. **1868** Eng. Cyclo. 826: The shock was always 60.

shocke, shoke, shokk. SHOCK

shyd, shyde, shyyd, sid. SHIDE

sieve—3 L chiphus; 8-9 sieve, sievf; ? L cipha (Prior), sife (Prior) [ME sive fr OE sife]. A m-c, a wicker basket, for dry products: apples and potatoes, approximately 1 bu (c3.52 dkl), cherries, 48 lb (21.772 kg), and plums, 56 lb (25.401 kg).—**1208** Bish. Winch. 24: In saccis lanæ, iij s. j d.... In chiphis, vj d. **1820** Second Rep. 32: Sievf, a flat basket for measuring or carrying fruit and vegetables...Kent: of apples and potatoes, about a bushel...of cherries, 48 lb. **1896** Wagstaff 35: A sieve of vegetables in West Ham = 1 bushel. **1956** Economist 51: Cherries: 1 sieve = 48 lb.... Plums: 1 sieve = 56 lb.

sievf, sife. SIEVE

sillyon. SELION

sirplithe. SARPLER

sistarius, sistern, sisterne. SESTER

six [ME six fr OE six]. An Imperial m-c for beer containing 6 gal (2.728 dkl); hence, the name (Economist 54).

skain, skaine, skane. SKEIN

skape. SKEP

skayn, skayne. SKEIN

skeb. SKEP

skef. SKIVE

skein—5 skayn (OED); 5-7 skayne, skeyne (OED); 6 scan (OED), scayne (OED), skane (OED); 6-7 skaine (OED); 6-9 skain; 7 scane (OED), schaine (OED); 7-9 skein; 8 scain (OED) [ME skayne, sheyne fr MF escaigne, of obscure origin]. A m-l for yarn: Hampshire, 480 yd (43.891 dkm), and Suffolk, 1600 or 2400 yd (146.304 or 219.456 dkm). Presently in the United Kingdom a skein of cotton yarn is 120 yd (10.973 dkm) and of woolen yarn, 256 yd (23.409 dkm).—**1612** Halyburton 323: Packthreid in skaynes the hundredth weght. **1664** Gouldman sv skain: A skain of gold or silver thread. **1778** Loch I.231: Scots Thread of twenty or thirty threads in the skein. **1820** Second Rep. 32: Skain or Skein, Hamphire: of yarn, 480 yards...Suffolk: of yarn reeled, 20 leas, each of 80 or 120 yards. **1840** Waterston 148: Yarn skein or rap of 80 threads, yds. 120. **1956** Economist 58: Cotton...1 skein = 120 yards.... Woollen...1

skein = 256 yards.

skeipp. SKEP

skep—1 sceppe (OED); 3-4 L eskippa; 3-7 L sceppa; 4 scep (OED), skipp (Prior); 4-6 skeppe (OED); 4, 7 scepp (OED), skippe (OED); 4-7 skepe (OED); 4-9 skep; 5 scappe (OED), schepp (Fab. Rolls), skype (OED); 5-6 skyppe; 5-7 skepp; 6 skeipp (OED); 7-9 skip (OED); 8 scape (OED), scepe (OED), schepe; 9 skape (OED), skeb (OED); ? L escheppa (Prior), L eschippa (Prior) [ME skep fr OE sceppe, skepful, fr ON skeppa, bushel]. A m-c for grain and other dry products, varying in size from 1 or 2 bu (c3.52 or c7.05 dkl) to approximately 1 or 2 SEAMS (c28.19 or c56.38 dkl).—**1200** Cur. Reg. 8.218: De ordeo xj. sceppas et j. quarterium et xxxj. summas avene. c**1320** Thorpe 11: Continet quinque eskippas de duro blado. **1490** Salzman 2.47: No one shall have nor kepe within hows eny bussell skepp, whych is the iiij[th] parte of a quarter. **1534** Fitzherbert 115: .xii. coffyns or skyppes of fragmentes. **1678** Du Cange sv sceppa: Mensura salis.... Aliorumque aridorum, puta farinæ. **1695** Kennett Glossary sv sceap: Hence a basket is call'd a Skip or Skep in the South parts of England. **1745** Fleetwood 142: For 28 Quarter[s] and one Schepe...of Wheat. **1777** Nicol. and Burn 613: Skep...a measure of uncertain quantity: In a survey of the forest of Englewood in 1619, it is defined to contain 12 bushels, and every bushel (Penrith measure) 16 gallons and upwards. **1880** Britten 158: Skep, a basket without a lid, and with short handles.

skepe, skepp, skeppe. SKEP

skevy. SKIVE

skeyne. SKEIN

skif. SKIVE

skin [ME skin fr OE scinn fr ON skinn]. A wt of 3 Cwt (146.964 kg) for cinnamon (c1800), originally the amount bound in animal´s skin (Second Rep. 32). See HUNDRED

skip, skipp, skippe. SKEP

skive—5 skef, skevy, skif (Southampton 1), skive (Southampton 1), skyve (Southampton 1) [perh fr shive, a thin piece or fragment; cf ON skīfa, a shaving, slice]. A m-q at Southampton for teasels, consisting of approximately 500 in number.—**1439** Southampton 2.28: 35 skevys tesellarum. **1443** Brokage II.57: Pro vi skevys tesell´; ibid 96: Cum 1 skef tesell. c**1475** Gras 1.193: Of a skef tasill´.

skoir, skor, skore, skowre. SCORE

skron [*]. A wt of 2 Cwt (97.976 kg) for almonds and 3 Cwt (152.406 kg) for barilla (c1800) (Second Rep. 32). See HUNDRED

skype, skyppe. SKEP

skyve. SKIVE

sleek [prob sleek (or slick) measure, a level or striked measure]. A m-c for apples and pears in Clydesdale (c1800-1900), containing 18 pt or 2 1/2 gal (c1.10 dkl) (Second Rep. 32 and Britten 176).

solin—1-4 L solinum (Baxter); 2 L solinus (Baxter); 6-9 solin [E solin fr MedL solinum, solinus, SULUNG]. Equivalent to SULUNG.—**1599** Richmond Appendix 2.9: And then I take it, that a Solin of Ground, after English

Account, containeth 216 Acres. If after Norman Tale, then Nine score Acres. **1888** Round 3.200-01: The carucata and the bovata (like the solin and the jugum) are both of them terms obviously derived from the team of oxen for the plough. **1895** Round 2.103-04: In the Domesday of Kent we find the form solin, or its Latin equivalent solinum, used for the unit of assessment, like the hide and the carucate in other countries. In the Kent monastic surveys it is found as sullung or suolinga. **1897** Maitland 395: Domesday Book shows us that in Kent the solin (sulung) is the fiscal unit that plays the part that is elsewhere played by the hide. **1904** Vinogradoff 283: Sulungs (solins) or aratra, with subdivisions termed yokes (iuga), at the ratio of four yokes to the sulung.

some, somme. SUM

soortt. SORT

sort—5-6 sort, sorte; 6 sortte; ? soortt (OED), sortt (OED) [ME sort fr MF sorte, prob fr MedL sors, sortis, sort, kind, fr LL sors, sortis, way, manner, fr L sors, sortis, a lot, share]. A m-c for fruit, generally containing 3 PIECES or 12 QUARTERNS (152.406 kg). Occasionally it was described as the equivalent of 3 FRAILS (40.824 to 102.057 kg).—**1439** Southampton 2.16: 2 sort´ fructui; ibid 17: 3 sort´ et 1 pecia fructui. **1443** Brokage II.100: Cum ix sortes fructui; ibid 111: iiii sortes fyges. **1507** Gras 1.697: Fygges the sortte that ys to saye iii frayles for the sortte.... Fygges the tunne that ys to say x sorte for the tunne. **1590** Rates 2.14: Figs the sort containing three peeces.

sorte, sortt, sortte. SORT

sowme. SUM

span—4-5 spane (OED); 4-7 spanne; 4-9 span; 5 spayn (OED); 8 spand (OED) [ME spanne fr OE spann]. A m-l which originated as a unit of body measurement equal to the distance from the tip of the smallest finger to the tip of the thumb on the outstretched hand and considered equal to 1/2 CUBIT. Based on the ft of 12 inches, the span was made equal to 9 inches (2.286 dm).—**1624** Huntar 8: 3. Inches is a palme. 3. Palmes is a spanne; ibid 10: A spanne containeth...Inches—9. **1639** Bedwell B1-B2: An Hand, or Hands breadth, a Spanne, a Foote, a Cubit, a Passe. **1717** Dict. Rus. sv: Span, a Measure from the Thumb´s end to the top of the little Finger, containing three Hands-breadth or 9 Inches. **1820** Second Rep. 32: Span, 9 inches. **1832** Edinburgh XII.569: 9 Inches = 1 Span = 0.2286 [m]. **1880** Courtney 168: 9 inches 1 span. **1956** Economist 8: Span...9 inches.

spand, spane, spanne, spayn. SPAN

spindle [ME spindel fr OE spinel; akin to OHG spinala, spindle, MHG spinel, spinle, OE spinnan, to spin]. A m-l for thread in Clydesdale (c1800) containing 14,400 yd (131.674 hm) or 48 cuts of 120 threads each on a reel 2 1/2 yd in circumference (Second Rep. 32). Presently in the United Kingdom a spindle of cotton yarn is 18 HANKS or 15,120 yd (138.257 hm); of linen yarn, 48 LEAS or 14,400 yd (131.674 hm); and of woolen yarn, 45 SKEINS or 11,520 yd (105.339 hm) (Economist 58).

square [ME squyre, square fr MF esquerre, esquarre fr (assumed) VL exquadra fr (assumed) VL exquadrare, to square]. A m-a (c1900) of 100 sq ft (9.290 sq m) for architectural roofing and flooring measurement (Thurston 23 and McConnell 13).

srone [*]. A m-c for oatmeal in some parts of Ireland (c1800) reckoned equal to 3 POTTLES totaling 12 lb (5.442 kg) in weight (Kelly 115).

sstakke, stac, staca, stacca. STACK

stack—2 L staca (Prior); 3, 6 stac (OED); 3-7 stak (OED); 4-9 stack; 5 sstakke (OED); 5-6 stake (OED), stakk (OED); 5-7 stacke (OED); 6 stayke (OED); 7 L stacca (Henllys) [ME stak, stack fr ON stakkr, haystack]. A m-c for dry products (c1600 to c1900): barley and wheat, Glamorganshire, 3 bu (c1.06 hl); coal, Derbyshire, 105 cu ft (2.973 cu m), and Shropshire, 4 cu yd (3.058 cu m) totaling 25 Cwt (1270.050 kg); oats, Glamorganshire, 6 bu (c2.11 hl); and wood, Bedfordshire, Middlesex, and Northamptonshire, 4 cu yd (3.058 cu m) (Henllys 138, Jessop 41, and Second Rep. 32).

stacke. STACK

stade. STADIUM

stadium—1-7 L stadium; 8-9 F stade [L stadium, a measure of length, a furlong, fr Gr stadion, a measure of length]. Equivalent to FURLONG.—c**1075** Hall 2: Digitus, uncia, palmus, sextas, pes, cubitus...passus...stadium, miliaria. c**1260** Bracton I.58: Stadium vero dicitur octava pars milliarii. c**1289** Bray 10: Quinque pedes passum faciunt; passus quoque centum viginti quinque stadium; si miliare des octo facet stadia. c**1300** Hall 7: 125 passus faciunt stadium.

c**1325** Rameseia I.76: Passus centum viginti et quinque unum stadium.... Et stadia quindecim unam leucam. **1395** York Mem. I.142: Octo stadia faciunt miliare Anglie. c**1400** Brit. Mus. 20.lv: Cxxv passus faciunt stadium. c**1425** Hall 9: Et sexdecim pedes et dimidia faciunt perticatam Regis. Et quadraginta perticate faciunt unum stadium. c**1440** Promp. Parv. 183: Furlonge. Stadium. **1664** Spelman 474: Quarentena...Stadium, Angl. a furlonge. **1665** Assize 6: Plinie Lib. 2. Cap. 23. deriveth Stadium to be a furlong. **1688** Bernardi 202: Pes Anglicus...1/660 Stadii aut Furlongi. **1780** Paucton 789: Furlong ou stade de 660 pieds Anglois...Mille légal de huit stades Anglois. **1784** Ricard II.155: Furlongs, ou stades.

stæn. STONE

staff [ME staf fr OE stæf; see WNID3]. A m-q for teasels (c1800-1900): Essex, 1250 or 50 GLEANS of 25 teasels each, and Gloucestershire, 500 or 25 gleans of 20 teasels each for middlings and 300 or 30 gleans of 10 teasels each for kings (Second Rep. 32 and Britten 176).

stain, staine. STONE

stak, stake, stakk. STACK

stan. STONE

stand—3-5 stonde (OED); 4-5 stoond (OED), stoonde (OED); 6-9 stand [ME stand, stond fr vb standen, stonden, to stand]. A wt of 1 1/4 to 3 Cwt (63.502 to 152.406 kg) for pitch.—**1701** Hatton 3.234: Stand...of Burgundy Pitch...2 1/2 C. to 3 C. weight. **1706** Phillips sv: Stand of Burgundy-Pitch...a quantity from two and a half to three Hundred Weight.

1840 Waterston 147: Burgundy pitch, stand...cwt. 1 1/4.

stane. STONE

stang—3, 7 stong; 3-9 stang; 4-5 L stanga; 4-7 stange; 6-7 stangue; 7 stonge (OED); 8 steng (OED); 8-9 stangell [ME stang, stange, stong fr ON stöng; akin to OE steng, pole]. A m-a for land in Wales commonly identified with the customary acre of 3240 sq yd (0.271 ha). In certain regions, however, it was much smaller for it was considered the equivalent of 1/4 ERW, or standard acre, of 4320 sq yd. In the East Riding of Yorkshire, 1/4 acre (c0.10 ha) was called a stang.—**1400** trans in Cal. Close 17.202: Five stangs (stanga) of meadow called ´Farthyngstanges´. **1603** Henllys 134: 8 poles in bredth, and xx in length, or 4 in bredth and 40 in length maketh a stange, w[hich] is Just in accompte (thoughe not in measure) w[ith] the statute acre, and the difference is onely in the length of the landpole; ibid 134-35: 4 of those stangues make the Penbrokeshire acre. **1610** Folkingham 59: Foure square Pearches make a Daiesworke, 10. Daie-workes a Roode or Stong, 4. Roodes an Acre. **1695** Kennett Glossary sv furendellus: Which fourth part of an acre is in the East riding of Yorkshire call´d a stang. **1820** Second Rep. 32: Stang, or Stangell, S. Wales: 1/4 Erw. **1895** Donisthorpe 214: PERCH...South Wales: of land...Sometimes 9 feet square, 160 making 1 stangell; 4 stangells 1 erw of 5,760 square yards; ibid 218: STANG or STANGELL...1/4 erw. See ACRE

stanga, stange, stangell, stangue. STANG

stayke. STACK

stayne, stean, steane, sten. STONE

steng. STANG

step [ME step fr OE stæpe, stepe]. A m-l used principally in the early modern period and generally equal to 1/2 PACE or approximately 2 1/2 ft (c0.76 m) (Harkness xliii and Eng. Cyclo. 817).

stetch [a var of stitch, a narrow ridge of land, a ridge between furrows]. A m-l of 8 ft 2 inches (2.489 m) for land in Suffolk and other areas of Eastern England; earliest mention being around 1100. It was a ridge or plowed land between two furrows (Britten 159 and Prior 150).

stica. STICK

stick—3 estik, L estika, estike, L sticka, stik; 3,7 L stika; 3-4, 7 L stica; 3-6 sticke; 4 styk, styke; 4-5 stike; 5 stikke, styck; 7-9 stick [ME sticke fr OE sticca; see WNID3]. A m-q for eels, numbering 25 and equaling 1/10 BIND or 1/10 GWYDE.—**1202** Feet 3.196: Tenendam de predicto abbate et successoribus suis sibi et heredibus suis in perpetuum per liberum seruitium xx solidorum et sex millariorum et duodecim stikarum anguillarum per annum pro omni seruitio saluo forinseco seruitio. **1208** Bish. Winch. 71: Idem reddit compotum de xv estikis anguillarum proventis de vivario et molindino hoc anno. In decima, j estika. In expensis constabularii et familiæ dum fuerunt ad mensam, xiiij estika. c**1220** Evesham 217: Et sexaginta sticas anguillarum annuatim.... Et duodecim sticæ anguillarum annuatim. c**1253** Hall 12: E checun estike de xxv anguilles. c**1272** Ibid 10: Bynda vero anguillarum constat ex decem stickes; et quelibet sticke ex viginti et quinque anguillis. **1289** Swinfield 3: j. estik´ ang´ll. **1290** Fleta 120: Et quelibet sticka ex xxx anguillis. c**1300** Brit. Mus. 5.151: Bind anguillarum constat ex decem stikes. Et quelibet sticke constat ex .xxv. anguillis. **1303** Report 1.414: Et quelibet stik ex viginti quinque anguillis. **1390** Henry Derby 20: Et per manus Thome Fyssher pro xlviij styks anguillarum ab ipso emptis apud Boston, le styk ad iiij d.; ibid 29: Et per manus eiusdem pro j styke di. anguillarum, xiiij d. c**1435** Amundesham II.317: Item, unum stikke anguillarum. c**1461** Hall 17: Also Elys be sold by the stike, that ys xxv elys; and x styckys make a gwyde. c**1590** Ibid 23: A bynd of eeles consistith 10 stikes. **1664** Spelman 524: Stica...Vel Stika anguillarum. Mensura numeralis 25 anguillas continens. **1717** Dict. Rus. sv bind: 10 strikes [sic], each 25 Eels. **1805** Macpherson I.471: 25 eels 1 stick, 10 sticks 1 bind. **1820** Second Rep. 32: Stick of eels, 25.

sticka, sticke, stik, stika, stike, stikke. STICK

stimpart [of obscure origin; possibly contraction of saxteenth, sixteenth, + -part (OED)]. A m-c for grain in Ayrshire (c1600-1800) containing 153.6 cu inches (c0.25 hl) and equal to 1/4 pk of 614.4 cu inches or 1/16 FIRLOT of 2457.6 cu inches (Swinton 58 and Second Rep. 32). It was synonymous with the LIPPY or FORPIT.

stoan, stoane. STONE

stoipe. STOUP

stoke [E stock, in the sense of a store or supply of goods]. A m-q for dinnerware, consisting of 60 pieces.—c**1461** Hall 17: Also there ys a numbyr that ys called a stoke, and yt conteynyth lx; thereby be sold Pruse trenchers, dysshes and platters and dyuers oder.

stolp. STOUP

ston. STONE

stonde. STAND

stone—1-3 stan (OED); 3 stæn (OED); 3-4 F pere; 3-5 ston (OED); 3-7 L petra; 4 F piere; 4-5 sten (OED); 4-6 stoon (OED); 4-9 stane, stone; 5 stayne (OED); 5-6 stoone (OED); 5-7 stonne; 6 steane (OED), stoan (OED); 6-7 stain (OED), staine (OED), stoane; 8-9 stean (OED) [ME stan, ston, stoon fr OE stān]. A wt for dry products in England, Scotland, Ireland, and Wales. It generally weighed 14 lb (6.350 kg), but there were many important exceptions, ranging from 4 to 32 lb (1.814 to 14.515 kg): almonds, 8 lb; alum, generally 8 lb but occasionally 13 1/2 lb; barley, 17 1/2 lb in Selkirkshire; beef, London, Essex, and Gloucestershire, 8 lb, Herefordshire and Wales, 12 lb, Cumberland and Westmorland, 14 or 16 lb, Liverpool and Westmorland, 20 lb, and Buteshire and Arran, 24 lb; butter, Westmorland, 20 lb, Aberdeenshire and Lanarkshire, 22 lb, Berwickshire, Dumbartonshire, and Peeblesshire, 23 lb, Selkirkshire, 23 1/2 lb, Argyllshire, Ayrshire, Buteshire, Inverness, Roxburghshire, and Sutherlandshire, 24 lb, Aberdeenshire, 26 lb, Wigtownshire, 26 1/4 lb, and Aberdeenshire, 28 lb; cheese, Aberdeenshire and Lanarkshire, 22 lb, Berwickshire and Dumbartonshire, 23 lb, Selkirkshire, 23 1/2 lb, Argyllshire, Ayrshire, Buteshire, Dumfriesshire, Inverness, Roxburghshire, Sutherlandshire, and Wigtownshire, 24 lb, Aberdeenshire,

26 lb, Wigtownshire, 26 1/4 lb, and Aberdeenshire, 28 lb; cinnamon, sometimes 13 1/2 lb; cumin, 8 lb; fish, Selkirkshire, 17 1/2 lb, Dumbartonshire, 23 lb, and Argyllshire, 24 lb; flax, generally 14 lb but Galloway, 16 lb, and Fifeshire, 22 lb; glass, 5 lb; hay, Cumberland, 16 lb, Hebrides, 17 1/2 lb, Liverpool, 20 lb, Inverness, 21 lb, Berwickshire, 21 7/8 lb, Peeblesshire, 22 lb, Renfrewshire, 22 1/2 lb, Peeblesshire, 23 lb, Selkirkshire, 23 1/2 lb, Argyllshire, Ayrshire, Banffshire, Berwickshire, Buteshire, Dumfriesshire, and Roxburghshire, 24 lb, and Wigtownshire, 26 lb; hemp, generally 16 lb but occasionally 14 1/2, 20, 24, and 32 lb; hog´s lard, Aberdeenshire, 28 lb; iron, Dumfriesshire and Selkirkshire, 17 1/2 lb; lead, generally 12 lb but occasionally 15 lb; meal, Kirkcudbrightshire and Selkirkshire, 17 1/2 lb; mutton, Buteshire and Arran, 24 lb; nutmeg, sometimes 13 1/2 lb; oatmeal, Dumfriesshire, 17 1/2 lb; pepper, 8 lb, but sometimes 13 1/2 lb; potatoes, Angus, 16, 20, or 24 lb; raw hides, Berwickshire and Peeblesshire, 23 lb, Berwickshire and Selkirkshire, 23 1/2 lb, Buteshire and Roxburghshire, 24 lb; salt, Kincardineshire, 16 lb; straw, Liverpool, 20 lb, and Buteshire and Arran, 24 lb; sugar, generally 8 lb, but sometimes 13 1/2 lb; tallow, Cumberland, 16 lb, Berwickshire and Peeblesshire, 23 lb, Berwickshire and Selkirkshire, 23 1/2 lb, Argyllshire, Buteshire, Dumfriesshire, Roxburghshire, and Sutherlandshire, 24 lb, and Aberdeenshire, 28 lb; wax, 8 lb; wheat, West Riding of Yorkshire, 22 lb; wool, generally 14 lb but Wales 4, 5, 6, 7, 11, 13, 14, 15, 17, 18, 21, 22, 24, and 26 lb, Herefordshire, 12 lb,

Gloucestershire, 12 1/2 and 15 lb, Cumberland and Dublin, 16 lb, Yorkshire, 16, 16 3/4, 17 1/2, 18, and 19 lb, Dumbartonshire, 17 lb, Durham, 18 lb, Northumberland, 18 or 24 lb, Liverpool, 20 lb, Lanarkshire, 22 lb, Peeblesshire, 23 lb, Selkirkshire, 23 1/2 lb, Argyllshire, Ayrshire, Berwickshire, Buteshire, Dumfriesshire, Inverness, Roxburghshire, and Sutherlandshire, 24 lb, Wigtownshire, 24 to 26 1/4 lb, Aberdeenshire, 28 lb, Guernsey or Jersey, 32 lb; and yarn, Cumberland, 16 lb. It was sometimes abbreviated st. and commonly called a half-quartern.—c**1220** Evesham 211: De ecclesia de Stowe quinque solidos vel duas petras ceræ. c**1253** Hall 11: Checun pere [de plum] est de xii lib.... La centeine de cire, sucre, peyuer, cumin, almand, et de alume, si est de xiii peris et di., et checune pere de viii li.; ibid 12: La sem de veyr est de xxiiii peris, e checune pere est de V li. c**1272** Ibid 9: Petra [plumbi] constat ex duodecim libris; ibid 10: Item centena zucari, cere, piperis, cimini, amigdalorum, et allume continet tresdecim petras et dimidiam; et quelibet petra continet octo libras. **1275** Gras 1.233: xl sackes V peres et demy de laine en xxxix sarpellers et i poke que poisa xvi peres apaie. **1290** Fleta 119: Et quelibet petra ponderat xij. libras in pondere plumbi.... Item summa vitri constat ex xx. petris, et quelibet petra ex quinque libris. c**1300** Hall 8: Duodecim libre et dimidia faciunt petram Londonie. **1304** Gras 1.169: Pro xxii petris canabi. **1341** Rot. Parl. 2.133: C'est assavoir, XIIII livres pur la piere. **1389** Rot. Parl. 3.272: Qatorsze livers al Pere. **1391** Henry Derby 68: Clerico speciarie per manus

Johannis Scorell pro xj stone cere. **1425** Acts Scotland 2.10: That thare be maid a stane for gudes saulde & bocht be wecht the quhilk sall wey xv lele troyes pundes Ande at the stane de diuidyt in xvj lele scottes pundes. c**1425** Account 3.529: Quodlibet blome [con]tinet xv petras et quaelibet petra continet xiii [sic] lb. c**1461** Hall 13: xiiij lb. make a stone. **1474** Cov. Leet 396: xij li. & halfe the halfe quartern, the wich was called of olde tyme...beyng Stone of London. **1540** Recorde 203: In woolle...the 14 pounde is not named halfe quarterne, but a Stone. **1575** Mer. Adven. 57: Every stone of the same flax shall conteign but onelye fowertene pounds, and no more. **1587** Acts Scotland 3.521: The stane contening sextene pund trois. c**1590** Hall 23: 14 poundes waightes haberdepoyse is the stonne of woole; ibid 25: Item waxe and spyce...euery stonne 8 [lb]. **1597** Skene 1.11: The stane to weygh iron, wooll, and vther Merchandice. **1603** Henllys 138: The Stone of wooll is in those partes of the Countrye [Wales] that haunteth the Sheere marketts aforesaid accompted xvij lb. **1606** Hall 38: A halfe quarterne is 14 pounde. c**1610** Lingelbach 122: Of a Sack of woolle aboue four stoane. **1616** Hopton 164: Of Wooll...a stone is 14 pound. **1635** Dalton 149: Beefe and other flesh are 16. ounces averdepois to the pound, and 8 of them pounds to make the stone.... Hemp, 20. li. weight maketh the stone. **1638** Bolton 273: Beefe, and other flesh are 16. ounces Averdepois to the pound, and eight of those pounds to make the stone, except where the usage of the Countrey [= district] requireth more pounds to the stone.... Of Sugar, Spices, and waxe 8. pounds maketh the stone. **1665** Assize 5: And every Stone [of lead] must consist of twelve pound Avoirdupois. **1677** Roberts 297: A Stone of 8 l. 7 l. 10 l. 14 l. 16 l. 20 l. **1678** Du Cange sv petra: Ponderis species, quod constat 12. libris et dimidia. **1707** Justice 7: But Allum, Cinnamon, Nutmegs, Pepper, and Sugar, have...13 Pound and an half to the Stone. **1708** Chamberlayne 207: Butchers commonly allow but eight Pounds to the Stone.... Iron and Shot are Weighed 14 Pounds to the Stone. **1717** Dict. Rus. sv: Stone, a certain Quantity or Weight of some Commodities. A Stone of Beef at London, is the quantity of 8 Pounds; in Herefordshire 12. A Stone of Glass is 5 l. Of Wax 8 l. **1804** Renton 9: Smithfield, of 8 lb. per stone and 16 oz. per lb.... Ayrshire, of 16 lb. per stone and 24 oz. per lb. **1805** Macpherson I.471: 12 1/2 pounds 1 stone of London.... 15 ounces of lead 1 pound, 12 pounds 1 stone. **1819** Cyclopædia sv weights: 8 Pounds...1 Stone of butcher´s meat.... 14 Pounds...1 Stone of horseman´s weight.... Wool [in Gloucestershire] is generally sold by the stone weight of 12 1/2 lbs.... A stone of wool in York market is sixteen pounds.... That at Ripon market, a stone of wool is 16 pounds 12 ounces. And a stone of wool in the Western Moorlands is 17 1/2 pounds.... But that at Darlington...the stone is 18 pounds. And that in the Eastern Moorlands, the weights...vary up to 19 pounds to the stone.... In Liverpool, 20 lbs. are the weight allowed for the several articles under that denomination, as beef, hay, straw, &c.... [In Cumberland] a stone of butcher´s meat 14 lbs., but in many places 16 lbs. **1819** Cooke Preface:

Stones of eight pounds are only used in and near London. **1820** Second Rep. 32-33: Stone, Formerly in London 12 1/2 lb. 31 Ed. 1; that is 1/8 of 100 lb...of alum, 13 1/2 lb...of glass, 5 lb...of hemp, 20 lb. 21 H. 8...Sometimes 32 lb...of hemp, or flax, 16 lb. 24 G. 2...of lead, 15 lbs, each 25 shillings in weight. 31 Ed. 1; that is, each of 6750 grains.... In modern times, 12 lb make a stone of lead. **1829** Palethorpe sv: The stone of glass is 5 lbs,; the stone of meat in London is 8 lbs; a stone of wax is also 8 lbs.... A stone of wool, according to the statute of Henry VII. is to weigh 14 lbs. yet in Gloucestershire it is 15 lbs. and in Herefordshire 12 lbs. **1834** Pasley 112: 1 Market Stone of Wool, in parts of Wales...4 [lb]...5...6...7...11...13...14. 1 Stone of Meat in Bedfordshire, north of the River Ouse...14 [lb]. 1 Market Stone of Wool, the largest of many used in North Wales...15 [lb]; ibid 113: 1 Market Stone of Wool...at Darlington, Durham, and in part of Northumberland...18 [lb].... 1 Market Stone of Wool, in part of S. Wales...21 [lb]...22. 1 Stone of Wheat, in the West Riding of Yorkshire...22 [lb]. 1 Stone of Wool in part of Northumberland, and in some markets of South Wales...24 [lb]. 1 Market Stone of Wool, the largest of many used in South Wales...26 [lb].... 1 Stone of Wool, Guernsey or Jersey...32 [lb]. 1 Stone of Hemp, sometimes...32 [lb]. **1880** Britten 171: In Liverpool, however, a stone of hay or straw is 20 lbs; ibid 176: Stone of hemp, or flax, 16 lbs.; of meat, 8 lbs.... (Cumb.), of hay, tallow, wool, or yarn, and sometimes of meat, 16 lbs. (Durh.), of wool, 18 lbs. (Ess.), of beef, 8 lbs.... (Glouc.), of beef, 8 lbs.; of wool, 12 1/2 lbs.... (Nhumb.), 24 or 18 lbs. (Suff.), of hemp, 14 1/2 lbs.... of meat...(Westm.), 14, 16, or 20 lbs.; of butter, 16 lbs. of 20 ounces each = 20 lbs. (Yks.), of wool, 16 lbs.... of wheat (W.R.), 22 lbs.... [Wales], provincial weights [for wool] of 4, 5, 6, 7, 11, 13, 14, 15, 17, 18, 21, 22, 24, and 26 lbs.; of butcher´s meat, commonly 12 lbs.... (Angus), of potatoes, 16, 20, or 24 lbs. av. (Argylesh.), of butter, cheese, hay, lint, tallow, and wool, 24 lbs. av. (Banffsh.), of hay, 24 lbs. av. (Berwicksh.), of hay, at Berwick, 24 lbs. av.... (Galloway), of flax, 16 lbs.... (Hebrides), of hay, 17 1/2 lbs. av.... (Dumfriessh.), of butter, hay, tallow, and wool, and of cheese sold wholesale, 24 lbs. av. (Invernesssh.), of hay, 20 lbs. Dutch, or about 21 lbs. av.... (Peeblessh.), of hay, 22 lbs. English av. (Renfrewsh.), of hay, 22 1/2 lbs. av.... (Sutherland)...of wool, 24 lbs. av. (Wigtonsh.), of hay, 26 lbs.; of cheese, 24 lbs.; of wool, 24 to 26 lbs. (Dublin), of wool, 16 lbs. av. **1880** Courtney 163: 7 pounds = 1 clove. 2 cloves = 1 stone common articles. **1907** Hatch 34: 1 stone = 6.35029 kilograms. **1951** Trade 28: Stone = 14 pounds.

stong, stonge. STANG

stonne. STONE

stook—5-6 stouk (OED), stouke (OED), stowk (OED); 5-7 stowke (OED); 6 stuk (OED); 6-9 stook; 9 stuck (OED) [ME stouk, stowke; akin to MLG stūke, pile; see WNID3]. A m-c for corn, generally containing 12 SHEAVES and equal to 1/2 THRAVE.—**1820** Second Rep. 34: Sometimes 12 sheaves make a stook.... Devonshire: of thresher's work, 10 sheaves from 7 to 10 inches through at the band. **1880** Britten 160: Stook...ten or twelve sheaves set upright in a double row. **1883** Simmonds sv: Stook, a name for 12 sheaves of corn.

stoon. STONE

stoond, stoonde. STAND

stoone. STONE

stoope, stop, stopa, stoppa. STOUP

stored [perh fr store, in the sense of a supply or provision for future use]. A m-c for corn in North Wales (c1800-1900) containing 2 bu (c7.05 dkl) (Second Rep. 34 and Donisthorpe 219).

stouk, stouke. STOOK

stoup—3 L stoppa (Prior); 4 L stopa; 4-6 stowpe (OED); 6 stoipe, stolp (OED), stop; 6-7 stoope (OED); 6-7, 9 stoupe (OED); 6-9 stoup, stowp [ME stowpe, prob of Scand origin; cf ON staup, cup, MLG stop, Du stoop]. A m-c for liquids, generally wine and honey, varying in size from 1 pt (c0.47 l) to 1 or more gal (c3.78 + l).—**1390** Henry Derby 9: Clerico speciarie per manus eiusdem pro j stopa et di. mellis per ipsum emptis ibidem, xix d.; ibid 14: Clerico buterie super vino per Yngram Northouer pro vij [X] xx [+] viij stopis vini Rochell ab ipso emptis ibidem, le stopa ad v d. lxj s. viij d. **1391** Ibid 39: Et eidem pro cccij stopis vini emptis per manus Johannis Payn, et expensis in hospicio domini ibidem per tempus predictum, le stopa j scot, in toto xij marc; ibid 47: Et pro iij barellis vini de Rynesch continentibus ccclxxvj stopas, le stopa xxxiiij d. **1573** Acts Scotland 3.82-83: That euerie Barrel of Hering & quhite fische contene nyne gallounis of the samin stop. **1587** Ibid 521: And in quantitie and wecht be the said stoipe of stirling. **1618** Ibid 4.587: And the Pinct Stowp/committed to the keiping of the Burgh of Sterline/conteineth the weght of Thrie Pounds seaven unces of frensh Troys weght. **1681** Ibid 8.400: Pynt choppin and mutchen stoups. **1895** Donisthorpe 39: The following came over from Holland: the last, the hogshead, the kilderkin, the firkin, the stoup; ibid 45: We have the pint stoup, the mutchkin stoup, and the gill stoup.

stoupe. STOUP

stowk, stowke. STOOK

stowp, stowpe. STOUP

strica, strik, strika. STRIKE

strike—3 L estrica, L strica (St. Paul´s); 3-4 L estarium (Prior), L esteria (Prior), L estricha (Prior), L estricum; 4 estrike, strikill, strykel; 4-6 strik (OED), stryk; 4-8 stryke; 4-9 strike; ? L strika (Prior), L strikum (Gras 2) [ME strik, strike fr vb striken, in the sense of leveling or scraping off with a straight instrument called a strike or streek]. A m-c for grain generally containing 2 bu (c7.05 dkl) and equal to 1/4 SEAM. In some of the shires, however, strikes of 1/2 to 4 bu (c1.76 to c14.10 dkl) were occasionally used. The strike was commonly called a half-coomb. Since the establishment of the Imperial system in 1824 the strike has been reckoned at 2 bu (7.274 dkl) and equal to 1/2 COOMB and 1/4 QUARTER. Occasionally it was abbreviated str.—**1208** Bish. Winch. 7: Summa, ccxlviij quarteria, dimidium, j estrica, j hopa.... ccxlviij quarteria, dimidium, j estricum, j hopa. **1350** Rot. Parl. 2.230: Et que les Estrikes soient auxi bien enseales, come Bussels & autres Mesures. **1393** York Mem. 2.13: Et j strykel ligni. **1395** Ibid 10: Et unum strikill ligni. **1434** Cov. Leet 151: That all the strikes of this Cite shuld be accordant to the standard made and delyuered vnto this Cite in the tyme of John. **1440** Palladius 21: A strike is for vi [X] xx con daies mete. **1534** Fitzherbert 21: Well sowen with two London busshelles of pease, the whyche is but two strykes in other places. **1540** Nottingham 378: We thengk that the brassen stryk be lefful acordyng to the 5 Kynges Standard, after viij gallans to the stryke. **1566** Recorde Kv: The bushell ther called a Stryke. **1616** Hopton 162: All kind of graine is measured by...strikes, or halfe coombs. **1628** Hunt C: Two Bushells in a Strike; [2] Strikes in a Combe. **1682** Hall 30: 1 Last conteynes: 10 Quarters...40 Strikes, 80 Bushels. **1756** Rolt sv: Or Stryke. A measure, containing two bushels. **1780** Paucton 810: Strike = 2 bushels = 8 pecks. **1820** J. Sheppard 45: 1 Bushel, (commonly called a Strike). **1820** Second Rep. 34: Strike...a measure of corn, varying in its contents from 1/2 to 1, 2, and 4 bushels. **1877** Leigh 202: Strike...A bushel. **1888** Fr. Clarke 37: A strike is 2 bushels. **1931** Naft 197: 1 Quarter = 2 coombs = 4 strikes. **1966** O´Keefe 671: 1 strike = 2 bu. = 72.735 l.

strikill, strikum. STRIKE

stroak [prob fr STRIKE]. A m-c for corn in northern England (_c_1800) containing 2 pk (_c_1.76 dkl) or 1/2 bu (Hunter 124).

stryk, stryke, strykel. STRIKE

stuck, stuk. STOOK

styck, styk, styke. STICK

suling, sulinga, sulinge, sullinga, sullung. SULUNG

sulung—1-? sulung (Select Doc.); 1-3 sullung (Round 2), L suolinga (Round 2); 3 L sullinga; 7 L sulinga, sulinge, L swulinga; 9 suling [OE _sulung_ fr _sulh_, a plow, an area of land]. A m-a for land in Kent generally equivalent to a HIDE. Sometimes it is described as the equivalent of 4 YOKES OF LAND.—**1204** Cur. Reg. 10.208: Kent.—Jurata inter priorem Roffensem petentem et Hugonem de Bosco tenentem utrum debet forinsecum servitium de tenemento suo de Brunhee de una sullinga vel de dimidia sullinga ponitur in respectum usque in octabas sancti Yllarii pro defectu juratorum. **1664** Spelman 530: Swulinga.... Rectiùs fortè _Sulinga_. Dici enim mihi videtur a Saxonico _sul_ vel _sulh_, id est _aratrum_; _idemque_ significare quod carucatus terræ; hoc est quantum sufficit ad annuum pensum unius aratri. **1695** Kennett Glossary sv selio: From the _Sax_. sul came the Lat. _Sulinga_, old Eng. _Sulinge_, a Plough-land. **1867** C. I. Elton 124-25: Occasionally the number of sulings and of carucates in a manor is the same.... Generally, however, the measurements disagree; _ibid_ 126: Instances of measuring by the hide in Kent are comparatively rare compared to those where the suling is taken as the standard of mensuration; _ibid_ 131: The yoke was the fourth part of the suling, and varied in size from forty to fifty of our acres, or a little more; ibid 132: As to the dimensions of the suling.... There are more reasons in favour of an estimate of 160 acres of arable than of any other. In some of the manors, however, of the see of Rochester it contained 180 acres, and in the Isle of Thanet and the neighbouring possessions of the Abbey of St. Augustine 200 and even 210 acres. **1888** Taylor 160-61: The Kentish ploughland or ´sulung´ is the English equivalent of the carucata of the Danish shires. Its area was the same, for if 2 1/2 sulungs contain 450 acres, one sulung would contain 180 acres. **1904** Vinogradoff 283: Kentish documents...keep a special reckoning in sulungs (solins) or aratra, with subdivisions termed yokes (iuga), at the ratio of four yokes to the sulung. See BOVATE; CARUCATE; KNIGHT´S FEE; VIRGATE

sum—5 somme (Brokage II), summe; 6 some, sum; 7 sowme [ME summe, somme fr OF summe, somme fr L summa fr summus, highest]. A wt of 10,000 lb (4535.900 kg) for nails.—**c1461** Hall 17: A summe...conteynythe X [X] M lbs.: and therby be sold paten nayle, sadelers´ naylys, cardemakers´ nalys, and dyuers odyr. **1545** Rates 1.31: Patten nayles the same. **1590** Rates 2.18: Harnesse nailes the sum containing x [X] M; ibid 25: Nailes called patten nailes the sum containing x. thousand; ibid 35: Sprigs the sum containing x [X] M. **1612** Halyburton 322: Copper naillis rose naillis and saidleris naillis the sowme contening ten thowsand. **1756** Rolt sv scavage: Nails...the sum containing 10,000. **1895** Donisthorpe 219: SUM: of nails, 10,000.

suma, summa, summagium. SEAM

summe. SUM

suolinga. SULUNG

swod [prob var of swad, pod, shell, prob fr OE swethian, to bind]. A m-c for fish in Sussex (c1850), a basket generally holding the equivalent of 1 bu (c3.52 dkl) (Cooper 81).

swulinga. SULUNG

systern, systerne. SESTER

szhide. SHIDE

T

taal, taale. TALE

tablespoonful [tablespoon + -ful]. A culinary m-c containing 1/2 fluid oz or 4 fluid dr (14.206 ml) and equal to 2 DESSERTSPOONSFUL or 4 TEASPOONSFUL (Stevens 3 and Economist 8).

tail, taile, taill, taille. TALE

tale—2-9 tale; 3-5 talle (OED); 3-6 tayle (OED); 4 taal (OED), taale (OED), tayl (OED); 4-5 taille (OED); 4-7 tail (OED); 5 tayll (OED), taylle (OED); 5-6 taill (OED); 5-7 taile; 6 tell (OED), telle (OED); 6-9 teale (OED) [ME tale, talk, narrative, list, fr OE talu; akin to ON tal, tala, number, speech, Sw tal, Dan tal, number]. A m-q for the number or count of fish or other products in a C, M, or LAST.—c**1461** Hall 18: Also there be odyr merchandyse that go by tale. **1566** Recorde Kiiij: Hearinges also are sold by the tale, 120 to the hundred, tenne thousande to the last. **1590** Rates 2.39: Whetstones the C. by tale. **1603** Henllys 139: Oysters are allso sold by tale, as by hundred and thousand and not by the bushell, as ys used in London. **1635** Dalton 149: Also Herrings are sold by taile, sc. six score herrings shall goe to the hundred, ten hundred to the thousand, and ten thousand to the Last. **1665** Sheppard 60: Also Herrings are sold by Tale, viz. Six score Herrings go to the 100 [one hundred]; 1000 [ten hundreds] to the Thousand and Ten thousand to the Last. **1699** Hatton 1.154: Other Commodities of Tale are bought and sold by the C. Five-score to the C. except headed ware, to wit, Cattel, Nails, and Fish. **1755** Postlethwayt II.190: Things that are not sold by weight and measure, are sold by tale. See COUNT

talle. TALE

talvett. TOVET

tancard, tancarde, tanckerd. TANKARD

tankard—4-5, 8 tancard (OED); 4-9 tankard; 5-7 tankerd (OED); 6 tancarde (OED), tanckerd (OED), tankarde (OED), tankert (OED); 7, 9 tanker (OED) [ME tankard, of obscure origin; cf MDu tanckaert]. A m-c for ale containing 1 qt (cl.15 1). It was originally a large, wooden tub, hooped with iron or leather staves, which was used principally for carrying water.—**1820** Second Rep. 34: Tankard of ale, a quart.

tankarde, tanker, tankerd, tankert. TANKARD

tapnet—6 tapnett, topnet (OED); 6-9 tapnet; 7 topnett [of obscure origin; possibly a measure whose contents were drawn from a tap (OED)]. A m-c, a basket made of rushes, used for importing figs, totaling 20 to 30 lb (9.072 to 13.608 kg).—**c1550** Welsh 128: 1 tapnet figs; ibid 191: 124 tapnetts figs. **1612** Halyburton 310: Figgs the topnett contening xxx pound. **1883** Simmonds sv: Tapnet, a frail or basket made of rushes, &c. in which figs are imported.

tapnett. TAPNET

tarcian. TERTIAN

tavort. TOVET

tayl, tayle, tayll, taylle. TALE

teacupful [teacup + -ful]. A culinary m-c containing 5 fluid oz (14.206 cl) or 1 GILL (Economist 8).

teale. TALE

tearce, tearse. TIERCE

teaspoonful [teaspoon + -ful]. A culinary m-c containing 1/8 fluid oz or 1 fluid dr (3.55153 ml) and equal to 1/2 DESSERTSPOONFUL or 1/4 TABLESPOONFUL (Stevens 3 and Economist 8).

teers, teirce, teirse. TIERCE

tell, telle. TALE

ten—6-8 tenn; 6-9 ten [ME ten, tene, tenn, ten, fr OE tēn, tēne, ten]. A m-c for coal in Newcastle generally considered equal to 10 Newcastle CHALDERS or to 1 KEEL of 21 TONS or 420 Cwt (21,336.840 kg).—**1603** Hostmen 46-47: Provyded that there be no staie maide of delyverie till that quarter which shalbe formost haue delyvered One Thousand Tens of Coles. **1675** Ibid 137: Twenty seaven thousand tenns of Coales.

tenn. TEN

terce. TIERCE

tercian, terciane, tercien, tercion, tercyan, tercyen. TERTIAN

ters. TIERCE

tersan. TERTIAN

terse. TIERCE

tertian—4-6 terciane, tercyan (OED), tertiane; 4-7, 9 tercian; 5 tercien (Southampton 1); 5-9 tertian; 6 tarcian (OED), tercyen (OED), tersan (OED), tertione; 7-8 tercion; 8 tiertian [ME tercian fr L tertianus, of the third]. A m-c for wine, oil, and honey containing 84 gal (c3.18 hl) and equal to 1/3 TUN of 252 gal. It was synonymous with the PUNCHEON and was double the TIERCE of 42 gal.—**1423** Rot. Parl. 4.256: Plese it to your wise discretions tendirly to consider, howe that of ald tyme ordined and trewly used Tonnes, Pipes, Tertians, Hoggeshedes of Wyn of Gascoign.... The Terciane XX [X] IIII [+] IIII galons.... The pipe XX [X] VI [+] VI galons, and so aftir the afferant the Tercian, and the Hoggeshede of Wyn of Gascoign. **1439** Rot. Parl. 5.30: Chescun Tercian XX [X] IIII [+] IIII galons; ibid 31: Toutz maners Tonels, Pipes, Tertians & Hoggeshedes, taunt de Vin, come de Oyle & Mele.... Ascune Tonelle, Pipe, Tertiane ou Hoggeshed, de vin, Oyle ou Mele. **1540** Recorde 206: Of wine and oyle the Tertian holdeth 84 Gallons. **1587** Stat. 267: The tercian lxxxiiii. galons. c**1590** Hall 21: The tertiane or punchione of a tunne, which is 1/3 part of a tunne, contenith 84 gallons; ibid 23: And so the kilderkynne, firkyn, and tertione. **1595** Powell C: Euery Tertian fourscore and foure gallons. **1615** Collect. Stat. 466: The pipe 126 gallons, and so after the rate, the tercian, & the hogshead of Gascoine wine; ibid 467: Euerie Tercian fourescore and foure gallons. **1656** Howes 4: 84 gallons, or 2 Terces, is a Tercion, or Punchion. **1665** Sheppard 58-59: The Tercian fourscore and four Gallons. **1731** Hatton 2.22: Ton Pipes Hogsh. Tiertian Gallons Quarts. **1805** Macpherson I.638: The tertian and hogshead in proportion. **1820** Second Rep. 35: Tertian of wine, 84 gallons. **1868** Eng. Cyclo. 825: The puncheon was anciently called the tercian (of a tun). See PUNCHEON

tertiane, tertione. TERTIAN

teyrse. TIERCE

thousand—1-3 thusend (OED); 2-3 thusennd (OED), thusent (OED); 3 thousunt (OED), thusand (OED), thusund (OED); 3-4 thousend (OED); 3-7 thousande; 4-6 thowsande; 4-7 thowsand; 4-9 thousand; 5 thouzand (OED) (see additional var in OED sv thous- and thows-) [ME thousend, thusend fr OE thūsend; see WNID3]. A m-q, the mil (M), generally 10 times larger than the C of 100, 106, 120, 124, 132, 160, and 225; and a wt, the M, generally 10 times larger than the Cwt of 100, 104, 108, 112, and 120 lb. Hence, for example, if the C for the product or the region numbered 120, the corresponding M numbered 1200; if the Cwt for the product or the region weighed 112 lb (50.802 kg), the corresponding M weighed 1120 lb (508.020 kg).—c**1461** Hall 12: And of thowsands be made mylyons; ibid 13: Also of this Weyght there goo v [X] xx [+] xii lb. to the C; and x [X] c make a M of ony weyght.... xxviij lb. [make a quarter Cwt]; lvj lb. make half a C; v [X] xx [+] xij lb. make a C...and x [X] c make a M. **1534** Fitzherbert 115: And howe he fed fyue thousande with two fysshes. **1545** Rates 1.2: Alhastes the thousande; ibid 4: Bawels the thowsande. **1549** Gras 1.627: Pro M waight rosen´. **1559** Fab. Rolls 353: In byllot or shydes, with ii thowsand lyeng on Lynts grene, iiij thowsand. **1578** Mer. Adven. 100: Yt is enacted, &c., That whosoever, broother or suster of this Feoloship aforesaid, that shall rate him or herselfe at twoo thowsande skinnes. **1601** Wheeler 25/151: At least sixtie thousande white Clothes, besides coloured Clothes of all sortes. **1603** Henllys 139: Oysters are allso sold by tale, as by hundred and thousand. c**1610** Lingelbach 61: Thowsand waight of Tallow, starch, alome, or suche lyke ware. **1615** Collect. Stat. 465: But a last of herrings containeth ten thousand, and euery thousand containeth ten hundred, and euery hundred sixe score. **1616** Hopton 162: Herring are counted by the hundreds, thousands, and Lasts. c**1634** Hall 51: Item hee hath seized and taken awaie many thowsands of those waightes. **1635** Dalton 149: Six score herrings shall goe to the hundred, ten hundred to the thousand, and ten thousand to the Last. **1678** Du Cange sv miliarium: Mille pondo librarum. **1708** Chamberlayne 205: Herrings 120 to the C...12 Hundred to the Thousand.

thousande, thousend, thousunt, thouzand, thowsand, thowsande. THOUSAND

thraf, thrafe, thraive, thraue, thrava. THRAVE

thrave—2-4 L trava; 3-4 L thrava; 4-6 threve (Fountains), L trefa (Fountains); 5 thraf (OED), threfe (OED); 5-6 thrafe (OED), thraue (OED), thrawe (OED); 5-9 thrave; 6 thravffe (OED), thrayf (OED), threafe (OED), threff (OED), threif (OED), threiff (OED); 7 thref (OED); 7-9 threave; 8 threive (OED); 9 thraive (OED), thrief (OED), thrieve (OED) [ME thrave, threve fr OE threfe, of Scand origin; cf ON threfi, thrave, Dan trave]. A m-c for grain and straw for thatching, generally containing 12 or 24 SHEAVES, but there were several exceptions: Derbyshire, corn, 2 KIVERS or SHOCKS equal to 24 sheaves in all; Durham, straw, 12 BATTENS equal presumably to 12 sheaves; Fifeshire, wheat, 20 sheaves; Gloucestershire, straw, 24 BOLTINGS or TRUSSES of 24 lb each or 576 lb (261.268 kg) in all; Kincardineshire, 2 STOOKS of 12 sheaves each; West Lothian, wheat, 14 sheaves; and Yorkshire, straw, 12 BUNDLES.—**1164** Malcolm 264: Et quatuor clauinos farine et decem trauas euene. c**1290** Worcester 59b: Unam thravam de frumento. **1621** Best 184: 40 threaves of mown rye in the lathe. **1820** Second Rep. 15: Thrave of corn, Derbyshire: 2 kivers...of straw, Gloucestershire: 24 boltings or trusses, of 24 lb. each = 576 lb.... Threave...Yorkshire, E.R. 12 bundles, not precisely limited in magnitude. **1849** Dinsdale 134: Threave...A bundle of straw equal to 12 battens. **1880** Britten 161: Threave...twenty-four sheaves. In Yorks., twelve loggins or bundles of straw is a threave. In West Lothian, fourteen sheaves of wheat is a threave. In Fifesh., twenty sheaves of wheat; ibid 177: Of corn in reaping (Kincardinesh.), 2 stooks of 12 sheaves each.

thravffe, thrawe, thrayf, threafe, threave. THRAVE

thredendel—5 thredendel; 6 thurdendell, thyrdendell; 6-7 thurdendel [ME threde, thurde, thyrde, third, + (n), + -del, -dell, part, fr OE dǣl, part]. A m-c used by innkeepers or taverners for the resale selling of ale and beer; often reckoned as 1/3 gal (c1.50 l) or sometimes as 3 pt.—**1423** Rot. Parl. 4.256: And howe nowe late by sotilte and ymagynation bene made of lesse mesure, in deceite of the peple; that is to say, Tonnes of Wyn of XX [X] XI galons or lesse, Pipes of Wyn of XX [X] V [+] XII galons or lesse, thredendels and hoggeshedes so aftur lesse mesure. **1517** Hall 50: As the kynges standard´ pynte hathe made all hother mesurys, he makythe the thyrdendell´ that inholderes & typlers sell stale ale bye, iii pyntes for a peny potte, that ys, accordyng to the thyrdendell´ in the kynges Escheker. Thys measure owghte to be usyd & alowyd. **1595** Powell F: The which are named or called, hooped quart and pinte measures, thurdendells, and halfe thurdendells, being a small quantitie some what bigger then the foresaide standerd: in respecte of the working and assending of the yeste and froth as aforesaide.... And by the same thurdendel, and halfe thurdendel, the Victulers shall retaile their drinke, being after the rate of three pence the gallon. **1665** Assize 12: The which are named and called hooped quart and pint measures, Thurdendels, and half Thurdendels, being a small quantity somewhat bigger then the aforesaid Standard; in respect of working and ascending of the Yest and Froth.

thref, threfe, threff, threif, threiff, threive, threve, thrief, thrieve. THRAVE

thurdendel, thurdendell. THREDENDEL

thusand, thusend, thusennd, thusent, thusund. THOUSAND

thymber. TIMBER

thyrdendell. THREDENDEL

tierce—4-8 terse; 4-9 tierce; 5 tyerce (OED), tyrse (OED); 5-9 terce; 6 teers (OED), teyrse (Dur. House); 6-7 tearce; 7 tearse (OED), teirse (OED), ters, tierse; 7-8 teirce [ME terce, tierce fr MF tierce fr fem of ters, tiers, third, fr L tertius, third]. A m-c for wine, oil, and honey containing 42 gal (cl.59 hl) and equal to 1/2 TERTIAN, or 1/3 PIPE of 126 gal or 1/6 TUN of 252 gal. The Irish tierce (c1800) contained 9139.2 cu inches (1.498 hl) or 42 Irish gal of 217.6 cu inches each

(Edinburgh XII.572). Since the establishment of the Imperial system in 1824 the tierce of beer has been reckoned at 42 gal (1.909 hl) throughout the United Kingdom. Occasionally it is abbreviated tc.—c**1590** Hall 21: The terce of a pipe or a butt, which is 1/6 part of a tunne, contenith 42 gallons. **1607** Cowell sv ters: Ters, is a certaine measure of liquide things, as wine, oyle, &c. conteining the sixth of a Tunne...or the third part of a pipe. **1616** Bullokar sv tearce: The sixt part of a tun, and the third part of a pipe. **1616** Hopton 161: 1 Tierce...42 Gallon. **1628** Hunt C: 3 Teirce in a Butt. **1665** Sheppard 14: A Terse or Tierse is but the 6th part of a Tun, or third part of a Pipe. **1677** Roberts 296: Each Hogshead to contain 63 Gallons, every Terce 84. **1682** Hall 29: 1 Tunne conteynes: 2 Pipes or Butts...6 Tierces. **1695** Wigan sv Liquid Measure: 42 Gallons...1 Teirce. **1708** Chamberlayne 210: Of these Gallons...a Tierce of Wine holds 42 Gallons. **1710** Harris 1. sv measures: The common Wine Gallon sealed at Guild-Hall in London...is supposed to contain 231 Cubick Inches; and from thence the Tierce will contain 9702 Cub. Inch. **1716** Ibid 2. sv measures: Terces...42 Galls. **1717** Dict. Rus. sv: Tierce or Terce, a Liquid Measure containing Forty two Gallons. **1756** Rolt sv: Or Teirce. A measure of liquid things, as cyder, wine, oil, or the like, containing the third part of a pipe, or 42 gallons. **1778** Diderot XXI.673: Tierce...42 Gallon. **1816** Kelly 87: 42 Gallons...1 Tierce...158,9673 [1]. **1820** Second Rep. 35: Tierce of wine, 42 gallons. **1880** Courtney 161: The barrel, hogshead, tierce, pipe, butt and tun, are the names of casks. **1956** Economist 54: Tierce = 42 gallons.

tierse. TIERCE

tiertian. TERTIAN

timber—3 L timbrium, L tymbra; 3-4 L tymbrium; 3-9 timber; 4 L timbra; 4-5 tymbre; 5-7 tymber; 6 thymber, tymbber (Gras 1); 7 L timbria; 9 timbre [OF timbre fr MLG timber, timmer, lumber; so called because the fur skins were packed and shipped between two heavy boards]. A m-q generally numbering 40 for the following fur skins: beaver, calaber (squirrel), cony (rabbit), ermine, ferret, fitch (fitchew), gray (badger), jennet, martin, miniver, mink, otter, and sable. The beaver, jennet, miniver, and otter furs sometimes were sold singly rather than as a timber. Most other fur skins were measured by the C.—c**1253** Hall 12: La timber de peus de cunnis et de gris, ou ver, est de xl peus. c**1272** Ibid 10: Tymbra vero de pellibus cuniculorum et grisonum constat ex quadraginta pellibus. **1290** Fleta 120: Lunda autem pellium continet xxxij timbria, et seuellio cuniculorum et de grises continet xl. pelles. c**1300** Brit. Mus. 5.151: Timber de pellibus cuniculorum...constat ex .xl. pellibus. **1303** Gras 1.166: Timbra squirellorum. **1323** Ibid 210: De quolibet tymbrio de grys. **1392** Henry Derby 92: Et per manus Johannis Dyndon pro j furrura de grys per ipsum empta ibidem de vj tymbre, et de ij tymbre de meniuer, xij nobles. **1406** trans in Cal. Close 19.33: Six ′tymbre′ of ′menyver′. c**1461** Hall 17: And xl fells make a Tymber. **1507** Gras 1.695: Bever wombys the tymber.... Armyns the tymber that ys to saye the xl skynes; ibid 696: Callabur rawe that ys to saye xl skynes the tymber; ibid 698: Gray tawyd the tymber...that ys to saye xl skynnes.... Gray ontawed [the] thymber; ibid 700: Marterns the tymber.... Mynkes the tymber; ibid 701: Otter the tymber that ys to saye xl skynnes to the tymber; ibid 703: Sablles the tymber. **1590** Rates 2.2: Armines the timber containing xl. skinnes; ibid 7: Callaber tawed the timber containing xl. skins; ibid 24: Marterons tawed the timber containing xl. skinnes. **1615** Collect. Stat. 465: A timber of conie skins and grayes consisteth of 40. skins. **1616** Hopton 164: Sables, Martins, Minkes, Jenits, Fitches, & Grayes, haue 4[0] skins in the Timber. **1664** Spelman 540: Timbria.... Pellium dicitur earum certus numerus puta 40. **1665** Sheppard 57: A Tymber of Cony-skins and Grayes, consisteth of 40 Skins. **1682** Hall 30: Sables, Martins, Minks, Ferrits, Fitches, and Grayes: 40 Skins in a Timber. **1708** Chamberlayne 205: Of Furrs, Fitches, Grays, Jennets, Martins, Minks, Sables, 40 Skins in a Timber. **1717** Dict. Rus. sv: Timber, of Furrs, i.e. Fitches, Genets, Grays, Marterns, Sables, &c. is forty Skins. **1805** Macpherson I.471: 40 skins of conies or grise 1 timber. **1883** Simmonds sv timbre: A legal quantity of 40 or 50 small skins, packed between two boards.

timbra, timbre, timbrium. TIMBER

tirrs. TRUSS

toad. TOD

tobit. TOVET

tod—5 tode (OED), toode; 5-7 todde; 5-7, 9 todd; 6 toad (OED); 6-7 tood; 7-9 tod [ME todd, todde; cf LG tod, todde, bundle, small load]. A wt for wool of 28 lb (12.700 kg), equal to 1/13 SACK or 1/26 SARPLER. In some of the shires, however, the following variations were found: Gloucestershire and Yorkshire, 28 1/2 lb (12.927 kg); Bedfordshire, sometimes 29 lb (13.154 kg); and Sussex, Guernsey, and Jersey, 32 lb (14.515 kg).—c**1461** Hall 13: Woll...is most used to be sold by the Stone and by the Todde; ibid 19: That ys to say...Todde content´ xxviij [lb].... The Todd amountythe in Poundes xxviij; ibid 20: The Sacke amountythe in...Toddes xiij. **1479** Cely 21: And bogwyt for me in Cottyswolde xxxvij sacke be the toode and sacke and halfe sacke. **1540** Recorde 203: In woolle, 28 pounde is not called a quarterne, but a Todde. c**1590** Hall 23: 28 poundes waight haberdepoyse is the tood of woole. **1603** Henllys 138: By the todde there is none sold except yt be to an Englishe buyer, that cometh of purpose, and maketh his bargaine by the todde, as a weight best knowne to himself. **1616** Hopton 164: A Todde is 28 pounds, or two stone. **1635** Dalton 149: 28 pounds goeth to the Tod. **1665** Sheppard 63: For Wooll, some say 14 pound goeth to the Stone; 28 pounds to the Todd; ibid 64: Each Tod 2 stone, each stone 14 pounds. **1678** Du Cange sv todde: Pondus 28. librarum, Angl. Tod. **1682** Hall 31: A Todd, 2 Stone; a Stone, 14 pound. **1695** Kennett Glossary sv todde: A tod of wooll is a parcel containing twenty eight pounds or two stone.... But in these parts the wooll-men buy in twenty nine pounds to the Todd, tho´ they sell out but twenty eight. **1701** Hatton 3.9: 7

Pound—1 Clove. 2 Cloves—1 Stone. 2 Stone—1 Tod. **1708** Chamberlayne 207: Wooll is Weigh'd by the...Tod, i.e. 28 Pounds; ibid 208: Tod 28 Pound, to 1 Sack 13 Tods. **1717** Dict. Rus. sv: Tod of Wooll, the quantity of 28 Pounds or 2 Stone. **1819** Cyclopædia sv weights: 2 Stone...1 Todd. **1820** Second Rep. 35: Tod of wool, 2 stone = 28 lb. **1834** Pasley 113: 1 Tod of Wool in Gloucestershire, and in Holderness, Yorkshire...28 1/2 [lb]...Bedfordshire, sometimes 28, and sometimes 1 lb. over. **1880** Courtney 163: 2 stone = 1 tod of wool. **1880** Britten 177: Tod, of wool, 2 stone = 28 lbs.... (Beds.), 28 lbs., and sometimes a pound over for pitch-marks, making 29 lbs. (Glouc.), 28 1/2 lbs. (Suss.), 32 lbs. (Yks., Holderness), 28 1/2 lbs. (Guernsey and Jersey), 32 lbs. **1956** Economist 58: 1 tod [of wool] = 28 lb.

todd, todde, tode. TOD

tofet, toffet, toflet. TOVET

toltrey [tolt, toll, + rey, king; so called because it was the fixed toll in kind on salt, paid by the men of Malden to the Bishop of London]. A m-c for salt (c1400) containing 2 bu (c7.05 dkl) (Prior 158).

tolvet, tolvett. TOVET

ton—4-7 tonne; 5 tone (Southampton 1), toun (OED); 5-7 tunne; 5-9 ton; 6 tune; 6-7 toonne, town (Halyburton); 6-9 tun; 7-8 tunn; 8 tonn [ME tonne, toun, unit of ship capacity or of weight; originally the same word as TUN]. A wt generally containing 20 Cwt or 2 M of 10 Cwt each. The total wt was 2000, 2160, 2240, or 2400 lb (907.180, 979.760, 1016.040, or 1088.620 kg) depending on whether the corresponding Cwt contained 100, 108, 112, or 120 lb. It is commonly abbreviated t. or tn.—**1440** Palladius 118: Or lette a tonne of barly him comprende. c**1461** Hall 13: v [X] xx [+] xij lb. make a C...and x [X] c make a M; and ij M make a Tunne. **1507** Gras 1.699: Iryne the tune. **1524** Ibid 197: Pro xx tonne de cave stonne. **1555** York Mer. 155: A tonne of iron, accomptinge sex score endes to the tonne, receyved on the shipe bord. c**1590** Hall 24: The tunne is 20 hundrid waight, conteninge 2240 poundes waight haberdepoyse; after the ratte 112 poundes to the 100; ibid 27: The hundred of irne is but 5 skore to the hundred and 20 hundrid to the tunne; ibid 28: Euery loade of wood ought to be 20 hundred waight, which is a tunne. **1603** Henllys 139: Iron is sold by the stone w[hich] consisteth of xvi haberdepoys, of which stones viij make the C. of Iron, and xx [X] c make a toonne. c**1610** Lingelbach 114: All other Iron by the Tonne. **1616** Hopton 163: Iron is counted by the pound, hundred, and Tun. **1657** Tower 547: That all owners of ships, during such time as they shall serve the King, may have 3 s. 4 d. for every Tonne over and above his fraight, according to the Custome. **1682** Hall 30: A Tunne of iron. **1689** Child 9: For Bantam 3 Ships, two of them 360 Tuns a peece, the third 600 Tuns. **1704** Mer. Adven. 244: Ffor carrying of every tonn of iron and weighing the same at the weighouse, belonging to a fforreigner; ibid 245: Every tunn of oak timber. **1708** Chamberlayne 206: 16 Drachms make an Ounce, 16 Ounces a Pound, 28 Pounds a Quarter, 4 Quarters an Hundred, 20 Hundred a Tun. **1728** Chambers 1.360: Tun...2240 Pound. **1740** Barlow 458: The word Ton is applied both to Weight and liquid Measure; viz. because the same Quantity of Liquor is a Ton both in Weight and Measure. **1778** Diderot XXVI.422: Tonn...2240 Livre. **1779** Swinton 31: Coals are also sold by the tun of 20 cwt. **1784** Ency. meth. 137: Londres. Le ton, tun...est de 20 hundreds, ou de 2240 l. **1789** Hawney 7: Let .43569 of a Ton be reduced to Hundreds, Quarters, and Pounds. **1817** Keith 1.11: An Avoirdupois tun of 2240 neat pounds. **1829** Palethorpe sv: TON, a weight containing 20 cwt. of 112 lbs. each, or 2240 lbs. of what is called avoirdupois weight. A ton of gold and silver weighs 2000 lbs. **1830** Crüger 157: 1 Stone (Stein) hat 14 lb, 8 solcher machen das Centweight, wovon 20 die Ton ausmachen. **1849** Murphy 54: Eight barrels of twenty stone each (a ton). **1880** Courtney 172: A ton of hay, or any other coarse bulky article usually sold by that measure, is 20 gross hundreds, that is 2240 lbs. **1951** Trade 28: Ton = 2,240 pounds. **1969** And. & Bigg 12: 1 ton = 1016.05 kg.

There were several tons, however, that did not contain 20 Cwt: 1344 lb (609.625 kg) or 12 Cwt for lime in North Wales; 1904 lb (863.635 kg) or 17 Cwt for culm in South Wales; 2520 lb (1143.047 kg) or 22 1/2 Cwt for potatoes in Essex; and 2688 lb (1219.250 kg) or 24 Cwt for coal in North Wales (Pasley 115).—**1851** H. Taylor 61: The ton...in some cases has been more than the usual 20 hundreds of weight. **1907** Hatch 20: 20 hundredweights = 1 ton = 80 quarters = 2240 pounds; ibid 34: 1 ´short´ ton of 2000 lbs. = 0.90718486 tonne...1 ´long´ ton of 2240 lbs. = 1.01604704 tonnes. **1956** Economist 4: Ton: (a) United Kingdom = 2,240 lb.

ton. TUN

tone. TON

toneal, tonel, tonell, tonelle, tonellum, tonellus. TUN

tonn. TON

tonne. TON; TUN

tonnel, tonnell. TUN

tood, toode. TOD

toonne. TON

topcliff [*]. A m-c for tin in Cornwall (c1600-1800) containing approximately 1/2 gal (c1.89 l) (Carew 2.45).

tope [of obscure origin; cf G topf, a pot]. A m-c of uncertain size used in Durham for dry products.—**c1530** Dur. House 263: Item 4 topez of pyese.

topette—5-6 topette; 6 toppet, topynett (Dur. House) [prob dim of TOPE]. A m-c of uncertain size, possibly 1/2 TOPE, in Durham and elsewhere for dry products.—**1443** Brokage 28: Et v topettes rasemorum. **c1550** Welsh 72: 20 toppets...10 pieces figs.

topnet, topnett. TAPNET

toppet. TOPETTE

topston [*]. A wt of 6 1/2 lb (2.948 kg) for wool in South Wales (c1800-1900), equal to 1/4 MAEN (Second Rep. 36 and Donisthorpe 220).

topynett. TOPETTE

toun. TON

tovet—7 talvett (OED), tovitt (OED); 7-8 tofet, toffet (OED); 7, 9 tovit; 7-9 tovet; 9 tavort (OED), tobit (OED), toflet (OED); ? tolvet (OED), tolvett (OED), tuffet (Jones) [of obscure origin; see OED]. A m-c for grain, generally containing 2 pk or 4 gal (cl.76 dkl) and equal to 1/2 bu. It arose as a local measure of Kent.—**1674** Ray 77: A Tovet or Tofet, half a bushel: Kent. **1696** Jeake 81: 1 Bushel 2 Tovits or Half Bushels, 1 Tovit 2 Pecks. **1853** Cooper 82: Tovet, a measure of two gallons. **1858** Shuttleworths 1094: 2 gallons = a peck; 2 pecks (4 gallons) = a tovet; 2 tovets, 4 pecks, or 8 gallons = a bushel. **1868** Eng. Cyclo. 824: There was anciently a dell, or half-bushel (also called a tovit).

tovit, tovitt. TOVET

town. TON

towne. TUN

trava. THRAVE

tray—3-7 trey; 4-? tray (OED); ? treye (Prior), treyy (Prior) [ME tray, trey fr OE trēg, trīg; akin to OSw trȫ, a wooden grain measure, OE trēow, tree, wood]. A m-c for dry products containing 2 SEAMS or 16 bu (5.638 hl).—**1270** Select Cases 2.7: Detinuerunt ei quinque marcas et quinque solidos...pro xj. treys ordei sibi venditis. **1317** Ibid 105: Cum simul emissent xx. treys carbonis maris.

trefa. THRAVE

trenda, trendal, trendel, trendell, trendelle, trendil, trendill. TRENDLE

trendle—1-5 trendel (OED); 4 L trenda; 4-6 trendil (OED); 4-? trendle; 5 trendill (OED), trendull (OED), trendyl (OED); 5-6 trendell, trendelle (OED), trendyll (OED); 7 trendal (OED) [ME trendle fr OE trendel, a circle, ring, dish]. A m-c for wax and other products. It was a round or oval tub of uncertain size.—**1393** trans in Cal. Close 15.173: One trendle (trenda) of wax found at Byrchelton. **1394** trans in ibid 218: One trendle (trenda) of wax found at Dyngemersshe. **1440** Scrope 230: A trendell with iij. bushell of whete malt.

trendull, trendyl, trendyll. TRENDLE

trev [W trev, a vill]. A m-a for land in Wales (c1300) containing 4 GAVAELS or 256 ERWS (c92.42 ha) (Laws Wales 1004 and Lloyd 5-7).

trey, treye, treyy. TRAY

troiss, tross, trosse. TRUSS

trousall, trousell. TRUSSELL

trug [prob a dial var of trough fr ME trough, trogh fr OE trog]. A m-c for grain (c1400) equal to 1/12 SEAM (c2.35 dkl) (Thor. Rogers 1.168).

trus, truse. TRUSS

truss—3 L trussa; 3-8 trusse; 4-6 trosse (OED), trus (OED); 5 troiss (OED), truse (OED), turss (OED); 5-7 turs (OED); 7 tirrs (OED), turse (OED); 7-9 truss; 9 tross (OED) [ME trusse fr OF trousse, bundle, pack, fr trousser, to pack]. A m-c for 1/36 LOAD of hay and generally weighing 56 lb (25.401 kg). In 1795 it was standardized at 56 lb (25.401 kg) for old hay and 60 lb (27.215 kg) for new hay. A truss of straw weighed 36 lb (16.329 kg) or 1/36 load of straw.—**1202** Feet 2.180:

Et in vigilia sancti Ædmundi mittent predictis infirmis sex rationabiles trussas straminis et in vigilia natale domini sex trussas et in vigilia pasche sex trussas. c**1590** Hall 27: 36 trusses makith a loade of haye, and euery trusse is 56 poundes waight haberdepoyse. **1665** Assize 13: Thirty six Trusses of Hay shall make the load, every Truss of Hay to weigh the full weight of fifty six pounds Avoirdupois. **1708** Chamberlayne 38: Hay is sold by the Truss 56 Pounds, and by the Load 36 Trusses. **1819** Cyclopædia sv weight: The truss of new hay is 60 lb. until the 1st of September. **1820** Second Rep. 36: Truss of hay, 56 lb. if old; 60 lb. if new...London: formerly 36 lb. **1850** Alexander 116: Truss; of hay...56.—pounds.... of new hay...60.—pounds. **1880** Courtney 56: A truss of hay, new, is 60 lbs.; old, 56 lbs.... A load of hay is 36 trusses. **1883** McConnell 14: Hay sold between 1st June and 31st Aug. is reckoned new hay, and must weigh 60 lbs. per truss. Hay sold between 31st August and the succeeding 1st June is reckoned old, and must weigh 56 lbs. per truss.... 36 lbs. Imp. of straw = 1 truss.... 36 trusses = 1 load. **1934** Int. Traders´ 86: Truss (new hay)...United Kingdom...60 pounds.

trussa, trusse. TRUSS

trussel. TRUSSELL

trussell—3-5 L trussellus; 4 trussel; 5 trusselle (OED); 5-9 trussell (OED); 6 tursall (OED); 6-7 trousall (Halyburton), trousell (Halyburton), tursell (OED) [ME trussell, bundle, trussell, fr MF troussel, trousel, dim of OF trousse, a bundle, TRUSS]. A m-q for the number of skins or the amount of cloth that formed a convenient bundle.—c**1243** Select Cases 3.lxxxvii: Item ij. trusselli de pellipatorio, qui custaverunt xv. marcas. **1323** Gras 1.209: De uno trussello panni cum cordis legatis. c**1360** Hale 137: De chescun trussel de quire en cares.... De chescun trussel de draps. c**1400** Ibid 214: De quolibet trussello de kerseys Walssh russet et mantell´ d´Irland.

trusselle, trussellus. TRUSSELL

tub [ME tubbe, tobbe; see WNID3]. A m-c (c1700-1850) for butter, 84 lb (38.102 kg); camphor, 56 to 86 lb (25.401 to 39.009 kg); corn for export, South Wales, 4 bu (c1.40 hl); tea, 60 lb (27.215 kg); and vermilion, 3 to 4 Cwt (152.406 to 203.208 kg) (Hatton 3.235, Second Rep. 36, Waterston 147, and Pasley 113).

tuffet. TOVET

tumblerful [tumbler + -ful]. A culinary m-c containing 10 fluid oz (28.412 cl) or 2 GILLS or 2 TEACUPSFUL (Economist 8).

tun—3 L tonnellus; 3-4 L tonellum, L tonellus, L tunellum; 4 toneal; 4-5 tonel, tonelle (York Mem. 1); 4-8 tonne; 5 tonell, tonnel (Nicholson), tonnell; 5-7 tunne; 6 towne (Dur. House), tune (OED), twn; 6-7 twne (Halyburton); 6-9 tun; 8 ton, tunn; ? L tunellum (Prior) [ME tonne, tunne, a tun, fr OE tunne, a tun, tub, a large vessel]. A m-c for wine, oil, honey, and other liquids, generally containing 252 gal (c9.54 hl), but occasionally 208, 240, and 250 gal capacities were used. It was also known as a DOLIUM. The Irish ale tun (c1800) contained 320 Irish gal (c11.42 hl) or 8 Irish ale bbl of 40 gal each and was equal to 16 Irish ale KILFERKINS or 32 Irish ale FIRKINS (Edinburgh XII.572). Since the establishment of the Imperial system in 1824 the tun for ale and beer throughout the United Kingdom has been reckoned at 216 gal (9.819 hl).—c**1205** Hoveden IV.99: Quod nullum tonellum vini...et nullum tunellum vini. **1208** Bish. Winch. 30: Et de xiij s. ix d. de j tonello et quarta parte j tonelli vini pessimi vendito; ibid 41: Et de l x. x s. viij d. ob. de v tonellis vini...unde iij tonella venerant de Wivesseia, ij de Waltham; ibid 55: Idem reddunt compotum de viij tonnellis siceræ proventis de Merewella.... In dono domini Episcopi H. balistario, j tonnellus. **1310** trans in Memorials 74: One aletonne, value 18 d. **1330** Rot. Parl. 2.39: C´est assavoir, de chescun Nief portant vintz tonels.... Issint que nule de eux porta vintz toneals. **1341** Gras 1.174: Pro xii tonellis vini. c**1350** Ibid 182: De ii tonellis oleo. **1390** Heales xci: Deux tonelles de vyn. **1396** Coopers 7: Et auxi facent tonelx vates kemelynes & autres vesseux. **1410** Hale 135: De chescun tonnell de vin. **1423** Rot. Parl. 4.256: The Tonne of Wyn xx [X] xii [+] XII galons. **1439** Rot. Parl. 5.30: De chescun Tonell...de Vin. **1444** Ibid 114: And more overe, wher as of old tyme euery Tonne...heeld the full gauge after the gauge of Englond.... That every Tonne contene XX [X] XII and XII Galons. c**1460** Capgrave 244: Thei took a hundred schippis, in whech thei had nyneteen thousand tunnes of wyn. c**1461** Hall 15: The tonne cont[aineth] ii [X] c [+] l galouns.... Who some euer schall retayle any tunne or pype...he schall rakyn the tunne but xii [X] xx gallounes. c**1517** Ibid 49: The tonne, xii [X] xx & xii galons. **1572** Mer. Adven. 97: By hoggeshed or hoggesheds, tonne or tonnes. **1573** Acts Scotland 3.82: That is to say the Twn of Burdeaux wine and Burdeaux bind for twentie four pundis. **1587** Stat. 267: The tunne of wine cc.lii galons. c**1590** Hall 21: The tunne contenith 252 gallons of liquor.... The tonne, xii [X] xx et xii galons. **1599** Mer. Adven. 57: That if anie person or persons...shall sell within this toune anie wyne, by hogshead or tunne. **1615** Collect. Stat. 466: Tunne of Wine...it containe of English measure 252. gallons. **1628** Hunt B: (252,) Gallons in a Tunne, of full Gage. So commonly there are...240. Gallons in a Tunne. **1635** Dalton 148: Wine, Oyle, and Honey...Tunne: 252. gallons. **1665** Sheppard 14: Two Pipes, a Tun, wherein are 252 Gallons. **1675** Mayne 151: The most Practical Way of GAUGING TUNNS. **1682** H. Coggeshall 2: A long Square Tun, whose length let be 6.2 F. the breadth 5.8 F. the depth 5 F. **1682** Hall 29: 1 Tunne conteynes: 2 Pipes or Butts, 3 Punchions, 4 Hogsheads, 6 Tierces, 8 Barrells, 14 Rundlets, 252 Gallons. **1701** Hatton 3.94: Of Wine Measure.... In 1 Ton, are...6 Tierce or 252 Gallons. **1704** Mer. Adven. 243: Ffor takeing forth of every tonne of wine from a keel or boat. **1708** Chamberlayne 210: A Tun 252 Gallons. **1783** Beawes 918: A Ton 252 Gallons, or 2016 Pints. **1787** Hale 170: Every tunn of wine. **1794** Martin 30: Wine ton is 252 gallons, which, at 231 cubic inches per gallon, is...58212 cubic inches. **1805** Macpherson I.637: A tun of wine...252 gallons. **1820** Second Rep. 36: Tun of wine, 2 pipes = 252 gallons. **1829** Palethorpe sv: TUN, a large vessel or cask, of an oblong form, biggest in the middle, and diminishing towards its two ends; girt round with hoops. **1850** Alexander 116: Tun; for wine; old measure...252.—gallons. **1882** Jackson 287: Tun = 2 butts. **1956** Economist 54: Tun = 216 gallons.

tun. TON

tune. TON; TUN

tunellum. TUN

tunn, tunne. TON; TUN

turs. TRUSS

tursall. TRUSSELL

turse. TRUSS

tursell. TRUSSELL

turss. TRUSS

twn, twne. TUN

tyddyn [W tyddyn, a tenement]. A m-a for land in Wales (c1300) containing 4 ERWS (c1.44 ha). Originally it referred to an area of ground encompassing one homestead (Laws Wales 1005 and Lloyd 7).

tyerce. TIERCE

tymbber, tymber, tymbra, tymbre, tymbrium. TIMBER

tyrse. TIERCE

U

uirga. VERGE

ulna. ELL; YARD

unc, unce. OUNCE

unch. INCH; OUNCE

unche. INCH

uncia. INCH; OUNCE

vadome. FATHOM

vaga. WEY

vathym. FATHOM

verge—3-7 L virga; 3-8 verge; 4-7 virge; ? L uirga (Prior) [ME verge fr MF verge fr L virga, a twig, rod]. Equivalent to YARD and occasionally to PERCH.—c**1300** Brit. Mus. 21.61v: .iij. pedes faciunt virgam.... Quinque virge & di...faciunt .j. perticam. **1308** Gras 1.365: Carcavit iiii [X] c ferri et xx virgas panni lanosi. **1390** Henry Derby 8: Eidem pro xvij virgis j quarterio de blanket ab eodem emptis pro eisdem garcionibus domini, pretium virge, xvj d.... Eidem pro j virga et j quarterio de blu fryse. c**1400** Henley 68: E pur ceo ke les acres ne sunt mye touz de une mesure kar en acon pays mesurent il par la verge de xviii peez e...de xx peez e...de xxii peez e...de xxiiij peez. **1410** Rot. Parl. 3.645: Qare l´ou le court Drap duist teigner la longure de XXVIII verges, il ne tient que XXIII verges, & l´ou le Dussein de Drap duist teigner XIIII verges, il ne tient que XI verges. **1454** Scrope 204: Continentem xij. virgas terræ et j. pedem, qualibet virgâ continente tres pedes. c**1500** Hall 7: iii pedes faciunt virgam; ibid 8: V virge dimidia faciunt perticam. **1580** Kytchin 13: Auxi si ascun ad, & use ascun measures de bushelles, galons, virge, ou aulnes...sont inquirable. **1664** Spelman 8: Pertica verò dimensionis virga, sexdecem pedes & dimid. habens in longitudine. **1688** Bernardi 12: Virga Anglica. 3 Pedes, 12 Palmi, 2 Cubiti, 4 Spithamæ, 4/5 Ulnæ Anglicæ. **1695** Kennett Glossary sv pertica: A Perch, which in the reign of King

John was the measure of twenty foot, and was the same as Virga,—Quælibet virga, unde quarantanæ mensurabuntur, erit viginti pedum. **1755** Postlethwayt II.191: The yard, or verge, being the ordinary measure for cloth, silks, and all other such goods.

vet. FATT

vetheym, vethym. FATHOM

virga. VERGE

virgat, virgata. VIRGATE

virgate—1-7 L virgata; 3-? virgate; 6, 8-9 virgat [MedL virgata fr L virga; see VERGE]. A m-a for land generally synonymous with the YARDLAND and in Sussex with the WISTA. Like the acreage of other superficial measures, its total acreage depended on local soil conditions, but virgates of 12, 15, 16, 20, 24, 28, 30, 32, 36, 38, 40, 44, 45, 48, 50, 60, 62, 64, 72, and 80 acres (c4.86 to c32.40 ha) were the most common. It was generally equal to 1/4 HIDE and was occasionally the sum of 2 or 3 BOVATES or 4 FARTHINGDALES. Occasionally, however, the number of virgates in the hide was as high as 7 (as in Barnwell, Hemington, and Gravele) or as low as 3 (as in Lawshall). It was frequently abbreviated v., vir., or virg. in medieval MSS.—**1086** Barfield Appendix V: De hac terra tenet Almær iij virgatas, Raynerus unam virgatam, Gislebertus j hidam et unam virgatam et dimidiam. c**1124** Malcolm 135: Scilicet .ix. virgatas terre pro .vij. virgatis in Magna Paxtona. c**1130** Slade 14: Ibidem Ricardus Basset iij car´ et j virg´. c**1157** Malcolm 203: Et quatuor virgatas terre in

Cameston´. **1200** Cur. Reg. 8.144: De dimidia virgata terre et de xvj. acris terre cum pertinenciis in Waleton´; ibid 145: De placito dimidie virgate terre. **1204** Cur. Reg. 10.220: Et ipse Gilbertus venit et reddidit Orenge matri sue x. virgatas terre sicut dotem suam; ibid 221: j virgate terre in Wihcthill´. **1212** Cur. Reg. 13.192: De ij. virgatis terre et de dimidia cotlanda cum pertinenciis in Wikefeld. **c1221** Clerkenwell 136: Dimidiam virgatam terre et duas acras cum mesuagio. **1222** St. Paul´s 147: Habet hæc ecclesia unam virgatam terræ liberam ab omni sæculari officio. **1266** Gray 464: Robert Abovetun tenet unam virgatam terre continentem quadraginta quatuor acras in utroque campo. **c1283** Battle xiii: Quatuor virgatæ seu wystæ faciunt unam hydam. **c1300** Bray 8: Henricus de Bray tenet tres virgatas terrae et continent de terra et prato ccvi acras et dimidiam. **c1300** Brit. Mus. 21.61v: Item .xxx. acræ...faciunt unam virgatam terre. Quatuor virgatæ faciunt hidam terre. **1304** Swinfield 221: Heredes Rogeri de la Sale de Hompton´ tenent .ij. virgatas terræ per militiam. **1322** Wellingborough 123: Et de ix. sol receptis de vna virgate pasture vendita. **1338** Langtoft 600-01: Decem acræ faciunt ferdellum. Quatuor fardella faciunt virgatam unam. Quatuor virgatæ faciunt hidam unam; ibid 601: Fardellum Acræ X./virgata XL./hida. CLX. **1454** Scrope 210: Quælibet virgata continens xxiv. acras terræ. **c1500** Hall 8: viii [X] xx pertice faciunt acram; duodecim acre faciunt bovatam; ii bovate faciunt virgatam. **c1500** Brit. Mus. 6.7: j hide...iiij virgat.... j virgat...iiij fferdalles. **1569** Ault 87: Great Horwood, Bucks....

Preceptum est quod quilibet custodiet pro virgata terre xl oves de propriis suis ovibus uel de vicinis suis. **1599** Richmond Appendix 2.11: VIRGATA...in Leverington...is LX Acres, in Fenton XXX Acr. in Tyd. XXXII Acr.... In Coln. Virgata, operabilis XV Acr. And in another Town not named...XX Acr. **1603** Henllys 135-36: There is allso a quantytie of land measure called a yard of land, in latin Virgata terræ. **1607** Gray 434: Johannes Bates, Clericus, Tenet...unum mesuagium, sive Tenementum et dimidiam virgatam terre cum omnibus pertinentibus. **1664** Spelman 558: Virgata terræ.... Aliàs enim 20, aliàs 24, aliàs 30, aliàs 40 acris æstimatur. **1665** Sheppard 22-23: A Yard Land (in Latine Virgata Terræ) is a quantity of Land called by this name, but it is no certain Quantity. For in some places it containeth 20 acres; in others 24 acres; in other places 30 acres, according to the estimation of the Country [= district]. **1688** Holme 137: Virgate of land is 20, in some places 24 Acres, or in some 30 Acres. **1695** Kennett Glossary sv virgata: VIRGATA terræ. A Yard-land.... Two virgates or yard-lands in Chesterton 24. Hen. III. contain'd fourscore and ten acres. **1710** Langtoft 600: The town, according to Domesday Book, consisted of VIII. virgats of Land.... Each virgat comprehending fourty acres. **1755** Willis 358: Virgata, or Yard-Land, whereof 4 make an Hide, was in different Counties 15, 20, 30 or 40 Acres. **1777** Nicol. and Burn 615: Virgate of land; a yard land consisting (as some say) of 24 acres, whereof four virgates make an hide. **1895** Round 2.37: But not only were there thus, in Domesday, four virgates to a hide; there were also in the Domesday virgate thirty Domesday acres. **1897** Maitland 395: The virgates on the Gloucestershire manors of Gloucester Abbey contain the following numbers of acres: 36, 40, 36, 38, 48, 48, 48, 48, 50, 48, 40, 64, 64, 64, 48, 50, 60, 48, 48, 64...44, 80, 48, 48, 72.

virge. VERGE

vunce. OUNCE

W

waga, wage. WEY

wagh. WAW

waghe, waigh. WEY

wal, wall. WAW

warp [ME warp fr OE wearp, a warp in weaving; akin to OHG warf, warp, ON varp, a casting, throwing, Sw varp, the draft of a net]. A m-q for herrings in Sussex and Kent (c1850). It consisted of a cast of 4 (Cooper 85).

wash [fr vb wash; see WNID3]. A m-c for oysters (c1500), probably originally the amount washed at one time. It contained approximately 1 gal (c4.40 l) or 1/8 bu (Prior 170).

waugh. WAW

waw—4 wagh (OED), waugh (OED), wawe (OED); 5 wal (OED); 5-6 wall (OED), waw [MLG and MDu wage (Du waag); see WEY]. A m-q for glass containing 40 BUNCHES of uncertain wt.—**1507** Gras 1.698: Glasse called Flemyche glass the waw that ys to saye xl bunchys. **1508** Fab. Rolls 359: A Waw of glasse.

wawe. WAW

way, waya, waye. WEY

web [ME web fr OE web; akin to OHG weppi, web, OE wefan, to weave]. A m-l in Fordingbridge (c1800) for ticking, a strong, usually twilled, cotton fabric, containing 70 yd (64.008 m) (Second Rep. 36), and a m-q in Scotland (c1600) for window glass, consisting of 60 BUNCHES (Halyburton 308).

wegh, weigh. WEY

weight [ME weght, wight fr OE wiht, gewiht]. Equivalent to WEY.—**1595** Powell C2: Two weights of wooll make a sacke, and xii. sackes make a last. **1615** Collect. Stat. 466: There is a weight aswel of lead as of wool, tallow and cheese, & weigheth foureteene stone. **1665** Sheppard 56: There is a Weight of Lead, of Wooll, Tallow, and Cheese, and weigheth 14 stone. **1880** Britten 177: Weight (Dors.), of hemp, 8 heads of 4 lbs., twisted and tied, making 32 lbs. (Som.), of hemp, 30 lbs.

werkhop [ME werk, work, + hop, hopper, receptacle]. A m-c for grain (c1300) containing approximately 2 1/2 bu (c8.81 dkl) and representing one day´s work in thrashing (Battle xxiv).

wey—1-3 L pondus; 2 L vaga; 3 L waya, L weya; 3-7 waye; 3-9 weye; 4 wage, waghe; 5 wegh (OED); 5-9 way, weigh, wey; 7 L waga, waigh [ME waye, weye fr OE wǣge, wǣg, weight, wey; see WEIGHT]. A m-c and a wt for dry products. It was originally called L pondus (weight), a name superseded in the thirteenth century by its ME translation, weye. Its size varied with the product as well as with the region: barley, corn, and malt, 40 bu (c14.09 hl before 1824 and 14.547 hl afterward) or 5 SEAMS or CHALDERS of 8 bu each; cheese, 180 lb (81.646 kg) or 15 STONE of 12 lb each, 182 lb (82.553 kg) or 14 stone of 13 lb each, 224 lb (101.604 kg) or 32 CLOVES of 7 lb each or 2 Cwt of 112 lb each, 256 lb (116.119 kg) or 32 cloves of 8 lb each, 336 lb (152.406 kg) or 42 cloves of 8 lb each, and 416 lb (188.693 kg); coal, South Wales, 8 tons 2 Cwt or 18,144 lb (8229.937 kg), refuse coal, Swansea, approximately 9 1/2 tons or

21,280 lb (9652.395 kg); culm, Swansea, 10 tons or 22,400 lb (10,160.416 kg); flax, 182 lb (82.553 kg) or 14 stone of 13 lb each; flour or meal, 36 Cwt or 4032 lb (1828.875 kg); glass, 60 BUNCHES or CASES of uncertain wt; hemp, Dorsetshire, 32 lb (14.515 kg), and Somersetshire, 30 lb (13.608 kg); lead, generally 182 lb (82.553 kg) or 14 stone of 13 lb each, but occasionally 175 lb (79.378 kg) or 1/12 FOTHER of 2100 lb; lime, Devonshire, 48 double Winchester bu (c33.83 hl); salt, generally 42 bu (c14.80 hl); tallow and wool, generally 182 lb (82.553 kg) or 14 stone of 13 lb each.—**c1150** Acts Scotland 1.309: Item lapis ad lanam et ad alias res ponderandas debet ponderare .xv. libras. Item vaga debet continere .xij. petras. **1208** Bish. Winch. 13: Et de xiiij s. iij d. de cxxiij velleribus agninis venditis quæ fecerunt vj petras, unde xiiij faciunt pondus. **c1253** Hall 11: La waye de plum, layne, sue, et de furmage peyse xiiii peris. **c1270** Report 1.419-20: Petra duodecim Libræ et dimid´ faciunt unam petram...xiv petræ faciunt unum pondus, quod Anglice dicitur weye; ibid 420: Unum pondus casei xv petræ, et una petra xii lib. **c1272** Hall 9: Charrus constat ex xii wayes; ibid 10: Waya enim tam plumbi quam lane...ponderat quatuordecim petras. Et due waye faciunt unum saccum. **c1300** Brit. Mus. 13.29: Constat la charre [plumbi] ex duodecim Wages.... Due Waghe lane faciunt unum saccum. **1303** Report 1.414: Weya enim tam plumbi quam lane, lini, sepi, casei, ponderabunt xiiii petras. **1430** Rot. Parl. 4.381: That the weight of a weigh of cheese shal containe xxxii. cloues, that is to say, euery cloue vii. lb.... Que le pois d´une weye de formage, puisse tener xxx & II cloues; c´est assavoir, chescun cloue VII li. c**1475** Gras 1.193: Of a way chese. **1507** Ibid 696: Chesse the waye; ibid 703: Sawllte the waye. **1540** Recorde 206: A cloue shoulde contayne 7 pounde: and a wey 32 cloues, that is 224 poundes. **1566** Ibid Kiiij: A Weye 32 Cloues, that is 224 poundes.... The common Wey is of 256 li. c**1590** Hall 23: 182 poundes waight haberdepoyse is a waye of woole.... The wey of cheesse is 32 cloves, conteninge 224 poundes waight haberdepoyse; ibid 28: The way of sault is 42 bushells: 10 wayes makith a last. **1600** Hill 67: 8. Bushels...1. Quarter. 5. Quarters...1. Way. c**1600** Brit. Mus. 16.70: So that a clove should contayne .7. pounde & a waye .32. cloves that is 224 poundes. **1603** Henllys 137-38: Neither ys the Cranoke or Wey measures used in selling thereof. **1613** Tap 1.63: Cheese. One Waigh contains Cloues. 32 Pounds. 256. **1615** Collect. Stat. 466: The weight of a weigh of cheese shal containe xxxij. cloues...euery cloue vij. l. **1616** Hopton 164: A Last of wooll is 4368 pounds, or 12 sackes; a sacke is 364 pounds, or 2 weyes. **1628** Hunt B2: Cheese Essex Weigh, 336 [lb]...Suffolke Weigh, 256 [lb]; ibid C: 182 Lb. in a Weigh of Wolle. **1635** Dalton 149: A weigh of cheese must containe 32 cloves, and every clove 8. l. of averdepois weight: although the statute 9. H. 6...seeme to make 7. l. to be a clove. And yet by the booke of assize, the weigh of Suff. cheese must containe 256. l.... But the weigh of Essex cheese...is 300 l. weight, after the rate of five score and xii. li. to the hundred, which is 336. l. **1664** Spelman 351: Waga tam plumbi quàm lanæ &c. pendeat 14 petras. **1665**

Sheppard 61-62: But Suffolk Cheese must be 256 pound...the Wey of Essex...Cheese must be...336, or sixteen score and 16 l. Averdepoys. **1677** Roberts 296: Two Waighs of Wooll make a Sack. **1701** Hatton 3.235: Weigh.... Of Glass = 60 Bunches; of Salt or Corn 40 Bushels. **1704** Mer. Adven. 244: Ffor bearing of a weigh of salt. **1707** Acts Scotland 11.407: All forreign Salt imported into Scotland...under a Weigh or fourtie Bushells. **1708** Chamberlayne 207: Wooll is Weigh'd by the...Way, 182 Pounds. **1716** Harris 2. sv weight: Weighs, or Weys, are commonly 165 Pound, or 180 Pound, or 200 Pound. **1717** Dict. Rus. sv weigh: Cheese or Wooll [sic], the Weight of 256 Pounds Aver-du-pois: Of Corn, 40 Bushels: Of Barley or Malt 6 [sic] Quarters or 40 Bushels. Of Glass 6[0] Bunches; ibid sv wey: The greatest Measure for dry things, containing 5 Chaldron. **1732** J. Owen 82: Wey, 40 bushels. **1755** Postlethwayt II.188: A weye of cheese 32 cloves, each clove 7 pounds. **1756** Rolt sv weigh: Way, or Wey. A weight of cheese, wool, or the like, containing 256 pounds avoirdupois. Of corn, the weigh contains 40 bushels; of barley, or malt, 6 quarters. **1805** Macpherson I.471: 12 1/2 pounds 1 stone of London, 14 stones 1 weye. **1820** Second Rep. 36: Weigh or Wey of cheese, flax, lead, tallow and wool, 14 stone...of window glass, 60 cases...Devonshire: of lime...sometimes 48 double Winchester bushels...Dorsetshire: of hemp, 8 heads of 4 lb. twisted and tied, making 32 lb...Somersetshire: of hemp, 30 lb. **1822** G. Gregory III. sv weigh: Way, or Wey, waga, a weight of cheese, wool, &c. containing 256 pounds avoirdupois. **1834** Pasley 114-15: 1 Wey of

Wool...182 [lb].... 1 Wey of Cheese...224 [lb].... 1 Wey of Cheese in Essex...256 [lb].... 1 Wey of Cheese in Suffolk (3 cwt.)...336 [lb].... 1 Wey of Cheese in Essex, sometimes...416 [lb].... 1 Wey of Meal or Flour, legal (36 cwt.)...4,032 [lb].... 1 Wey of Coals, South Wales (8 Tons 2 cwt.) 18,144 [lb]...refuse Coal, at Swansea (about 9 1/2 Tons)...21,280 [lb]...of Culm at Swansea (10 Tons)...22,400 [lb]. **1888** Fr. Clarke 37: 2 weys make 1 last, = 80 bushels. **1956** Economist 8: Wey or weigh...5 quarters or 40 bushels. **1966** O'Keefe 671: 1 load or wey = 40 bu. = 14.547 hl. See WEIGHT

weya, weye. WEY

windle [ME windle fr OE windel, basket]. A m-c for grain in Lancashire (c1800) containing 3 bu (c1.08 hl) for corn and 3 1/2 bu (c1.23 hl) for barley, beans, and wheat (Second Rep. 36 and Cyclopædia sv weights).

wineglassful [wineglass + -ful]. A culinary m-c containing 2 1/2 fluid oz (7.103 cl) or 1/2 GILL and equal to 1/2 TEACUPFUL or 1/4 TUMBLERFUL (Economist 8).

wista—3-4 L wista (Battle); 3-4, 8 L wysta [MedL wista, prob fr OE wist, food, sustenance]. Equivalent to VIRGATE.—c**1283** Battle xiii: Quatuor virgatæ seu wystæ faciunt unam hydam. **1722** Richmond 257: Virgata itaque & Wysta sunt una eademque quantitas terræ.... At si Wysta & Virgata sint idem, Wysta ex 40 tantùm acris constare debet.

wrap [ME wrappe fr wrappen, to cover, wrap]. A m-q for worsted yarn (c1950) containing 80 yd (73.152 m) or 1/7 HANK of 560 yd (Economist 58).

wysta. WISTA

X

xx. SCORE

Y

yacker [dial var of ACRE in Durham, Northamptonshire, and Wiltshire]. Equivalent to ACRE.—**1842** Akerman 59: Yacker...An acre. **1849** Dinsdale 150: Yacker...an acre. **1851** Sternberg 125: Yacker...An acre. Fields, also, of much larger extent than an acre are called by this name, generally in composition with some other word, as Green's Yacker.

yard—3 yeorde (OED), yherde (OED); 3-5 L ulna; 3-6 yerd, yerde; 4 yeird (OED); 4-7 yarde; 5 yeerde, yerdd, yerede (OED); 5-7 yeard, yearde; 5-9 yard; 6 yerdde (OED) [ME yarde, yerde fr OE gierd, geard, a rod, stick, a measure, a yard]. A m-l of 36 inches (0.914 m) or 3 ft for land and sometimes by custom 37 inches (0.940 m) for cloth; the latter resulting from marking the end of each yd by placing the thumb on the cloth and starting the next yd from the other side of the thumb. The yd generally was equal to 4/5 ELL, and was synonymous with the VERGE. However, the ell (L ulna) was occasionally equated with the yd, and the YARDLAND (yard of land, F verge de terre, VIRGATE) was sometimes called a "yard" although it retained its own proper dimensions.—c**1272** Hall 7: Et xij pollices faciunt pedem; et tres pedes faciunt ulnam. c**1400** Ibid 41: Nota quod tres pedes regii faciunt ulnam Regis. **1433** Rot. Parl. 4.451: Clothe of colour shold conteigne in lenght XXVIII yerdes.... Clothes called Streytes, holdyng XIIII yerdes in lenght, and yeerde brode unwette. c**1450** Common 167: For ij yerddys...of tawny clothe. c**1461** Hall 14: Off the length of the yerd and off odyr mesures conteynynge lengthis.... And xii ynchis make a fote; and iij fote make a yard. **1474** Cov. Leet 396: And xij Inches makith a foote, and iij fote makith a yarde. **1507** Gras 1.703: Sattyn crymsen cunterfett the yard...Sattyn ryght purled with goold the yarde. **1519** Mer. Adven. 57: It is assented, &c., for our imposicyons beyonde see, to pay of every yerde of canves, one halffpeny. **1541** Mag. Carta 25: Two yardes within the lystes. **1566** Recorde Kv: 3 Foote make a Yearde. c**1590** Hall 27: The yard in lenght is 3 foott. **1592** Masterson 139: Note that 4 elles at London is 5 yeardes. **1603** Henllys 137: Yet doeth yt agree in the ynche, foote, and yard. **1615** Collect. Stat. 464: xii ynches make a foot, three feet make a yard. **1616** Hopton 165: 3 foote a yard. **1628** Hunt B2: The Ell of 20. Neyles, and the Yeard of 16. Neyles. **1635** Dalton 150: Three foot make a yard. **1647** Digges 1: Three Foote a Yard. **1665** Sheppard 16: 3 foot a yard. **1682** Hall 28: A Yard is two Cubits, or three feete. **1708** Chamberlayne 207: 2 Cubits a Yard. **1717** Dict. Rus. sv: Yard, a well known long Measure that consists of 3 Foot. **1805** Macpherson 1.656: An abuse had crept in of measuring cloths, not by the yard and full inch, but by the yard and full hand. **1820** Second Rep. 36: Yard, 3 feet = 36 inches. But by custom, the legal yard for cloth has become 37 inches in many cases. **1882** Beck 1.377: Goods for export are measured by what is called the short stitch—that is, a yard of "35 inches and a thumb," that is, 36 inches; goods for the home market are measured by "long stitch," a yard of "36 inches and a thumb," 37 inches. **1893** Mendenhall 145: 1 yard = 3600/3937 meter. **1951** Trade 27: YARD = 0.9144 metre.

yarde. YARD

yardland [YARD + land]. Equivalent to VIRGATE.—**1550** Ault 85: Great Horwood, Bucks.... Hit is ordeyned at this court that every man schall kepe for a yardelande xl shepe x vaccas et x bestias et no more. **1600** Ibid 94: Newton Longville, Bucks.... Inprimus wee all agreed to keepe xxx sheepe for a yeard land. **1603** Henllys 135-36: There is allso a quantytie of land measure called a yard of land, in latin Virgata terræ. **1608** Gray 439: Editha Reade...I yearde of land.... Thos. Hudd...1/2 yearde of land. **1610** Norden 59: Foure yard land, which in latine is called quatronaterræ euery yard land thirty acres. **1610** Folkingham 60: The Yard-land (Virgata terræ siue quatrona terræ) varies from 20, 24, 30 acres. **1616** Bullokar sv yardland: In some places, it is 20. Acres of land: in some, 24. and in some 30. **1635** Dalton 71: Every plow land or Carve, is foure yard land...every yard land, containing 30. Acres. **1665** Sheppard 22-23: A Yard Land (in Latine Virgata Terræ) is a quantity of Land called by this name, but it is no certain Quantity. For in some places it containeth 20 acres; in others 24 acres; in other places 30 acres. **1695** Kennett Glossary sv virgata: A Gird-land or Yard-land, was originally no more than a certain extent or compass of ground surrounded with such bounds and limits. And therefore the quantity was uncertain according to the difference of place and custom. They reckoned in some parts fourty, in other thirty, twenty, and at Wimbleton in Surry but fifteen acres. **1708** Chamberlayne 208: 30 Acres ordinary make a Yard-Land. **1717** Dict. Rus. sv yard-land: A certain quantity of Land; which at Wimbleton in Surrey, is only 15 Acres; but in other counties it contains 20, in some 24, in some 30, and in others 40. **1755** Willis 358: Virgata, or Yard-Land, whereof 4 made an Hide, was in different Counties 15, 20, 30 or 40 Acres. **1777** Nicol. and Burn 615: Virgate of land; a yard land consisting (as some say) of 24 acres, whereof four virgates make an hide. **1816** Kelly 86: 30 Acres is called a Yard of Land. **1855** Jessop 35: The yard of land = 30 acres. **1892** Andrews 161: In the majority of cases...the yard-land would consist of about thirty acres, though...it cannot be considered a uniform measure; in the thirteenth century virgates of 15, 16, 18, 24, 40, 48, 50, 62 and 80 acres were known.

yeard, yearde, yeerde, yeird, yeorde, yerd, yerdd, yerdde, yerde, yerede, yherde. YARD

ynce, ynch, ynche, ynsh. INCH

yoke of land—1-7 L jugum terre (terrae) [yoke of land, trans of L jugum terre (terrae)]. A m-a for land in Kent, sometimes described as the equivalent of 4 VIRGATES, sometimes as 1/2 HIDE, and sometimes as a 40 to 50 acre strip of which 4 made a SULUNG.—**1202** Cur. Reg. 9.121: Assisa inter de Cusinton´ petentem et Johannem Hanin tenentem de j. jugo terre cum pertinenciis in Seling´. **1220** Cur. Reg. 2.322: Mabilia filia Gaufridi petit versus Willelmum...dimidium jugum terre cum pertinentiis in Aynesford´. c**1320** Thorpe 2: Frendesberia habet XXI jugum terre de Gaveland unius servicii et unius redditus. **1599** Richmond Appendix 2.11: Domesdei Kant. In villa de Hadone, quæ fuit Episcopi Baioc. Odo tenet de Episcopo unum Jugum terræ, & est dim. Car.

1867 C. I. Elton 131: The Yoke-land or jugum.... The Yoke was the fourth part of the suling, and varied in size from forty to fifty of our acres, or a little more. **1872** Robertson 94: They reckoned in sulings and juga, or in plough-lands and yoke-lands; for the jugum was evidently the "gioc ærtheslondes," or the amount allotted in early days to the yoke of oxen,—the quarter-ploughland.... The latter [jugum] evidently contained forty acres, giving a hundred and sixty to the suling.

zame, zeam, zeme. SEAM

BIBLIOGRAPHY

Abba "Glossarium Abba," in M. Inguanez and C. J. Fordyce (eds.), Glossaria Latina, V, 9-143. Paris, 1931. (There are separate entries for stadium and dragma.)

Abingdon Accounts of the Obedientiars of Abingdon Abbey, ed. R. E. G. Kirk. (Camden Society Publication, New Series, Vol. 51.) Westminster, 1892.

Account 1 "An Account of a Comparison Lately Made by Some Gentlemen of the Royal Society, of the Standard of a Yard, and the Several Weights Lately Made for Their Use, etc." Philosophical Transactions, 42 (1742-43), 544-56. (This article is concerned with the yard and ell, and with the troy and avoirdupois weight standards.)

_____ **2** "An Account of the Proportions of the English and French Measures and Weights, from the Standards of the Same, Kept at the Royal Society." Ibid., 42 (1742-43), 185-88. (This article dwells on the similarities among the following weights and measures: Paris half-toise and English yard; Paris dimark and English troy pound; Paris foot and English foot.)

_____ **3** "Account Roll of a Fifteenth-Century Iron Master," ed. G. T. Lapsley. English Historical Review, 14 (1899), 509-29.

Acts A Collection of Acts and Ordinances of General Use, Made in the Parliament Begun and Held at Westminster the Third Day of November, Anno 1640, ed. Henry Scobell. London, 1658.

Acts Scotland 1 The Acts of the Parliaments of Scotland. (Great Britain Record Commission Publications). A.D. MCXXIV-MCCCCXXIII. London, 1814.

_____ **2** _____. A.D. MCCCCXXIV-MDLXVII. London, 1814.

_____ **3** _____. A.D. MDLXVII-MDXCII. London, 1814.

_____ **4** _____. A.D. MDXCIII-MDCXXV. London, 1814.

_____ **5** _____. A.D. MDCXXV-MDCXLI. London, 1817.

_____ **6** _____. A.D. MDCXLI-MDCLXI. London, 1817.

_____ **7** _____. A.D. MDCLXI-MDCLXIX. London, 1820.

_____ **8** _____. A.D. MDCLXX-MDCLXXXVI. London, 1820.

_____ **9** _____. A.D. MDCLXXXIX-MDCXCV. London, 1822.

_____ **10** _____. A.D. MDCXCVI-MDCCI. London, 1823.

_____ **11** _____. A.D. MDCCII-MDCCVII. London, 1824.

Adames Adames, Jonas. The Order of Keeping a Court Leete and Court Baron. London, 1593.

Adams Adams, John Quincy. Report of the Secretary of State upon Weights and Measures Prepared in Obedience to a Resolution of the House of Representatives of the Fourteenth of December, 1819. (16th Congress, 2nd Session, H.R. Document No. 109.) Washington, 1821. (This excellent report deals generally with the simplification and standardization of weights and measures in the early nineteenth century. Adams quotes freely from many English statutes, and he includes a short, but effective, history of English measuring units.)

Agricola Agricola, Georgi. Medici libri quinque de mensuris et ponderibus. Basil, 1533. (Agricola concentrates almost exclusively on the Greek and Roman systems of weights and measures and pays very little attention to the medieval.)

Airy Airy, George B. "Account of the Construction of the New National Standard of Length, and of its Principal Copies." Philosophical Transactions, 147 (1857), 621-702.

Akerman Akerman, John Yonge. A Glossary of Provincial Words and Phrases in Use in Wiltshire. London, 1842.

Alexander Alexander, J. H. Universal Dictionary of Weights and Measures. Baltimore, 1850. (This is one of the best compilations of weights and measures to appear in the nineteenth century. Not only does Alexander provide detailed descriptions of individual units, but he defines the standards upon which they were based.)

Americana The Americana, ed. Frederick Converse Beach. Vol. 22. New York, 1912. (See article entitled Weights and Measures.)

Amundesham Annales monasterii S. Albani, a Johanne Amundesham, monacho, ut videtur, conscripti, (A.D. 1421-1440). Quibus præfigitur Chronicon rerum gestarum in monasterio S. Albani, (A.D. 1422-1431) a quodam auctore ignoto compilatum. (Rerum Britannicarum Medii Aevi Scriptores, 2 vols.) London, 1870-71.

Anc. Char. Ancient Charters Royal and Private prior to A.D. 1200: Part I. (Pipe Roll Society.) London, 1888.

Anc. Laws Ancient Laws and Institutes of England, ed. Commissioners of the Public Records. London, 1840.

And. & Bigg Anderton, Pamela and Bigg, P. H. Changing to the Metric System: Conversion Factors, Symbols, and Definitions. London, 1969.

Andrews Andrews, Charles McLean. The Old English Manor. Baltimore,

1892.

Ansileubus Ansileubus. "Glossarium," in W. M. Lindsay, J. F. Mountford, and J. Whatmough (eds.), Glossaria Latina, I, 1-604. Paris, 1926. (There is an entry for libra.)

Arbuthnot Arbuthnot, John. Tables of Ancient Coins, Weights and Measures, Explain'd and Exemplify'd in Several Dissertations. London, 1727. (Tables of weights and measures follow p. 327.)

Arnold Arnold, Richard. Chronicle. London, 1502.

Arnoult Collection des décrets de l'Assemblée nationale constituante, ed M. Arnoult. Vols. 2 and 6. Dijon, 1792. (Mainly important for French metrology, these volumes contain references to some English weights and measures.)

Assize The Assize of Bread with Sundry Good and Needful Ordinances. London, 1665. (This book contains translations into English of the most important assizes, among which are those for fuel and tile.)

Astle Astle, Thomas. An Account of the Tenures, Customs, &c. of the Manor of Great Tey, in the County of Essex. London, 1795.

Ault Ault, Warren O. "Open-Field Husbandry and the Village Community." Transactions of the American Philosophical Society, 55, Part 7 (1965), 1-102. (The appendix contains by-laws taken from the rolls of 37 manors in 12 counties, and date from 1270 to 1608.)

Avery Avery, W. and T. Suggestions for the Amendment of the Law Relating to Weights and Measures. London, 1888.

B. J. B. J. The Merchants Avizo. London, 1607. (The author was a merchant; a handbook for regional and local commercial dealings.)

Badcock Badcock, Benjamin. Tables Exhibiting the Prices of Wheat, from the Year 1100 to 1830. London, 1832.

Bailey Bailey, Nathan. An Universal Etymological English Dictionary. London, 1721.

Baker Baker, Anne Elizabeth. Glossary of Northamptonshire Words and Phrases. 2 vols. London, 1854.

Bald Bald, Alexander. The Farmer and Corn-Dealer's Assistant. Edinburgh, 1780.

Barbon Barbon, Nicholas. A Discourse of Trade. London, 1690. (Jabob H. Hollander, ed., 1905.)

Barfield Barfield, Samuel. Thatcham, Berks, and Its Manors. Oxford and London, 1901.

Barlow Barlow, William. "An Account of the Analogy betwixt English Weights and Measures of Capacity." Philosophical Transactions, 41 (1740), 457-59.

Barrington Barrington, Daines. Observations on the More Ancient Statutes from Magna Carta to the Twenty-First of James I. London, 1796.

Bartlett-Amati Bartlett-Amati, L. Weights, Measures, Moneys and Interest Tables. 6th ed. Rome, 1891.

Battle Custumals of Battle Abbey in the Reigns of Edward I and Edward II (1283-1312) from MSS in the Public Record Office, ed S. R. Scargill-Bird. (Camden Society Publication, New Series, Vol. 41.) Westminster, 1887. (The documents provide information on the wista and other superficial measures.)

Baudouin Collection complète des lois, décrets, ordonnances, réglemens, avis du Conseil d'Etat, publiée sur les éditions officielles du Louvre; de l'Imprimerie nationale, par Boudouin; et du Bulletin des Lois; de 1788 à 1830 inclusivement, ed. J. B. Duvergier. 30 vols. Paris, 1834. (Although most of the metrological information contained in these volumes is French, there are references to English weights and measures, especially for the sake of comparison.)

Baxter Baxter, J. H., and Johnson, Charles. Medieval Latin Word-List from British and Irish Sources. London, 1962.

Beamont Beamont, William. An Account of the Rolls of the Honour of Halton. Warrington, 1879.

Beawes Beawes, Wyndham. Lex Mercatoria Rediviva: or, the Merchant's Directory. London, 1783. (Discussions and tables of weights and measures.)

Beck 1 Beck, S. William. The Draper's Dictionary: A Manual of Textile Fabrics, Their History and Applications. London, 1882.

_____ **2** _____. Gloves, Their Annals and Associations: A Chapter of Trade and Social History. London, 1883. (Beck includes many documents illustrating the various metrological units found in glove manufacture and trade.)

Bedwell Bedwell, William. Mesolabivm architectonicvm, That is, A Most Rare, and Singular Instrument, for the Easie, Speedy, and Most Certaine Measuring of Plaines and Solids by the Foote. London, 1639.

Beilby Beilby, John. Several Useful and Necessary Tables for the Gauging of Casks. London, 1694.

Bello Chronicon Monasterii de Bello. (Anglia Christiana Society.) London, 1846. (This is a particularly valuable source for superficial measures.)

Bellot Bellot, James. "The Booke of Thrift, Containing a Perfite Order, and Right Methode to Profite Lands, and Other Things Belonging to Husbandry," in Francis Cripps-Day, ed., The Manor Farm, pp. 115 ff. London, 1931. (Reprint of original 1589 edition.)

Benedict Benedict of Peterborough. The Chronicle of the Reigns of Henry II. and Richard I: A.D. 1169-1192, ed. William Stubbs. (Rerum Britannicarum Medii Aevi Scriptores, 2 vols.) London, 1867.

Benese Benese, Rycharde. This Boke Sheweth the Maner of Measurynge of All Maner of Lande. Southwarke, 1537. (Benese defines those measures that pertain to land.)

Bernardi Bernardi, Edvardi. De mensuris et ponderibus antiquis. Oxford, 1688. (Bernardi discusses the weights and measures of Greece, Rome, and medieval England in great detail. Some of his computations, however, are incorrect.)

Berriman Berriman, A. E. Historical Metrology. London, 1953. (Berriman discusses the various pound weights used in medieval England and the standards for the gallon found at the Exchequer. He also comments on the historical importance of seals used in authenticating local and state standards.)

Berthelot La grande Encyclopédie: Inventaire raisonné des sciences, des lettres et des arts, ed. MM. Berthelot and Laurent. Vol. 26. Paris, n.d. (This volume contains one small section on the weights and measures of the Middle Ages.)

Best Rural Economy in Yorkshire in 1641, Being the Farming and Account Books of Henry Best, of Elmswell, in the East Riding of the County of York. (Surtees Society Publication, Vol. 33.) Durham, 1857. (The glossary contains descriptions of the leap and maund.)

Beverini Beverini, Bartholomæo. Syntagma de ponderibus et mensuris. Lucca, 1711.

Bish. Winch. The Pipe Roll of the Bishopric of Winchester for the Fourth Year of the Pontificate of Peter des Roches, 1208-1209, ed. Hubert Hall. London, 1903.

Black Prince "Palatinate of Chester: 1351-1365." Register of Edward the Black Prince, Part 3. London, 1932. (Several documents are concerned with the standardization of Cheshire's weights and measures.)

Blind Blind, August. Mass-, Münz- und Gewichtswesen. Leipzig, 1906. (Blind describes rather superficially several medieval English units.)

Boissonnade Boissonnade, P. Life and Work in Medieval Europe. New York, 1950. (Boissonnade mentions the hogshead and the pound.)

Bolton Bolton, Richard. A Justice of Peace for Ireland, Consisting of Two Bookes. Dublin, 1638.

Bonwick Bonwick, James. Romance of the Wool Trade. London, 1887.

Bourquelot Bourquelot, M. Felix. Etudes sur les Foires de Champagne.

Paris, 1865. (Bourquelot devotes one entire chapter to the weights and measures used by merchants at the fairs of Champagne.)

Bowring Bowring, John. The Decimal System in Numbers, Coins, and Accounts. London, 1854.

Bracton Henrici de Bracton. De Legibus et consuetudinibus Angliæ, ed. Travers Twiss. (Rerum Britannicarum Medii Aevi Scriptores, 6 vols.) London, 1878-83.

Bradley Bradley, Richard. Chomel's Dictionaire Aeconomique, or the Family Dictionary. Translated and revised by R. Bradley. London, 1725.

Brashear Brashear, John A. "Evolution of Standard Measurements," American Manufacturer and Iron World. March 29, 1900, pp. 256-58.

Bray The Estate Book of Henry de Bray of Harleston, Co. Northants (c. 1289-1340), ed. Dorothy Willis. (Camden Third Series, Vol. 27.) London, 1916. (There is a table of land measures among the documents.)

Breed Breed, W. Roger. The Weights and Measures Act: 1963. London, 1964. (A thorough treatment with attention to previous laws).

Brehaut Brehaut, Ernest. An Encyclopedist of the Dark Ages: Isidore of Seville. New York, 1912. (Brehaut includes a translation of Isidore's short treatise on weights and measures.)

Bridbury Bridbury, A. R. England and the Salt Trade in the Later Middle Ages. Oxford, 1955.

Bridges Bridges, Noah. Lux Mercatoria: Arithmetick Natural and Decimal. London, 1660.

Brisson Brisson, Mathurin. Réduction des mesures et poids anciens en mesures et poids nouveaux. Paris, 1798. (Metric conversions)

Britannica The Encyclopædia Britannica, or Dictionary of Arts, Sciences, and General Literature. Vol. 21. Edinburgh, 1860. (The eighth edition; see article entitled Weights and Measures.)

Brit. Mus. 1 British Museum Manuscript Collections. Add. 6159. Register of Christchurch, Canterbury. Folios 148-148v (Fourteenth century.)

_____ **2** _____. Add. 6666. Derbyshire Collections Analecta. Folio 299 (Seventeenth century.)

_____ **3** _____. Add. 14252. Ranulphi de Glanville. Tractatus. Folio 118v (Twelfth century.)

_____ **4** _____. Add. 17512. S. Gregorii Dialogorum Libri. Folios 106-106v (Eleventh century.)

_____ **5** _____. Add. 32085. Statuta tractatus varii registrum brevium. Folios 150v-151 (Fourteenth century.)

_____ **6** _____. Add. 36542. Stafford Family Evidences. Folios 5v, 7, 159-159v (Sixteenth century.)

_____ **7** _____. Calig. A. XV. Calendar. Folios 107v-108 (Eleventh century futile calendar attempt.)

_____ **8** _____. Cladius D. II. Judicium pillorie, etc. Folios 255v-256 (Fourteenth century penal and other laws.)

_____ **9** _____. Cotton Cleo. A. III. Glossaries of Latin and Anglo-Saxon. Folio 10 (Eleventh century source of legal and bureaucratic terminology.)

_____ **10** _____. Cotton Tiberius E. IV. Annales de Winchcombe Bede de temporibus. Folio 135 (Eleventh century.)

_____ **11** _____. Cotton Vesp. B. VI. Bede de compoto. Folios 105-109 (Eighth century document of valuable computations relating, in part, to weights and measures.)

_____ **12** _____. Egerton 1925. Traité des monnaies anciennes. (Sixteenth century source for ancient coins and coinage systems.)

_____ **13** _____. Galba E. IV. Composition of Weights and Measures. Folio 29 (Fourteenth century metrological lists.)

_____ **14** _____. Hargrave 313. De Scarrario. Folio 95 (Fourteenth century numismatic materials; some metrological items.)

_____ **15** _____. Harley 13. Astronomical Treatises. Folios 132v-134 (Fourteenth century; some metrological linear information.)

_____ **16** _____. Harley 660. Papers on Coins, etc. Folios 70-70v (Seventeenth century; valuable numismatic and metrological materials.)

_____ **17** _____. Harley 921. Phrasæolog Latina. Folio 1 (Seventeenth century word list.)

_____ **18** _____. Harley 1033. Ancient Statutes, etc. Folios 135v-136 (Modus amensurandi terram; a valuable source for superficial measures.)

_____ **19** _____. Harley 1712. Pierr Comestor Sermons. Folios 162-163 (Twelfth century; some metrological references.)

_____ **20** _____. Harley 3205. Assize of Inch, Ell, Perch, etc. Folios 1v-2v (Fifteenth century measures of length and area.)

_____ **21** _____. Harley 5394. Eada de figuris verborum. Folios 60v-61v

(Fourteenth century.)

_____ **22** _____. Lansdowne 48. Burghley Papers 1586. Folio 142 (Star Chamber case concerning the Assize of Bread.)

_____ **23** _____. Lansdowne 52. Burghley Papers 1587. Folio 16 (Comparative table of English and Classical units.)

_____ **24** _____. Otho. E. X. Papers Relating to Mines, Coinage, Weights and Measures. Folios 14-18v (Sixteenth century.)

_____ **25** _____. Royal 2 B. V. Psalterium Cantica, etc. Folios 188-189 (Anglo-Saxon; some metrological references.)

_____ **26** _____. Royal 7 B. X. Johan Borough. Pupilia Occult, etc. Folios 251-251v (Fifteenth century; some metrological references.)

_____ **27** _____. Royal 7 DXXV. Chronological and Other Collections. Folios 44-46 (Seventh century.)

_____ **28** _____. Royal 18 CXIV. Irish Accompts 1495-1496. Folios 154-157 (Composition of weights and measures.)

_____ **29** _____. Sloane 513. Tract on Weights and Measures by a Monk of Buckfast Abbey. Folio 25v (Fifteenth century.)

_____ **30** _____. Sloane 747. Regist. Cartar. Abb: De Missenden. Folio 52v (Fifteenth century.)

_____ **31** _____. Sloane 904. Politica and Other Tracts. Folios 212-213 (Seventeenth century; some metrological references.)

_____ **32** _____. Vitellus F. XII. Collectanea. Folios 182-183 (Seventeenth century; mention of several metrological units.)

Britten Britten, James. Old Country and Farming Words: Gleaned from

Agricultural Books. (English Dialect Society, Vol. 30.) London, 1880.

Brockett Brockett, John Trotter. A Glossary of North Country Words in Use; with Their Etymology, and Affinity to Other Languages; and Occasional Notices of Local Customs and Popular Superstitions. Newcastle Upon Tyne, 1829. (Brockett includes entries for 15 weights and measures, and he discusses the characteristics of heaped, striked, and shallow capacity measures.)

Brokage The Brokage Book of Southampton: 1443-1444, ed. Olive Coleman. 2 vols. Southampton, 1960-61. (Volume 2 is valuable for its lists of products.)

Brown Brown & Jackson. The British Calculator. London, 1814.

Browne Browne, W. A. The Money, Weights and Measures of the Chief Commercial Nations in the World with the British Equivalents. London, 1899. (Browne treats the systems of metrology in use before the nineteenth century rather superficially.)

Buckhurst Buckhurst, Helen McM. "An Anglo-Saxon Index," in W. M. Lindsay (ed.), The Corpus Glossary, pp. 267-91. Cambridge, 1921. (Buckhurst does not define the weights and measures included in her list but only gives their declensions in Anglo-Saxon.)

Budé Budé, Guillaume. Annotationes Gulielmi Budæi Parisiensis, secretarii regii, in quatuor et viginti pandectarum libros, ad Io annem deganaium cancellarium Franciæ. Paris, 1535. (The weights and measures discussed are mainly Greek and Roman.)

Bullokar Bullokar, John. An English Expositor: Teaching the

Interpretation of the Hardest Words Used in Our Language. London, 1616. (No. 11 in the Collection of Facsimile Reprints of English Linguistics, 1500-1800; published by the Scolar Press Limited, 1967.)

Burton "Annales de Burton (A.D. 1004-1263)," in Henry Richards Luard (ed.), Annales monastici. (Rerum Britannicarum Medii Aevi Scriptores.) London, 1864. (This manuscript is important for its table of unusual land measures.)

Caernarvon Registrum Vulgariter Nuncupatum: The Record of Caernarvon. London, 1838. (This collection of documents contains the famous description of the tower pound so often found in other medieval manuscripts. There is also a version of the Assisa Panis.)

Cal. Char. 1 Calendar of the Charter Rolls Preserved in the Public Record Office. Henry III: 1216-57. London, 1904. (Most of the volumes of the various Rolls included in this bibliography contain information dealing with infractions of statutory standards and with the punishments imposed for violations. Seldom are there actual tables or definitions of individual units.)

_____ **2** _____. Henry III-Edward I: 1257-1300. London, 1906.

_____ **3** _____. Edward I-Edward II: 1300-26. London, 1908.

_____ **4** _____. 15 Edward III-5 Henry V: 1341-1417. London, 1916.

_____ **5** _____. 5 Henry VI-8 Henry VIII: 1427-1516. London, 1927.

Cal. Close 1 Calendar of the Close Rolls Preserved in the Public Record Office. Edward II: 1302-07. London, 1908.

_____ **2** _____. Edward II: 1313-18. London, 1893.

_____ **3** _____. Edward II: 1318-23. London, 1895.

_____ **4** _____. Edward II: 1323-27. London, 1898.

_____ **5** _____. Edward III: 1330-33. London, 1898.

_____ **6** _____. Edward III: 1339-41. London, 1901.

_____ **7** _____. Edward III: 1343-46. London, 1904.

_____ **8** _____. Edward III: 1349-54. London, 1906.

_____ **9** _____. Edward III: 1354-60. London, 1908.

_____ **10** _____. Edward III: 1360-64. London, 1909.

_____ **11** _____. Edward III: 1369-74. London, 1911.

_____ **12** _____. Edward III: 1374-77. London, 1911.

_____ **13** _____. Richard II: 1377-81. London, 1914.

_____ **14** _____. Richard II: 1381-85. London, 1920.

_____ **15** _____. Richard II: 1392-96. London, 1925.

_____ **16** _____. Richard II: 1396-99. London, 1927.

_____ **17** _____. Henry IV: 1399-1402. London, 1927.

_____ **18** _____. Henry IV: 1402-05. London, 1929.

_____ **19** _____. Henry IV: 1405-09. London, 1931.

_____ **20** _____. Henry IV: 1409-13. London, 1932.

_____ **21** _____. Henry V: 1413-19. London, 1929.

_____ **22** _____. Henry V: 1419-22. London, 1932.

_____ **23** _____. Henry VI: 1435-41. London, 1937.

_____ **24** _____. Henry VI: 1441-47. London, 1937.

_____ **25** _____. Henry VI: 1447-54. London, 1947.

_____ **26** _____. Henry VII: 1500-09. London, 1963.

Cal. Fine 1 Calendar of the Fine Rolls Preserved in the Public Record Office. Edward I: 1272-1307. London, 1911.

_____ **2** _____. Edward II: 1319-27. London, 1921.

_____ **3** _____. Edward III: 1337-47. London, 1915.

Cal. Just. Calendar of the Justiciary Rolls or Proceedings in the Court of the Justiciar of Ireland, ed. James Mills. 2 vols. London, 1914.

Cal. Lib. 1 Calendar of the Liberate Rolls Preserved in the Public Record Office. Henry III: 1226-40. London, 1916.

_____ **2** _____. Henry III: 1240-45. London, 1930.

_____ **3** _____. Henry III: 1245-51. London, 1937.

_____ **4** _____. Henry III: 1251-60. London, 1959.

_____ **5** _____. Henry III: 1267-72. London, 1959.

Cal. Pat. 1 Calendar of the Patent Rolls Preserved in the Public Record Office. Henry III: 1266-72. London, 1913.

_____ **2** _____. Edward I: 1272-81. London, 1901.

_____ **3** _____. Edward II: 1307-13. London, 1894.

_____ **4** _____. Edward II: 1317-21. London, 1903.

_____ **5** _____. Edward III: 1327-30. London, 1891.

_____ **6** _____. Edward III: 1338-40. London, 1898.

_____ **7** _____. Edward III: 1340-43. London, 1900.

_____ **8** _____. Edward III: 1343-45. London, 1902.

_____ **9** _____. Edward III: 1345-48. London, 1903.

_____ **10** _____. Edward III: 1348-50. London, 1905.

_____ **11** _____. Edward III: 1350-54. London, 1907.

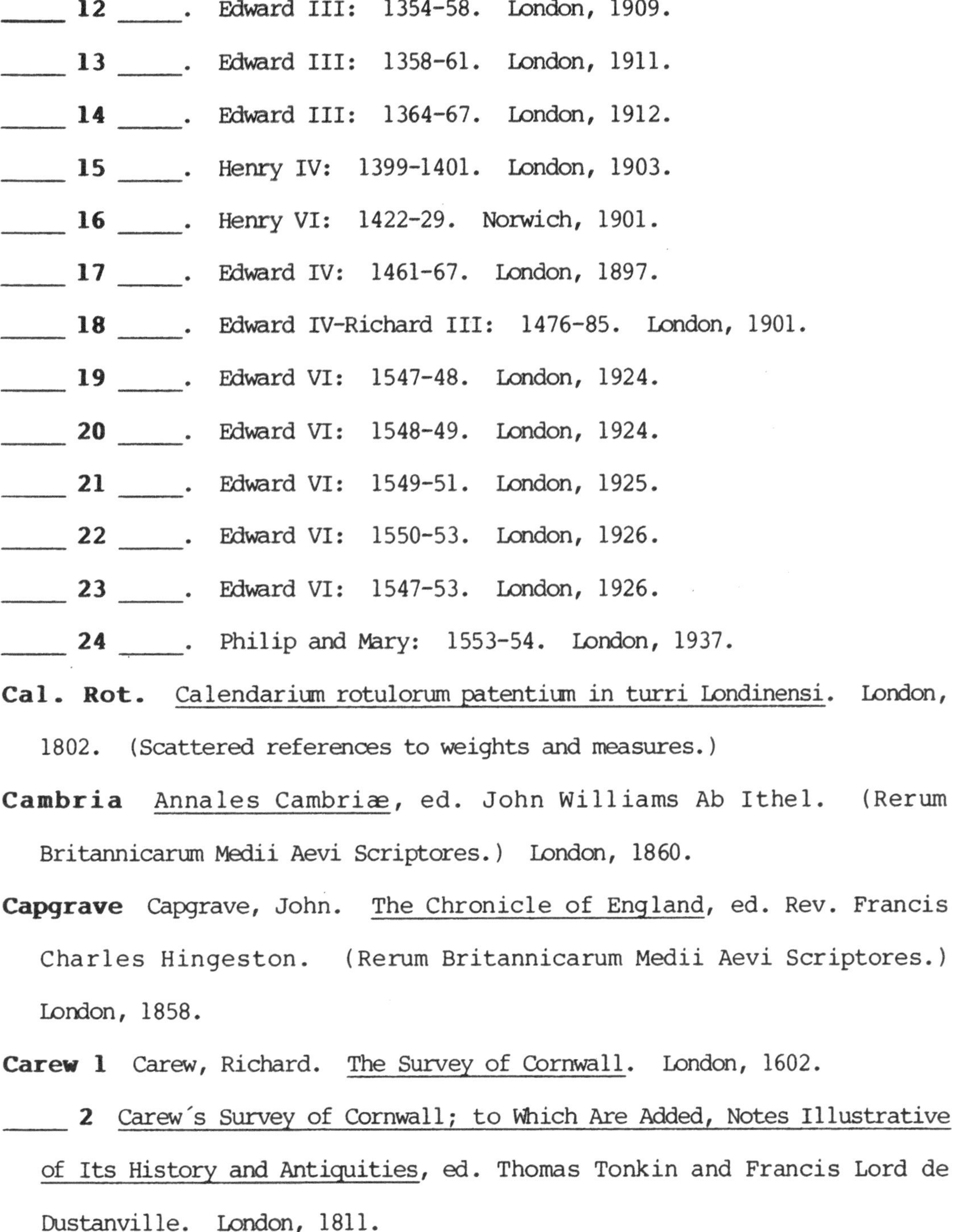

_____ **12** _____. Edward III: 1354-58. London, 1909.

_____ **13** _____. Edward III: 1358-61. London, 1911.

_____ **14** _____. Edward III: 1364-67. London, 1912.

_____ **15** _____. Henry IV: 1399-1401. London, 1903.

_____ **16** _____. Henry VI: 1422-29. Norwich, 1901.

_____ **17** _____. Edward IV: 1461-67. London, 1897.

_____ **18** _____. Edward IV-Richard III: 1476-85. London, 1901.

_____ **19** _____. Edward VI: 1547-48. London, 1924.

_____ **20** _____. Edward VI: 1548-49. London, 1924.

_____ **21** _____. Edward VI: 1549-51. London, 1925.

_____ **22** _____. Edward VI: 1550-53. London, 1926.

_____ **23** _____. Edward VI: 1547-53. London, 1926.

_____ **24** _____. Philip and Mary: 1553-54. London, 1937.

Cal. Rot. Calendarium rotulorum patentium in turri Londinensi. London, 1802. (Scattered references to weights and measures.)

Cambria Annales Cambriæ, ed. John Williams Ab Ithel. (Rerum Britannicarum Medii Aevi Scriptores.) London, 1860.

Capgrave Capgrave, John. The Chronicle of England, ed. Rev. Francis Charles Hingeston. (Rerum Britannicarum Medii Aevi Scriptores.) London, 1858.

Carew 1 Carew, Richard. The Survey of Cornwall. London, 1602.

_____ **2** Carew´s Survey of Cornwall; to Which Are Added, Notes Illustrative of Its History and Antiquities, ed. Thomas Tonkin and Francis Lord de Dustanville. London, 1811.

Cawdrey Cawdrey, Robert. A Table Alphabeticall, Conteyning and Teaching the True Writing, and Understanding of Hard Usuall English Wordes, Borrowed from the Hebrew, Greeke, Latine, or French. London, 1604. (Reproduced, with an introduction, by Robert A. Peters.)

Celsus Celsus, Aulus Corn. Medicina libri octo. Lipsiae, 1766.

Cely The Cely Papers: Selections from the Correspondence and Memoranda of the Cely Family, Merchants of the Staple, A.D. 1475-1488, ed. Henry Elliot Malden. (Camden Society Publication, Third Series, Vol. I.) London, 1900.

Chadwick Chadwick, Hector Munro. Studies on Anglo-Saxon Institutions. New York, 1963. (Chadwick discusses the pound; the book is a reissue of the original 1905 edition.)

Chamberlayne Chamberlayne, John. Magna Britannia Notitia: or, The Present State of Great-Britain with Divers Remarks upon the Ancient State Thereof. London, 1708. (One chapter is devoted to weights and measures. Chamberlayne is rather repetitious and sometimes quotes materials without indicating his sources. However, he provides many useful tables illustrating the dimensions of superficial and capacity measures.)

Chambers 1 Chambers, Ephraim. Cyclopædia: or, An Universal Dictionary of Arts and Sciences, Vol. 2. London, 1728.

_____ **2** Chambers´s Encyclopædia (Revised ed.), Vol. 10. London, 1874. (The sections on weights and measures in this edition are inferior to those in the earlier one.)

Chaney Chaney, H. J. Our Weights and Measures: A Practical Treatise on the Standard Weights and Measures in Use in the British Empire with Some Account of the Metric System. London, 1897. (Chaney outlines some of the duties of the clerks of the market in addition to defining very briefly several medieval English weights and measures.)

Chester Calendar of County Court, City Court and Eyre Rolls of Chester, 1259-1297, ed. R. Stewart-Brown. (Chetham Society Publication, Vol. 84.) Manchester, 1925.

Child Child, Josiah. A Discourse Concerning Trade. London, 1689.

Chisholm Chisholm, H. W. "On the Science of Weighing and Measuring, and the Standards of Weight and Measure." Nature, 8 (1873), 1-192. (Two discussions by Chisholm are of special value: the toise de Perou and the Imperial standard gallon and bushel.)

Chomel Chomel, M. Noel. Dictionnaire Œconomique. 2 vols. Paris, 1740. (Separate entries for mesures and poids in volume 2.)

Chron. Abing. Chronicon Monasterii de Abingdon, ed. Rev. Joseph Stevenson. (Rerum Britannicarum Medii Aevi Scriptores.) London, 1858.

Chron. Joh. Chronica Johannis de Oxenedes, ed. Sir Henry Ellis. (Rerum Britannicarum Medii Aevi Scriptores.) London, 1859.

Chron. Lon. Chronicles of London, ed. Charles Lethbridge. Oxford, 1905.

C. I. Elton Elton, Charles Isaac. The Tenures of Kent. London, 1867. (Chapter 6—The Domesday Survey—is valuable for land measures.)

Cinque Ports Charters of the Cinque Ports, Two Ancient Towns, and Their Members, ed. Samuel Jeake. London, 1728.

Clark Clark, George Thomas. "The Custumary of the Manor and Soke of Rothley, in the County of Leicester." Archaeologia, 47 (1882), 89-130.

Clarke Clarke, A. R. "Results of the Comparisons of the Standards of Length of England, Austria, Spain, United States, Cape of Good Hope, and of a Second Russian Standard, Made at the Ordnance Survey Office, Southampton." Philosophical Transactions, 163 (1873), 445-69.

Clerkenwell Cartulary of St. Mary Clerkenwell, ed. W. O. Hassall. (Camden Third Series, Vol. 71.) London, 1949.

Clode Clode, Charles M. The Early History of the Guild of Merchant Taylors of the Fraternity of St. John the Baptist, London, with Notices of the Lives of Some of Its Eminent Members. Part 1 (The History). London, 1888. (Some valuable documents.)

Close 1 Close Rolls of the Reign of Henry III Preserved in the Public Record Office. Henry III: 1227-34. London, 1902.

_____ **2** _____. Henry III: 1234-37. London, 1908.

_____ **3** _____. Henry III: 1237-42. London, 1911.

_____ **4** _____. Henry III: 1242-47. London, 1916.

_____ **5** _____. Henry III: 1247-51. London, 1922.

_____ **6** _____. Henry III: 1251-53. London, 1927.

_____ **7** _____. Henry III: 1254-56. London, 1931.

_____ **8** _____. Henry III: 1261-64. London, 1936.

_____ **9** _____. Henry III: 1268-72. London, 1938.

Cocker Cocker, Edward. Accomplish´d School-Master. London, 1696. (No. 33 of A Collection of Facsimile Reprints of English Linguistics,

1500-1800; published by The Scolar Press Limited, 1967.)

Coggeshall Radulphi de Coggeshall Chronicon Anglicanum, ed. Josephus Stevenson. (Rerum Britannicarum Medii Aevi Scriptores.) London, 1875.

Coles Coles, E. An English Dictionary, Explaining the Difficult Terms That Are Used in Divinity, Husbandry, Physick, Philosophy, Law, Navigation, Mathematicks, and Other Arts and Sciences. London, 1724. (Edition of 1732 used also.)

Collect. Stat. A Collection in English of the Statutes Now in Force, Continued from the Beginning of Magna Charta, Made in the 9. Yere of the Raigne of King H. 3. until the End of the Parliament Holden in the 7. Yere of the Raigne of Our Soveraigne Lord King James. London, 1615. (This collection provides some valuable commentaries on the statutes, and it is especially important as a source of variant spellings.)

Colles Colles, George W. The Metric Versus the Duodecimal System. Boston, 1896. (Originally a paper presented to the American Society of Mechanical Engineers.)

Common A Common-place Book of the Fifteenth Century, ed. Lucy Toulmin Smith. London, 1886.

Cooke Cooke, Layton. The Grazier's Manual. London, 1819. (Second edition; the first edition, entitled Tables Adapted to the Use of Farmers and Graziers, was published in 1813 and contains more metrological information, especially on the Scots units.)

Cooper Cooper, William Durrant. A Glossary of the Provincialisms in Use in the County of Sussex. London, 1853. (Cooper's glossary contains some rather detailed accounts of unusual weights and measures such as the draught, leap, meal, swod, tovet, warp, and wint.)

Coopers Coopers Company, London. Historical Memoranda, Charters, Documents, and Extracts, from the Records of the Corporation and the Books of the Company, 1396-1848. London, 1848. (Compiled by James F. Firth, a member of the Company.)

Cottenham "Common Rights at Cottenham and Stretham in Cambridgeshire," in W. Cunningham (ed.), Camden Miscellany. Camden Third Series, 12 (1910), 169-290.

Cotton Bartholomaei de Cotton. Historia Anglicana (A.D. 449-1298), ed. Henry Richards Luard. (Rerum Britannicarum Medii Aevi Scriptores.) London, 1859. (Bartholomew mentions several of Richard I´s decrees dealing with weights and measures.)

Coulton Coulton, G. G. Medieval Village, Manor and Monastery. New York, 1960. (Coulton discusses rather briefly the perch, sheaf, thrave, and yardland and lists some of the problems resulting from variations in these measures.)

Courtney Courtney, W. S. The Farmers´ and Mechanics´ Manual. New York, 1880. (Some information on local measures.)

Cov. Leet The Coventry Leet Book: or Mayor´s Register, Containing the Records of the City Court Leet or View of Frankpledge: 1420-1555, ed. Mary Dormer Harris. (Early English Text Society.) London, 1913. (This is one of the most important sources for information on capacity measures.)

Cowell Cowell, John. The Interpreter: or Books Containing the Signification of Words. Cambridge, 1607.

Crüger Crüger, Carl. Contorist. Eine Handels-Münz-Maass-und Gewichtskunde. Hamburg, 1830.

C. Sandys Sandys, Charles. Consuetudines Kanciae: A History of Gavelkind and Other Remarkable Customs in the County of Kent. London, 1851.

Cullyer Cullyer, John. The Gentleman & Farmer's Assistant. London, 1798. (One of many such manuals containing superficial measures.)

Cumberland Cumberland, Richard. An Essay Towards the Recovery of the Jewish Measures and Weights, Comprehending Their Monies; by Help of Ancient Standards, Compared with Ours of England. London, 1686.

Cunningham Cunningham, William. The Growth of English Industry and Commerce during the Early and Middle Ages. Cambridge 1927. (Cunningham defines briefly twelve weights and measures.)

Cur. Reg. 1 Introduction to the Curia Regis Rolls, 1199-1230 A.D.., ed. Cyril Thomas Flower. (Selden Society Publication, Vol. 62.) London, 1944. (This work contains several references to land measures.)

_____ **2** Curia Regis Rolls of the Reign of Henry III Preserved in the Public Record Office. 3-4 Henry III. London, 1938.

_____ **3** _____. 4-5 Henry III. London, 1952.

_____ **4** _____. 5-6 Henry III. London, 1949.

_____ **5** _____. 7-9 Henry III. London, 1955.

_____ **6** _____. 11-14 Henry III. London, 1959.

_____ **7** _____. 14-17 Henry III. London, 1961.

_____ **8** Curia Regis Rolls of the Reigns of Richard I and John Preserved in the Public Record Office. Richard I-2 John. London, 1922.

_____ **9** _____. 3-5 John. London, 1925.

_____ **10** _____. 5-7 John. London, 1926.

_____ **11** _____. 7-8 John. London, 1929.

_____ **12** _____. 8-10 John. London, 1931.

_____ **13** _____. 11-14 John. London, 1932.

_____ **14** _____. 15-16 John. London, 1935.

Curtler Curtler, W. H. R. A Short History of English Agriculture. Oxford, 1909.

Cyclopædia The Cyclopædia; or, Universal Dictionary of Arts, Sciences, and Literature, ed. Abraham Rees. Vol. 38. London, 1819. (See article entitled Weight.)

Daire Daire, M. Eugene. Oeuvres de Turgot. 2 vols. Paris, 1844. (Turgot´s works include discussions of weights and measures, principally French.)

Dalton Dalton, Michael. The Countrey Justice. London, 1635. (One entire chapter is devoted to weights and measures. Especially valuable is the discussion of the duties and responsibilities of the justices of the peace in regard to verification and enforcement of Crown standards.)

Davenport Davenport, Frances Gardiner. The Economic Development of a Norfolk Manor: 1086-1565. Cambridge, 1906. (Especially valuable are the appendixes of documents.)

D. Digges Digges, Dudley. The Defence of Trade. In A Letter to Sir Thomas Smith Knight, Gouernour of the East-India Companie. London, 1615.

Delambre 1 Delambre, Jean Baptiste Joseph. Grandeur et figure de la terre. Paris, 1912.

_____ **2** _____. Grundlagen des dezimalen metrischen Systems oder Messung des Meridianbogens zwischen den Breiten von Dünkirchen und Barcelona, ed. Walter Block. Leipzig, 1911.

Devizes "The Chronicle of Richard of Devizes," in Richard Howlett (ed.), Chronicles of the Reigns of Stephen, Henry II, and Richard I. (Rerum Britannicarum Medii Aevi Scriptores, Vol. 3, pp. 381-454.) London, 1886.

Dickinson Dickinson, William. A Glossary of Words and Phrases Pertaining to the Dialect of Cumberland. London, 1878.

Dict. Rus. Dictionarium rusticum, urbanicum et botanicum: or, A Dictionary of Husbandry, Gardening, Trade, Commerce, and All Sorts of Country-Affairs. London, 1717. (This dictionary contains approximately 100 entries for weights and measures. In several instances there are errors in computation and quite possibly a few printing errors.)

Dict. Univ. Dictionnaire universel Français et Latin. Paris, 1752.

Diderot Encyclopédie ou dictionnaire raisonné des sciences, des arts et des métiers, par une société de gens de lettres, ed. M. Diderot. Vols. 21 and 26. Geneva, 1778. (Information on measures is in Volume 21 and on weights in Volume 26.)

Digest Metric A Digest of "The Metric Versus the English System of Weights and Measures" from Research Report No. 42. (National Industrial Conference Board.) Special Report No. 20. New York, 1921.

Digges Digges, Leonard. A Booke Named Tectonicon. London, 1647. (Digges provides very little information on individual units aside from defining the inch, yard, and perch when setting up specific arithmetical problems for the reader to solve.)

Dinsdale Dinsdale, Frederick T. A Glossary of Provincial Words Used in Teesdale in the County of Durham. London, 1849. (Dinsdale has good definitions of noggin and score.)

Domesday Domesday Tables for the Counties of Surrey, Berkshire, Middlesex, Hertford, Buckingham and Bedford and for the New Forest, ed. Francis Henry Baring. London, 1909.

Domesday Book The Domesday Book of Kent, ed. Rev. Lambert Blackwell Larking. London, 1869.

Donisthorpe Donisthorpe, Wordsworth. A System of Measures of Length, Area, Bulk, Weight, Value, Force, &c. London, 1895.

Du Cange Du Cange, Charles du Fresne. Glossarium mediæ et infimæ Latinitatis. 10 vols. Paris, 1937. (These volumes contain a wealth of information on medieval English weights and measures. Not only are there definitions for some of them, but Du Cange includes ample documentation.)

Dugdale Dugdale, Sir William. "Monasticon Anglicanum," in William Harrison Douglas (ed.), The Old Historians of the Isle of Man, pp. 1-77.

Isle of Man, 1871. (There is some treatment of superficial measures in the entry for Rushen Abbey on p. 75.)

Durham The Inventories and Account Rolls of the Benedictine Houses or Cells, of Jarrow and Monk-Wearmouth, in the County of Durham. (Surtees Society Publication, Vol. 29.) Durham, 1854. (There is information on weights and measures in the glossary.)

Dur. House The Durham Household Book; or, the Accounts of the Bursar of the Monastery of Durham from Pentecost 1530 to Pentecost 1534. (Surtees Society Publication, Vol. 18.) London, 1844. (The glossary contains descriptions of some capacity measures and of several types of cloth.)

Economist The Economist Guide to Weights & Measures. London, 1956. (Compiled by the Statistical Department of The Economist.)

Eden Eden, Richard (trans.). Cortes´ (Martin) Arte of Nauigation. N. p., 1561.

Edinburgh The Edinburgh Encyclopædia, ed. by David Brewster. Vols. 12 and 18. Philadelphia, 1832. (The first American edition; volume 12 has an article on measures; volume 18 on weights.)

Edler Edler, Florence. Glossary of Medieval Terms of Business: Italian Series 1200-1600. Cambridge, 1934. (Although the weights and measures are Italian, there are references to their English equivalents in several instances.)

Eliot Eliot, F. Perceval. Letters on the Political and Financial Situation of the Country in the Year 1814; Addressed to the Earl of Liverpool. London, 1814. (Author signs name of Falkland; Pamphlet no.

7, vol. 4.)

Elton Elton Manorial Records: 1279-1351, ed. S. C. Ratcliff. (The Roxburghe Club.) Cambridge, 1946.

Ency. meth. Encyclopédie methodique: Commerce, Vol. 3. Paris, 1784. (There are many tables and charts comparing the metrological units of one country with another.)

Eng. Cyclo. Arts and Sciences or Fourth Division of The English Cyclopædia, ed. Charles Knight. Vol. 8. London, 1868. (See article entitled Weights and Measures.)

Eng. Gilds English Gilds: The Original Ordinances of More Than One Hundred Early English Gilds, ed. Lucy Toulmin Smith. (Early English Text Society.) London, 1870. (There are occasional references to weights and measures and to the verification and enforcement of gild standards.)

Evans Evans, John. The Palace of Profitable Pleasure. London, 1621. (No. 32 of A Collection of Facsimile Reprints of English Linguistics, 1500-1800; published by The Scolar Press Limited, 1967.)

Evesham Chronicon abbatiæ de Evesham, ad annum 1418, ed. William Dunn MacRay. (Rerum Britannicarum Medii Aevi Scriptores.) London, 1863.

Ewart Ewart, John. The Land Drainer's Calculator. London, 1862.

Exchequer Issues of the Exchequer; Being Payments Made Out of His Majesty's Revenue during the Reign of King James I, ed. Frederick Devon. London, 1836.

Eyre Rolls of the Justices in Eyre Being the Rolls of Pleas and Assizes for Lincolnshire 1218-9 and Worcestershire 1221, ed. Doris Mary Stenton. (Selden Society Publication, Vol. 53.) London, 1934.

Eyton Eyton, Rev. Robert William. Domesday Studies: An Analysis and Digest of the Somerset Survey (According to the Exon Codex), and of the Somerset Gheld Inquest of A.D. 1084. 2 vols. London, 1880. (Eyton discusses the hide as a unit of superficial measurement.)

Fab. Rolls The Fabric Rolls of York Minster. (Surtees Society Publication, Vol. 35.) Durham, 1859. (The glossary contains definitions of several capacity and superficial measures along with documentary materials illustrating their use.)

Fabyan Fabyan, Robert. The Newe Chronycles of Englande and of Fraunce. London, 1516.

Falkirk Scotland in 1298: Documents Relating to the Campaign of King Edward the First in That Year, and Especially to the Battle of Falkirk, ed. Henry Gough. London, 1888.

Fauve Fauve, Adrien. Les Origines du système métrique. Paris, n.d.

Feet 1 Feet of Fines for the County of Lincoln for the Reign of King John (1199-1216), ed. Margaret S. Walker. London, 1954.

____ **2** Feet of Fines for the County of Norfolk for the Reign of King John (1201-1215) and for the County of Suffolk for the Reign of King John (1199-1214), ed. Barbara Dodwell. London, 1958. (This work is important for its excellent descriptions of superficial measures such as the bovate, virgate, and knight's fee. The subject index also contains some valuable information.)

_____ **3** Feet of Fines for the County of Norfolk for the Tenth Year of the Reign of King Richard the First (1198-1199) and for the First Four Years of the Reign of King John (1199-1202), ed. Barbara Dodwell. London, 1952.

Finchale The Charters of Endowment, Inventories, and Account Rolls, of the Priory of Finchale, in the County of Durham. (Surtees Society Publication, Vol. 6.) London, 1837.

Fitzherbert Fitzherbert, Anthony. The Boke of Hysbandry. London, 1534. (In 1882 it was reprinted by the English Dialect Society and edited by Rev. Walter W. Skeat under the title The Book of Husbandry by Master Fitzherbert.)

Fleetwood Fleetwood, Bishop. Chronicon preciosum: or, An Account of English Gold and Silver Money; the Price of Corn and Other Commodities; and of Stipends, Salaries, Wages, Jointures, Portions, Day-Labour, etc. in England, for Six Hundred Years Last Past. London, 1745. (Fleetwood discusses 14 separate units of measurement, but his treatment is very superficial and he confuses the troy with the tower pound. He does have some timely quotations from medieval and early modern manuscripts, however.)

Fleta Fleta, ed. H. G. Richardson and G. O. Sayles. (Selden Society Publication.) London, 1955. (Fleta is a valuable source for many capacity measures and for information pertaining to the construction of the tower and mercantile pounds.)

Flores Flores Historiarum, ed. Henry Richards Luard. (Rerum

Britannicarum Medii Aevi Scriptores.) London, 1890.

Folkingham Folkingham, W. Fevdigraphia: the Synopsis or Epitome of Svrveying Methodized. London, 1610.

Forbes Forbes, William. The Duty and Powers of Justices of Peace, in This Part of Great-Britain Called Scotland; with an Appendix Concerning Weights and Measures. Edinburgh, 1707.

Fountains Memorials of the Abbey of St. Mary of Fountains, ed. Joseph Thomas Fowler. (Surtees Society Publication, Vol. 130.) Durham, 1918. (The glossary has definitions for several dry and liquid capacity measures.)

Fox Fox, Francis F. Some Account of the Ancient Fraternity of Merchant Taylors of Bristol. London, 1880.

Francis Francis, Sidney. Tables, Memoranda, and Calculated Results for Farmers, Graziers, Agricultural Students, Surveyors, Land Agents, Auctioneers, etc. London, 1889. (1890 and 1894 editions also used.)

Fr. Clarke Clarke, Frank Wigglesworth. Weights, Measures and Money of All Nations. New York, 1888. (Clarke lists the weights and measures of the nineteenth century individually and by country and he includes the location and United States-English equivalent for each unit.)

G. Gregory Gregory, G. A Dictionary of Arts and Sciences. 3 vols. New York, 1822. (Volume 2 has an article on measures, and volume 3 on weights.)

Glazebrook Glazebrook, Sir Richard. "Standards of Measurement: Their History and Development." Nature, 128 (1931), Supplement, pp. 17-28.

(Glazebrook concentrates on metrological standardization under Elizabeth.)

Gore Gore, J. Howard. "The Decimal System of Measures of the Seventeenth Century." American Journal of Science, 41 (1891), 241-46.

Gouldman Gouldman, Francis. A Copious Dictionary. London, 1664.

Granger Granger, Allan. Our Weights and Measures. London, 1917.

Grantham The Royal Charters of Grantham: 1463-1688, ed. G. H. Martin. Leicester, 1963.

Gras 1 Gras, Norman Scott Brien. The Early English Customs Systems. Cambridge, 1918. (Occasionally a certain capacity measure is defined, but generally only the price for its contents is given.)

____ **2** ____. The Economic and Social History of an English Village (Crawley, Hampshire) A.D. 909-1928. Cambridge, 1930.

____ **3** ____. The Evolution of the English Corn Market from the Twelfth to the Eighteenth Century. Cambridge, 1915. (There is a short description of the seam.)

Gray Gray, Howard Levi. English Field Systems. Cambridge, 1915. (The appendixes contain some excellent documents.)

Greaves Greaves, John. A Discourse of the Roman Foot and Denarius. London, 1737.

Greenstreet Greenstreet, James. Assessments in Kent for the Aid to Knight the Black Prince: Anno 20 Edward III. London, 1878.

Greenwood Greenwood, William. The Authority, Jurisdiction and Method of Keeping County-Courts, Courts-Leet, and Courts-Baron. London, 1730.

Gregory "William Gregory's Chronicle of London," in James Gairdner (ed.), The Historical Collections of a Citizen of London in the Fifteenth Century. (Camden Society Publication, pp. 55-239.) London, 1876.

Gross Gross, Charles. The Gild Merchant. 2 vols. Oxford, 1890. (A rich source for documentation.)

Grote Grote, George. "On Ancient Weights, Coins, and Measures." The Minor Works of George Grote, ed. Alexander Bain, pp. 135-74. London, 1873.

Guilhiermoz 1 Guilhiermoz, P. "Note sur les Poids du moyen age." Bibliothèque de l'Ecole des Chartres, 67 (1906), 161-233 and 402-50. (Guilhiermoz discusses the weights used in most European countries, and he includes tables comparing the various pounds, which he also converts into Paris grains and metric grams.)

_____ **2** _____. "Remarques diverses sur les poids et mesures du moyen age." Ibid., 80 (1919), 5-100. (The focus is again European-wide.)

Guyot Guyot, Arnold. Tables, Meteorological and Physical. (Smithsonian Miscellaneous Collections.) 4th ed. Washington, 1884.

Hale Hale, Lord Chief Justice. "A Treatise, in Three Parts. Pars Prima. De Jure Maris et Brachiorum ejusdem. Pars Secunda. De Portibus Maris. Pars Tertia. Concerning the Customs of Goods Imported and Exported," in Francis Hargrave, ed., A Collection of Tracts Relative to the Law of England, from Manuscripts; Now First Edited. Vol. 1. Dublin, 1787. (Especially important is Part 3, pp. 115-248.)

Hall Hall, Hubert, and Nicholas, Frieda J. "Select Tracts and Table

Books Relating to English Weights and Measures (1100-1742)." Camden Third Series, 41 (1929), 1-53. (This is the most complete single collection dealing specifically with medieval English weights and measures. The documents cover all five major divisions of measurement as well as cloth regulations. The Cottonian manuscripts have been used in addition to other valuable collections. With few exceptions, the editing is well done; important information is contained in the footnotes.)

Hallock Hallock, William. Outlines of the Evolution of Weights and Measures and the Metric System. New York, 1906. (Among the subjects discussed by Hallock are early standards; primary and defined standards; the metrological systems of the Babylonians, Egyptians, Greeks, Romans, and Moslems; Anglo-Saxon influences; and medieval weights and measures. His remarks on the latter are rather brief, and he tends to exaggerate the influence that ancient systems of metrology had on the development of medieval English and French units.)

Halyburton Ledger of Andrew Halyburton, Conservator of the Privileges of the Scotch Nation in the Netherlands: 1492-1503, Together with The Book of Customs and Valuation of Merchandises in Scotland: 1612. Edinburgh, 1867.

Hardwicke Hardwicke, Robert E. The Oilman's Barrel. Norman, Oklahoma, 1958. (Pages 3-46 contain some excellent discussions of capacity measures and their historical evolution.)

Harkness Harkness, William. "The Progress of Science as Exemplified in the Art of Weighing and Measuring." Smithsonian: Miscellaneous Collection, 33 (1888), XLIII-LX. (Harkness dwells on English and French measures of length before 1600; the mercantile and avoirdupois pounds; and the poids de marc and the pile de Charlemagne.)

Harpur Harpur, John. The Jewell of Arithmetick. London, 1617.

Harris 1 Harris, John. Lexicon technicum, or, An Universal English Dictionary of Arts and Sciences, Vol. 2. London, 1710. (Information is included under the entries "measures" and "weights.")

____ **2** ____. Lexicon technicum, or, An Universal English Dictionary of Arts and Sciences (3rd ed.). London, 1716. (This work is much more detailed than the earlier edition and contains some excellent tables comparing English weights and measures with ancient and contemporary systems.)

Hartmann Hartmann, Carl. Die Waagen und ihre Construction. Weimar, 1856. (Many detailed discussions.)

Hassler Hassler, Ferdinand Rudolph. Report upon the Comparison of Weights and Measures of Length and Capacity, Made at the City of Washington, in 1831, under the Direction of the Treasury Department, in Compliance with a Resolution of the Senate of the United States of the 29th May, 1830. (22nd Congress, 1st Session, H.R. Document No. 299.) Washington, 1832. (There is some information dealing with the composition of English gallons and bushels.)

Hatch Hatch, F. H. and Vallentine, E. J. Mining Tables: Being a Comparison of the Units of Weight, Measure, Currency, Mining Area, etc., of Different Countries; Together with Tables, Constants & Other Data Useful to Mining Engineers and Surveyors. London, 1907.

Hatfield Bishop Hatfield's Survey: A Record of the Possessions of the See of Durham, Made by Order of Thomas de Hatfield, Bishop of Durham, ed. Rev. William Greenwell. (Surtees Society Publication, Vol. 32). Durham, 1857.

Hatton 1 Hatton, Edward. Arithmetick; or, the Ground of Arts: Teaching that Science, Both in Whole Numbers and Fractions. London, 1699. (Originally written by Robert Recorde.)

_____ **2** _____. An Intire System of Arithmetic: or Arithmetic in All Its Parts. London, 1731.

_____ **3** _____. The Merchant's Magazine: or, Trades-Man's Treasury. London, 1701. (Mainly capacity measures.)

Hauy Hauy, René Just Abbe. Instruction sur les mesures déduites de la grandeur de la terre, uniformes pour toute la république, et sur les calculs relatifs à leur division decimale. Paris, 1795.

Hawney Hawney, William. The Complete Measurer: or, The Whole Art of Measuring. London, 1789. (This is basically an arithmetic book that contains occasional descriptions of linear and superficial measures to be used for the solution of the various problems.)

Hazlitt Hazlitt, W. Carew. Tenures of Land & Customs of Manors. London, 1874. (Originally written by Thomas Blount; re-arranged, corrected, and enlarged from the copies of Blount (1679), Beckwith (1784), and the third edition of 1815.)

H. Baker Baker, Humfrey. The Well-Spring of Sciences. London, 1646. (Tables of weights and measures.)

H. Coggeshall 1 Coggeshall, Henry. Timber Measure by a Line. London, 1677. (Mainly linear surveying measurements.)

____ **2** ____. A Treatise of Measures. London, 1682.

Heales Heales, Alfred. The Records of Merton Priory in the County of Surrey. London, 1898.

Henderson Henderson, David. Tables for Calculating the Price of Any Quantity of Grain. Edinburgh, 1838.

Henley Walter of Henley's Husbandry: Together with an Anonymous Husbandry, Seneschaucie and Robert Grosseteste's Rules, ed. Elizabeth Lamond. London, 1890. (There is information on the furlong, perch, and league in Walter's work, while in the Anonymous Husbandry there is a discussion of the perch, acre, and rood.)

Henllys Owen of Henllys, George. The Description of Penbrokshire, ed. Henry Owen. 3 vols. London, 1892. (Owen's book is valuable for Welsh linear, superficial, and capacity measures, and his description of the perch is especially important.)

Henrici Henrici Archidiaconi Huntendunensis Historia Anglorum, ed. Thomas Arnold. (Rerum Britannicarum Medii Aevi Scriptores.) London, 1879.

Henry Derby Expeditions to Prussia and the Holy Land Made by Henry Earl of Derby (Afterwards King Henry IV) in the Years 1390-1 and 1392-3 Being the Accounts Kept by His Treasurer during Two Years, ed. Lucy Toulmin Smith. (Camden Society Publication, New Series, Vol. 52.) London,

1894. (This account of expenditures contains some valuable information on capacity measures.)

Herschel Herschel, John F. W. "The Yard, the Pendulum, and the Metre," in Familiar Lectures on Scientific Subjects. Article 10 (419-51). New York, 1872.

Hewitt Hewitt, H. J. Medieval Cheshire: An Economic and Social History of Cheshire in the Reigns of the Three Edwards. (Chetham Society Publication, Vol. 88.) Manchester, 1929. (Appendix G describes some of the linear, superficial, and capacity measures that were found in Cheshire.)

Higden Polychronicon Ranulphi Higden monachi cestrensis, ed. Churchill Babington, et al. (Rerum Britannicarum Medii Aevi Scriptores, 9 vols.) London, 1865-86. (The volumes also contain translations of the Polychronicon by John Trevisa (A.D. 1387) and by an anonymous author of the fifteenth century.)

Hilderbrand Hilderbrand, Clifton. Metric Literature Clues: A List of References to Books, Pamphlets, Documents and Magazine Articles on Metric Standardization of Weights and Measures. San Francisco, 1921.

Hill Hill, Thomas. The Art of Vulgar Arithmeticke, both in Integers and Fractions, Deuided into Two Bookes. London, 1600.

Hodder Hodder, James. Hodder´s Arithmetick: or, That Necessary Art Made Most Easie. London, 1661.

Hofmann Hofmann, Joh. Jacob. Lexicon universale historiam sacram et profanam. Lugduni Batavorum, 1698.

Hogg Hogg, John. Answer to a Small Treatise Call'd Just Measures. London, 1693.

Holme Holme, Randle. The Academy of Armory, or a Storehouse of Armory and Blazon. London, 1688.

Holroyd The Letter Books of Joseph Holroyd (cloth-factor) and Sam Hill (clothier): Documents Illustrating the Organization of the Yorkshire Textile Industry in the Early 18th Century, ed. Herbert Heaton. Halifax, 1914. (County Borough of Halifax, Bankfield Museum Notes, Second Series, no. 3.)

Hooper Hooper, George. "An Inquiry into the State of the Ancient Measures, the Attic, the Roman, and especially the Jewish," in The Works of the Right Reverend George Hooper, D.D. Sometime Bishop of Bath and Wells. Vol. 2. Oxford, 1855. (Originally published in London in 1721.)

Hopton Hopton, Arthur. A Concordancy of Yeares. London, 1616. (Chapter 43 deals with weights and measures.)

Horwood Horwood, Alfred J. "A Manuscript in the Diocesan Library of Derry, in Ireland." Great Britain Historical Manuscripts Commission, pp. 639-40. London, 1881. (Horwood describes and presents selected citations from a treatise on weights and measures, predominantly ancient, written by William Harrison sometime around 1590.)

Hostmen Extracts from the Records of the Company of Hostmen of Newcastle-Upon-Tyne. (Surtees Society Publication, Vol. 105.) Durham, 1901.

Hoveden Chronica Rogeri de Hoveden, ed. William Stubbs. (Rerum Britannicarum Medii Aevi Scriptores, Vols. 3 and 4.) London, 1870-71.

Howes Howes, Edward. Short Arithmetick: or, The Old and Tedious Way of Numbering, Reduced to a New and Briefe Method. London, 1656.

H. Taylor Taylor, Henry. The Decimal System, as Applied to the Coinage & Weights & Measures of Great Britain. London, 1851. (The fourth edition.)

Hultsch Hultsch, Fridericus. Metrologicorum scriptorum reliquiæ. 2 vols. Lipsiae, 1864. (Hultsch includes the writings of Isidore of Seville on weights and measures. His book is especially valuable for late Roman and early medieval tracts dealing with linear measures.)

Hunt Hunt, Nicholas. The Merchants Iewell. London, 1628.

Huntar Huntar, Alexander. A Treatise of Weights, Mets and Measures of Scotland. Edinburgh, 1624. (The best early modern treatment.)

Hunter Hunter, Rev. Joseph. The Hallamshire Glossary. London, 1824. (Some valuable metrological entries.)

Huntley Huntley, Rev. Richard Webster. A Glossary of the Cotswold (Gloucestershire) Dialect. London, 1868. (Huntley discusses the lug.)

Hyda Liber monasterii de Hyda; Comprising a Chronicle of the Affairs of England, from the Settlement of the Saxons to the Reign of King Cnut; and a Chartulary of the Abbey of Hyde, in Hampshire: A.D. 455-1023, ed. Edward Edwards. (Rerum Britannicarum Medii Aevi Scriptores.) London, 1866.

Hylles Hylles, Thomas. The Arte of Vulgar Arithmeticke. London, 1600.

Ingalls Ingalls, Walter Renton. Systems of Weights and Measures. New York, 1945.

Int. Traders´ International Traders´ Handbook. Philadelphia, 1934.

Ireland Historic and Municipal Documents of Ireland, A.D. 1172-1320, ed. J. T. Gilbert. (Rerum Britannicarum Medii Aevi Scriptores.) London, 1870.

Jackson Jackson, Lowis D´A. Modern Metrology: A Manual of the Metrical Units and Systems of the Present Century. London, 1882.

Jacobus Excerpta e Libris Domicilii domini Jacobi quinti regis Scotorum, ed. The Bannatyne Club. Edinburgh, 1836.

Jager Jager, Robert. Artificial Arithmetick in Decimals. London, 1651. (One of the earliest treatments of decimals for the general public.)

Jeake Jeake, S. Arithmetic. London, 1696.

Jefferson 1 Jefferson, Thomas. "Plan for Establishing Uniformity in the Coinage, Weights and Measures of the United States: Communicated to the House of Representatives, July 13, 1790," in Saul K. Padover (ed.), The Complete Jefferson, pp. 974-95. New York, 1943. (Jefferson discusses many different linear, superficial, capacity, and quantity measures in addition to several types of weights.)

_____ **2** _____. "Standards of Measures, Weights and Coins," in Saul K. Padover (ed.), Ibid., pp. 1004-11.

Jenkins Jenkins, David. Pacis Consultum: A Directory to the Publick Peace. London, 1657.

Jessop Jessop, William H. R. A Complete Decimal System of Money and

Measures. Cambridge, 1855.

J. Eyre Eyre, J. The Exact Surveyor: or, the Whole Art of Surveying of Land. London, 1654.

Jocelinus Cronica Jocelini de Brakelonda, de rebus gestis Samsonis abbatis monasterii Sancti Edmundi, ed. Johanne Gage Rokewode. (Camden Society Publication, Vol. 13.) London, 1840.

Johnson Johnson, Samuel. A Dictionary of the English Language. 2 vols. London, 1773. (Although there are many entries for weights and measures, there are occasional errors in both their size and composition.)

John. Univ. Cyclo. Johnson's Universal Cyclopædia, ed. Charles Kendall Adams. Vol. 8. New York, 1895. (See article entitled Weights and Measures.)

Jolly Jolly, Alexander. Conversion of Weights Tables Showing the Live & Dead Weight of Cattle, Sheep, & Pigs. London, 1888.

Jones Jones, Stacy V. Weights and Measures: An Informal Guide. Washington, 1963. (Jones makes relatively few references to medieval or early modern English weights and measures.)

Jourdan Recueil général des anciennes Lois françaises, depuis l'an 420 jusqu'à la révolution de 1789, ed. MM. Jourdan, Decrusy, and Isambert. 30 vols. Paris, 1830. (Information on weights and measures is contained in volumes 1, 3, 9, 11, 13, 14, 18, 20, 22, 24, 25, 26, and 27.)

J. Owen Owen, John. The Youth's Instructor in the English Tongue.

London, 1732. (No. 14 of A Collection of Facsimile Reprints of English Linguistics, 1500-1800; published by The Scolar Press Limited, 1967.)

J. Sheppard Sheppard, James. The British Corn Merchant´s and Farmer´s Manual. Derby, 1820.

Judson Judson, Lewis Van Hagen. "Weights and Measures." Encyclopædia Britannica, 23 (1964), 479-88.

Justice Justice, Alexander. A General Treatise of Monies and Exchanges. London, 1707.

Kater 1 Kater, Henry. "An Account of the Comparison of Various British Standards of Linear Measure." Philosophical Transactions, 111 (1821), 75-94.

_____ **2** _____. "An Account of the Construction and Adjustment of the New Standards of Weights and Measures of the United Kingdom of Great Britain and Ireland." Ibid., 116 (1826), 1-52. (Kater lists the standards found in the cities of Edinburgh, London, and Westminster.)

_____ **3** _____. "On the Error in Standards of Linear Measure, Arising from the Thickness of the Bar on which They Are Traced." Ibid., 120 (1830), 359-81.

Keith 1 Keith, George Skene. Different Methods of Establishing an Uniformity of Weights and Measures. London, 1817.

_____ **2** _____. Tracts on Weights, Measures, and Coins. London, 1791. (Many valuable documents.)

Kelly Kelly, P. Metrology; or, an Exposition of Weights and Measures, Chiefly Those of Great Britain and France: Comprising Tables of

Comparison, and Views of Various Standards; with an Account of Laws and Local Customs, Parliamentary Reports, & Other Important Documents. London, 1816.

Kennelly Kennelly, Arthur E. Vestiges of Pre-Metric Weights and Measures Persisting in Metric-System Europe. New York, 1928. (Kennelly is interested principally in French weights and measures.)

Kennett Kennett, White. Parochial Antiquities Attempted in the History of Ambrosden, Burcester, and Other Adjacent Parts in the Counties of Oxford and Bucks. Oxford, 1695. (An important glossary.)

King King, Charles. The British Merchant; or, Commerce Preserv'd. Vol. 1. London, 1721.

Kisch Kisch, Bruno. Scales and Weights: A Historical Outline. New Haven, 1965. (Kisch concentrates chiefly on the ancient and modern periods. His most important contributions are his excellent descriptions of scales and his index of important weights used in the world today.)

Klimpert Klimpert, Richard. Lexicon der Munzen, Mässe, Gewichte: Zählarten und Zeitgrössen aller Länder der Erde. Berlin, 1896. (Klimpert is concerned primarily with French and German metrology.)

Kytchin Kytchin, John. Le Covrt Leete et Court Baron. London, 1580.

Labbe Labbe, Philippe. Bibliotheca bibliothecarum curis secundis auctior. Rothomagi, 1672.

Lambart Lambart, James. The Countrymans Treasure. London, 1683.

Langtoft Langtoft, Peter. Chronicle, ed. Thomas Hearne. London, 1810.

(The glossary contains descriptions of the larger superficial measures such as the farthingdale, virgate, and hide.)

Lavoisier 1 Lavoisier, Antoine Laurent. Oeuvres, ed. René Fric. 2 vols. Paris, 1955. (Lavoisier was one of the pioneers in the construction of the metric system, which put an end to the complexity and confusion of French weights and measures. These two volumes contain some of his writings on early modern measuring units.)

_____ **2** _____. Statistique agricole et projets de réformes, ed. Edouard Grimaux. Paris, 1888.

Laws The Laws of the Earliest English Kings, ed. F. L. Attenborough. Cambridge, 1922.

Laws Wales Ancient Laws and Institutes of Wales; Comprising Laws Supposed to be Enacted by Howel the Good, Modified by Subsequent Regulations Under the Native Princes Prior to the Conquest by Edward the First, etc. (Great Britain Record Commission Publications.) London, 1841.

Leake Leake, Stephen Martin. An Historical Account of English Money from the Conquest to the Present Time. London, 1793. (Leake discusses the troy and tower systems of weight.)

Leet Continuation of the Court Leet Records of the Manor of Manchester: A.D. 1586-1602, ed. John Harland. (Chetham Society Publication, Vol. 65.) Manchester, 1865.

Leet Juris. Leet Jurisdiction in the City of Norwich during the XIIIth and XIVth Centuries, ed. William Hudson. (Selden Society Publication,

Vol. 5.) London, 1892. (Some of the cases deal with infractions of metrological regulations and with the various fines and amercements levied as penalties.)

Leigh Leigh, Egerton. A Glossary of Words Used in the Dialect of Cheshire. London, 1877. (Leigh includes separate entries for 9 weights and measures.)

Leland The Itinerary in Wales of John Leland in or about the Years 1536-1539, ed. Lucy Toulmin Smith. London, 1906.

Letter 1 "Letter from the Secretary of the Interior Transmitting in Response to a Resolution of the House of Representatives, Reports Concerning the Adoption of the Metric System of Weights and Measures." Metric System Pamphlets, 1 (1878), No. 7.

_____ **2** "Letter from the Secretary of the Treasury Transmitting to the House of Representatives Certain Reports in Reference to the Adoption of the Metric System." Ibid., No. 2.

_____ **3** "Letter from the Secretary of War Transmitting Reports of Chiefs of Bureaus upon the Adoption of the Metrical System, in Response to a Resolution of the House of Representatives." Ibid., No. 6.

Levi Levi, Leone. The History of British Commerce and of the Economic Progress of the British Nation: 1763-1878. London, 1880.

Leybourn 1 Leybourn, William. The Compleat Surveyor. London, 1653.

_____ **2** _____. Planometria: or, The Whole Art of Svrveying of Land. London, 1650.

Liber Liber quotidianus contraolulatoris garderobæ: Anno regni regis

Edwardi primi: A.D. MCCXCIX and MCCC. London, 1787. (There are several descriptions of heaped, striked, and shallow capacity measures together with a discussion of the crannock.)

Lightwood Lightwood, John M. A Treatise on Possession of Land. London, 1894.

Lingelbach Lingelbach, W. E. The Merchant Adventurers of England: Their Laws and Ordinances with Other Documents. Philadelphia, 1902.

Lipson Lipson, E. The Economic History of England, Vol. 1. London, 1949. (Lipson discusses the supervision of weights and measures by state and local officials.)

Lloyd Lloyd, J. E. Early Welsh Agriculture. Bangor, 1894.

Loch Loch, David. Essays on the Trade, Commerce, Manufactures, and Fisheries of Scotland. 3 vols. Edinburgh, 1778-79.

London Croniques de London, depuis l'an 44 Hen. III. jusqu'à l'an 17 Edw. III, ed. George James Aungier. (Camden Society Publication, Vol. 28.) London, 1844.

Lyte Lyte, Henry. The Art of Tens, or Decimall Arithmeticke. London, 1619. (Another early decimal presentation to the general public.)

McCaw McCaw, G. T. "Linear Units Old and New." Empire Survey Review, 5 (1939-40), 236-59. (McCaw discusses the primitive measures of length, the foot and cubit, the Greek foot, the "natural" or Olympic foot, and the Gallic leuca.)

McConnell McConnell, Primrose. Note-Book of Agricultural Facts & Figures for Farmers and Farm Students. London, 1883.

Macpherson Macpherson, David. Annals of Commerce, Manufactures, Fisheries, and Navigation with Brief Notices of the Arts and Sciences Connected with Them. 4 vols. London, 1805. (Volume 1 has some excellent tables of medieval English measuring units.)

Mag. Carta The Great Charter Called I[n] Latyn Magna Carta with Divers Olde Statutes. London, 1541.

Maitland Maitland, Frederic William. Domesday Book and Beyond: Three Essays in the Early History of England. New York, 1966. (First published in 1897 in Cambridge, England and Boston, Massachusetts.)

Malcolm Regesta regum Scottorum: The Acts of Malcolm IV King of Scots 1153-1165, ed. G. W. S. Barrow. Edinburgh, 1960.

Malmesbury The Register of Malmesbury Abbey, eds. J. S. Brewer and Charles Trice Martin. (Rerum Britannicarum Medii Aevi Scriptores, 2 vols.) London, 1886.

Mann Mann, W. Wilberforce. A New System of Measures, Weights, and Money; Entitled The Linn-Base Decimal System. New York, 1871. (One of the worst reform proposals ever conceived.)

Manydown The Manor of Manydown Hampshire, ed. G. W. Kitchin. London, 1895. (There is a brief discussion of superficial measurement in the introduction.)

Margan "Annales de Margan (A.D. 1066-1232)," in Henry Richards Luard (ed.), Annales monastici. (Rerum Britannicarum Medii Aevi Scriptores.) London, 1864.

Marianae Marianae, Joannis. De rege et regis institutione. Types

Wechelianis, apud Laeredes, 1611.

Marshall Marshall, William H. The Rural Economy of Yorkshire. London, 1788. (Some information on superficial measurement.)

Martin Martin, William. An Attempt to Establish Throughout His Majesty's Dominions an Universal Weight and Measure, Dependant on Each Other, and Capable of Being Applied to Every Necessary Purpose Whatever. London, 1794.

Mason Mason, R. A Mirrovr for Merchants. London, 1609.

Masterson Masterson, Thomas. Thomas Masterson His First (-Third) Booke of Arithmeticke. London, 1592.

Matthaei 1 Matthaei Parisiensis. Chronica Majora, ed. Henry Richards Luard. (Rerum Britannicarum Medii Aevi Scriptores.) London, 1874. (Matthew refers to Richard I's attempt to standardize weights and measures in 1189 in a section entitled De persecutione Judæorum.)

_____ **2** _____. Historia Anglorum, ed. Sir Frederic Madden. (Rerum Britannicarum Medii Aevi Scriptores.) London, 1865-69.

Mayne Mayne, John. Arithmetick: Vulgar, Decimal, & Algebraical. London, 1675.

Melsa Chronica monasterii de Melsa, a fundatione usque ad annum 1396, auctore Thoma de Burton, abbate, ed. Edward A. Bond. (Rerum Britannicarum Medii Aevi Scriptores, 3 vols.) London, 1866.

Memorials Memorials of London and London Life in the XIIIth, XIVth and XVth Centuries, ed. Henry Thomas Riley. London, 1868. (There are several descriptions of capacity measures in addition to a number of inventories. The footnotes occasionally give information on the types of scales used by merchants.)

Mem. Roll The Memoranda Roll for the Michaelmas Term of the First Year of the Reign of King John (1199-1200), ed. H. G. Richardson. (Pipe Roll Society.) London, 1943.

Mendenhall Mendenhall, Thomas C. "Fundamental Units of Measure." Smithsonian Institution Annual Report, pp. 135-49. Washington, 1893. (Originally a paper read before the International Engineering Congress of the Columbian Exposition, Chicago, 1893.)

Mer. Adven. Extracts from the Records of the Merchant Adventurers of Newcastle-Upon-Tyne. (Surtees Society Publication, Vol. 93.) Durham, 1895. (There is a detailed discussion of the chalder and the keel.)

Metric The Metric Versus the English System of Weights and Measures. (National Industrial Conference Board.) Research Report No. 42. New York, 1921.

Met. Univ. "Metrology Universalized; or, A Proposal to Really Equalize and Universalize the Hitherto Unequalized and Arbitrary Weights and Measures of Great Britain and America." Metric System Pamphlets, 1 (1828), No. 8. (This article deals primarily with English linear measurement.)

Miller Miller, Sir John Riggs. "Equalization of Weights and Measures." The Parliamentary History of England from the Earliest Period to the Year 1803, 28 (1789-91), 639-650. London, 1816. (Also used 1790 monographic version entitled Speeches in the House of Commons Upon the

Equalization of the Weights and Measures of Great Britain, etc.)

Mon. Fran. Monumenta Franciscana: Being a Further Collection of Original Documents Respecting the Franciscan Order in England, ed. Richard Howlett. (Rerum Britannicarum Medii Aevi Scriptores, Vol. 2.) London, 1882.

Mon. Jur. Monumenta Juridica: The Black Book of the Admiralty, ed. Sir Travers Twiss. (Rerum Britannicarum Medii Aevi Scriptores.) London, 1871 and 1873.

Montgomery Montgomery, William Ernest. The History of Land Tenure in Ireland. Cambridge, 1889.

More More, Richard. The Carpenters Rule to Measure Ordinarie Timber, etc. London, 1602.

Morton Morton, John C. A Cyclopedia of Agriculture Practical and Scientific. London, 1855.

Mun. acad. Munimenta academica, or Documents Illustrative of Academical Life and Studies at Oxford, ed. Rev. Henry Anstey. (Rerum Britannicarum Medii Aevi Scriptores.) London, 1868. (The duties of the Chancellor of Oxford University in regard to the maintenance of Crown standards are described in several documents.)

Mun. gild. Munimenta gildhallæ Londoniensis: Liber Albus, Liber Custumarum et Liber Horn, ed. Henry Thomas Riley. (Rerum Britannicarum Medii Aevi Scriptores.) London, 1859-62. (The Liber Albus and the Liber Custumarum are the most important for information on weights and measures.)

Murphy Murphy, Edmund. The Agricultural Instructor or Farmer´s Class-Book. Dublin, 1849.

Naft Naft, Stephen. Conversion Equivalents in International Trade. Philadelphia, 1931.

Neilson Neilson, Nellie. Economic Conditions on the Manors of Ramsey Abbey. Philadelphia, 1898. (Valuable documents in the appendix.)

Nicholson Nicholson, Edward. Men and Measures: A History of Weights and Measures: Ancient and Modern. London, 1912. (There are tables of some medieval English, Irish, Scots, and Welsh weights and measures as well as some valuable discussions dealing with the ancient systems of metrology. Unfortunately, Nicholson exaggerates the influence which many of the ancient systems had on English metrological development. In addition, he seldom indicates the sources of his data.)

Nicol. and Burn Nicolson, Joseph, and Burn, Richard. The History and Antiquities of the Counties of Westmorland and Cumberland, Vol. 2. London, 1777. (The glossary has definitions for several superficial measures.)

Norden Norden, John. The Surveiors Dialogue. London, 1610.

Nottingham Records of the Borough of Nottingham: Being a Series of Extracts from the Archives of the Corporation of Nottingham. 2 vols. London, 1883.

Nourse Nourse, T. Campania Foelix: or, a Discourse of the Benefits and Improvements of Husbandry. London, 1700.

Noy Noy, R. Complete Lawyer. London, 1634. (1651 edition is entitled

The Compleat Lawyer, or a Treatise Concerning Tenvres and Estates.)

O´Keefe O´Keefe, John A. The Law of Weights and Measures. London, 1966. (Supplement published in 1967.)

Oldberg Oldberg, Oscar. A Manual of Weights and Measures. Chicago, n.d. (Weights and measures for everyday use.)

Oldfield Oldfield, Thomas. A Table of Silver Weight. London, 1696.

Osmund The Register of S. Osmund, ed. W. H. Rich Jones. (Rerum Britannicarum Medii Aevi Scriptores, 2 vols.) London, 1883-84.

Owen Owen, George A. A Treatise on Weighing Machines. London, 1922.

Oxford Medieval Archives of the University of Oxford, ed. Rev. H. E. Salter. Oxford, 1921. (There are several descriptions of the seals used by the Chancellor of Oxford University in authenticating weights and measures under his jurisdiction.)

Palethorpe Palethorpe, Joseph. A Commercial Dictionary of the Names of All the Coins, Weights and Measures in the World. Derby, 1829. (The title is an exaggeration.)

Palladius Palladius. On Husbondrie, ed. Rev. Barton Lodge. (Early English Text Society.) London, 1873.

Pasley Pasley, C. W. Observations on the Expediency and Practicability of Simplifying and Improving the Measures, Weights and Money, Used in This Country, without Materially Altering the Present Standards. London, 1834. (Very important for all measurement divisions.)

Paucton Paucton, Alexis Jean Pierre. Métrologie, ou traité des mesures, poids et monnoies des anciens peuples & des modernes. Paris, 1780.

Pauli Drei volkswirthschaftliche Denkschriften aus der Zeit Heinrichs VIII. von England, ed. Reinhold Pauli. Göttingen, 1878.

Pearman Pearman, M. T. A History of the Manor of Bensington. London, 1896. (Some remarks on superficial measures.)

Pegolotti Pegolotti, Francesco Balducci. La Pratica della mercatura, ed. Allan Evans. Cambridge, 1936. (Pegolotti describes the clove, hundred, and stone in addition to many non-English units.)

Pell Pell, O. C. "Summary of a New View of the Geldable Unit of Assessment of Domesday," in Domesday Studies, ed. P. Edward Dove. Vol. 2, pp. 561-619. London, 1888.

Penkethman Penkethman, John. A Perfect Table, Declaring the Assize or Weight of Bread. London, 1640.

People´s Cyclo. The People´s Cyclopedia of Universal Knowledge, with Numerous Appendixes Invaluable for Reference in All Departments of Industrial Life, ed. William Harrison De Puy. Vol. 3. New York, 1885.

Percy Northumberland Household Book (The Regulations and Establishment of the Household of Henry Algernon Percy, the Fifth Earl of Northumberland) 1512-25. London, 1770.

Perkin Perkin, F. Mollwo. The Metric and British Systems of Weights, Measures and Coinage. New York, 1907.

Perry Perry, John. The Story of Standards. New York, 1955.

Petrie 1 Petrie, Sir William M. Flinders. Measures and Weights. London, 1934. (Petrie was a famous Egyptologist.)

_____ **2** _____. "The Old English Mile." Proceedings of the Royal Society of Edinburgh, 12 (1882-84), 254-66.

Phillips Phillips, Edward. The New World of English Words: or, A General Dictionary. London, 1696. (Edition of 1706 used also.)

Pigott Pigott, I. The Canadian Mechanic's Ready Reckoner. Three Rivers, Canada, 1832.

Pipe The Great Rolls of the Pipe for the Seventeenth Year of the Reign of King Henry the Second: A.D. 1170-1. (Pipe Roll Society.) London, 1893.

Pope Pope, Charles. The Merchant, Ship-Owner, and Ship-Master's Import and Export Guide. London, 1831.

Postlethwayt Postlethwayt, Malachy. The Universal Dictionary of Trade and Commerce. 2 vols. London, 1755. (Weights and measures information is in volume 2, pp. 186-97.)

Powell Powell, John. The Assize of Bread. London, 1595.

Prior Prior, W. H. "Notes on the Weights and Measures of Medieval England." Bulletin du Cange: Archivvm Latinitatis medii ævi, 1 (1924), 77-170. (Prior defines briefly over 100 weights and measures. He makes occasional errors in computation, but, on the whole, his work is valuable for metrological study.)

Promp. Parv. Promptorium Parvulorum, ed. Albertus Way. (Camden Society Publication.) London, 1843.

Rameseia Cartularium Monasterii de Rameseia, ed. William Henry Hart and Rev. Ponsonby A. Lyons. (Rerum Britannicarum Medii Aevi Scriptores.) London, 1884-93.

Rastell Rastell, John. An Exposition of Certaine Difficult and Obscure Words and Termes of the Lawes of This Realme. London, 1579.

Rates 1 The Rates of the Custome House Bothe Inwarde and Outwarde the Difference of Measures and Weyghts and Other Commodities Very Necessarye for All Marchantes to Knowe Newly Correctyd and Imprynted. London, 1545.

_____ **2** The Rates of the Custome House Reduced into a Much Better Order for the Redier Finding of Any Thing Therin Contained. London, 1590.

Rathborne Rathborne, Aaron. The Svrveyor in Foure Bookes. London, 1616. (A pioneer of the "chain" for land measurement surveying.)

Rawlyns Rawlyns, Richard. Practical Arithmetick. London, 1656.

Ray Ray, John. A Collection of English Words not Generally Used. London, 1674. (Scant metrological information.)

Recorde Recorde, Robert. The Ground of Artes, Teachyng the Worke and Practice of Arithmetike. London, 1540. (Edition of 1566 used also.)

Records Records of the Coinage of Scotland from the Earliest Period to the Union, ed. R. W. Cochran-Patrick. 2 vols. Edinburgh, 1876. (There are several remarks on the avoirdupois and troy pounds.)

Red Book The Red Book of the Exchequer, ed. Hubert Hall. (Rerum Britannicarum Medii Aevi Scriptores, 3 vols.) London, 1896.

Relation A Relation, or Rather a True Account, of the Island of England; with Sundry Particulars of the Customs of These People, and of the Royal Revenues under King Henry the Seventh, about the Year 1500, ed. Charlotte Augusta Sneyd. (Camden Society Publication, Vol. 37.)

London, 1847. (This collection contains several lists of products and the capacity measures by which they were sold.)

Remembrance The Third Book of Remembrance of Southampton: 1514-1602, ed. A. L. Merson. Southampton, 1955.

Renton Renton, George. The Graziers´ Ready Reckoner. London, 1804.

Report 1 "Report from the Committee Appointed to Inquire into the Original Standards of Weights and Measures in This Kingdom, and to Consider the Laws Relating Thereto." Report from Committees of the House of Commons, 2 (1737-65), 411-51. (This report contains some excellent discussions of medieval English weights and measures. Ample documentation, especially from Fleta, and concise summaries of metrological laws make this an important source.)

_____ **2** "Report from the Committee Appointed (upon the 1st Day of December, 1758) to Inquire into the Original Standards of Weights and Measures in This Kingdom; and to Consider the Laws Relating Thereto." Ibid., 455-63.

_____ **3** "Report from the Select Committee of the House of Lords Appointed to Consider the Petition of the Directors of the Chamber of Commerce and Manufactures, Established by Royal Charter in the City of Glasgow Taking Notice of the Bill Entitled ´An Act for Ascertaining and Establishing Uniformity of Weights and Measures etc.´" Reports from Committees, 7 (1824), 1-35.

_____ **4** "Report from the Select Committee on the Weights and Measures Act; Together with the Minutes of Evidence." Ibid., 18 (1835), 1-60.

_____ **5** "Report of the Committee Appointed to Superintend the Construction of the New Parliamentary Standards of Length and Weight." Ibid., 19 (1854), 1-23.

Reynardson Reynardson, Samuel. A State of the English Weights and Measures of Capacity. London, 1750.

Ricard Ricard, Samuel. Traité général du Commerce. 2 vols. Amsterdam, 1781. (Vol. 2, pp. 150-56, discusses English weights and measures.)

Ricart The Maire of Bristowe Is Kalendar by Robert Ricart, Town Clerk of Bristol 18 Edward IV, ed. Lucy Toulmin Smith. (Camden Society Publication, New Series, Vol. 5.) Westminster, 1872. (There is a description of the crannock in this calendar.)

Rich Rich, E. E. The Ordinance Book of the Merchants of the Staple. Cambridge, 1937. (The Ordinance Book of 1565 covers pages 103-200.)

Richmond Registrum Honoris de Richmond, ed. Roger Gale. London, 1722. (Especially important are the discussions of land measures in the appendixes; appendix 2 on land dimensions was compiled by Arthur Agard in 1599 during his tenure as Deputy Chamberlain of the Exchequer from 1570 to 1615.)

Rider Rider, John. Riders Dictionarie, Corrected and Augmented with the Addition of Many Hundred Words Both Out of the Law, and Out of the Latine, French, and Other Languages. London, 1640.

Ridgeway Ridgeway, William. The Origin of Metallic Currency and Weight Standards. Cambridge, 1892.

Roberts Roberts, Lewes. The Merchants Map of Commerce. London, 1677.

Robertson Robertson, E. William. Historical Essays in Connexion with the Land, the Church, &c. Edinburgh, 1872.

Robinson Robinson, Francis Kildale. A Glossary of Yorkshire Words and Phrases Collected in Whitby and the Neighbourhood. London, 1855.

Rogers Rogers, W. A. "On the Present State of the Question of Standards of Length." Proceedings of the American Academy of Arts and Sciences, 15 (1879-80), 273-312.

Rolls Three Rolls of the King´s Court in the Reign of King Richard the First: A.D. 1194-1195. (Pipe Roll Society.) London, 1891.

Rolt Rolt, Richard. A New Dictionary of Trade and Commerce. London, 1756. (Local and regional comparisons.)

Romé de L´Isle Romé de L´Isle, M. de. Métrologie, ou tables pour servir à l´intelligence des poids et mesures des anciens. Paris, 1789.

Rördansz Rördansz, C. W. European Commerce: or, Complete Mercantile Guide to the Continent of Europe. London, 1818.

Rot. Parl. 1 Rotuli parliamentorum ut et petitiones, et placita in parliamento tempore Edwardi R. I, ed. Rev. John Strachey et al. Vol. 1 (1278-1325). London, 1832. (The Rotuli parliamentorum are especially valuable for information on capacity measures and for descriptions of cloth measurements.)

_____ **2** Rotuli parliamentorum ut et petitiones, et placita in parliamento tempore Edwardi R. III, ed. Rev. John Strachey et al. Vol. 2 (1326-77). London, 1832.

_____ **3** Rotuli parliamentorum ut et petitiones, et placita in parliamento tempore Ricardi R. II, ed. Rev. John Strachey et al. Vol. 3 (1377-1411). London, 1832.

_____ **4** Rotuli parliamentorum ut et petitiones, et placita in parliamento tempore Henrici R. V, ed. Rev. John Strachey et al. Vol. 4 (1413-37). London, 1832.

_____ **5** Rotuli parliamentorum ut et petitiones, et placita in parliamento ab anno decimo octavo R. Henrici sexti ad finem ejusdem regni, ed. Rev. John Strachey et al. Vol. 5 (1439-68). London, 1832.

_____ **6** Rotuli parliamentorum ut et petitiones, et placita in parliamento ab anno duodecimo R. Edwardi IV ad finem ejusdem regni, ed. Rev. John Strachey et al. Vol. 6 (1472-1503). London, 1832.

Rot. parl. Ang. Rotuli parliamentorum Anglie hactenus inediti MCCLXXIX-MCCCLXXIII, ed. H. G. Richardson and George Sayles. (Camden Third Series, Vol. 51.) London, 1935.

Round 1 Round, J. H. "Barons and Knights in the Great Charter," in Magna Carta Commemoration Essays. London, 1917.

_____ **2** _____. Feudal England. London, 1895. (The first part of the book consists of studies of the Domesday Book and of similar but less important surveys; the second part, of essays embodying the results of the author's researches in the history of this period.)

_____ **3** _____. "Notes on Domesday Measures of Land," in Domesday Studies, ed. P. Edward Dove. Vol. 1, pp. 189-225. London, 1888.

R. Powell Powell, Robert. A Treatise of the Antiquity, Authority, Vses and Jurisdiction of the Ancient Courts of Leet. London, 1642.

Ruding Ruding, Rogers. Annals of the Coinage of Great Britain and Its Dependencies; from the Earliest Period of Authentic History to the Reign of Victoria. 2 vols. London, 1840.

St. Edmunds Feudal Documents from the Abbey of Bury St. Edmunds, ed. D. C. Douglas. London, 1932.

St. Gile´s Memorials of St. Gile´s, Durham, Being Grassmen´s Accounts and Other Parish Records, Together with Documents Relating to the Hospital of Kepier and St. Mary Magdalene. (Surtees Society Publication, Vol. 95.) Durham, 1896. (This volume contains information on the standard weights for bread.)

St. Mary´s Chartularies of St. Mary´s Abbey, Dublin: with the Register of Its House at Dunbrody, and Annals of Ireland, ed. John T. Gilbert. (Rerum Britannicarum Medii Aevi Scriptores, Vol. 2.) London, 1884.

St. Paul´s The Domesday of St. Paul´s of the Year MCCXXII; or, Registrum de visitatione maneriorum per Robertum Decanum, ed. William Hale. (Camden Society Publication, Vol. 69.) Westminster, 1858. (This is a particularly important source for capacity measures and for large superficial measures such as the hide and virgate.)

St. Peter´s Historia et Cartularium Monasterii Sancti Petri Gloucestriæ, ed. William Henry Hart. (Rerum Britannicarum Medii Aevi Scriptores, Vol. 3.) London, 1867.

Salignacus Salignacus, Bernardus. The Principles of Arithmeticke, trans W. Bedwell. London, 1616.

Salisbury Charters and Documents Illustrating the History of the

Cathedral, City, and Diocese of Salisbury, in the Twelfth and Thirteenth Centuries, ed. Rev. W. Dunn Macray. (Rerum Britannicarum Medii Aevi Scriptores.) London, 1891.

Salzman 1 Salzman, L. F. English Industries of the Middle Ages. Oxford, 1923. (Salzman discusses the various capacity measures that were used for coal, iron, fish, and malted beverages.)

_____ **2** _____. English Trade in the Middle Ages. Oxford, 1931. (One chapter is devoted to English weights and measures. Most of Salzman's remarks are really too brief to be effective, and his most important contribution is his section on Roman and Arabic methods of computation.)

_____ **3** _____. "Hides and Virgates in Sussex," in English Historical Review, 19 (1904), 92-96.

Samson The Kalendar of Abbot Samson of Bury St. Edmunds and Related Documents, ed. R. H. C. Davis. (Camden Third Series, Vol. 84.) London, 1954. (This book contains information on socage land and socage dues in addition to several descriptions of unusual capacity measures found at St. Edmunds's.)

Sandys Sandys, Sir John Edwin. A Companion to Latin Studies. (3rd ed.) Cambridge, 1921. (Sandys discusses some aspects of Roman metrology.)

Scot. Lawes Scotland: The Lawes and Actes of Parliament. Edinburgh, 1597. (A valuable source, especially for variant spellings.)

Scott Scott, George W. The Decimal System of Weights and Measures As Authorized by Act of Congress. Albany, 1867.

Scrope Scrope, G. Poulett. History of the Manor and Ancient Barony of

Castle Combe in the County of Wilts. London, 1852. (Many important rental rolls and court records.)

Sebbenhuth Hereafter Ensueth the Auncient Customes of the Mannors of the Sebbenhuth. London, 1610.

Second Rep. "Second Report of the Commissioners Appointed by His Majesty to Consider the Subject of Weights and Measures." Reports from Commissioners, 7 (1820), 1-40. (This report is extremely valuable for it includes a 40-page listing of the state and local units of measurement that were common in the late 1700s and early 1800s.)

Seebohm Seebohm, Frederic. Customary Acres and Their Historical Importance. London, 1914.

Select Cases 1 Select Cases before the King's Council in the Star Chamber Commonly Called the Court of Star Chamber, ed. I. S. Leadam. (Selden Society Publication, Vol. 26.) London, 1911. (This book contains information on the wey, firkin, butter barrel, and virgate.)

_____ **2** Select Cases Concerning the Law Merchant: A.D. 1270-1638, ed. Charles Gross. (Selden Society Publication, Vol. 23 [Local Courts].) London, 1908. (There are several descriptions of infractions of weights and measures legislation together with a glossary which defines such capacity measures as the trey and ring.)

_____ **3** Select Cases Concerning the Law Merchant: A.D. 1239-1633, ed. Hubert Hall. (Selden Society Publication, Vol. 46 [Central Courts].) London, 1930.

_____ **4** Select Cases in the Council of Henry VII, ed. C. G. Bayne and

William Huse Dunham, Jr. (Selden Society Publication, Vol. 75.) London, 1958.

Select Col. A Select Collection of Early English Tracts on Commerce from the Originals of Mun, Roberts, North and Others, ed. John Ramsay McCulloch. Cambridge, 1952.

Select Com. "The Select Committee Appointed to Consider the Several Reports Which Have Been Laid Before This House, Relating to Weights and Measures." Reports from Committees, 4 (1821), 1-7.

Select Doc. Select English Historical Documents of the Ninth and Tenth Centuries, ed. F. E. Harmer. Cambridge, 1914. (There are relatively few documents dealing specifically with weights and measures.)

Select Pleas 1 Select Pleas in Manorial and Other Seignorial Courts, ed. F. W. Maitland. (Selden Society Publication, Vol. 28.) London, 1889.

_____ **2** Select Pleas of the Crown, ed. F. W. Maitland. (Selden Society Publication, Vol. 27.) London, 1888.

Seventh Rep. "Seventh Annual Report of the Warden of the Standards on the Proceedings and Business of the Standard Weights and Measures Department of the Board of Trade for 1872-73." Reports from Commissioners, 38 (1873), 1-105. (This long report deals primarily with the laws relating to the inspection and verification of weights and measures. There are some excellent metrological tables in addition to several photographs of common capacity measures.)

Sheppard Sheppard, W. Of the Office of the Clerk of the Market, of

Weights and Measures, and of the Laws of Provision for Man and Beast. London, 1665. (Sheppard's account of the responsibilities and duties of the clerk of the market is detailed and based on the statutes and ordinances and on his own personal observations. His tables of weights and measures are not always accurate, however, and there are too many repetitions, which are occasionally contradictory.)

Shipley Shipley, Joseph T. Dictionary of Early English. New York, 1955. (Shipley defines some unusual units such as the fust, fardel, and seron.)

Shuttleworths The House and Farm Accounts of the Shuttleworths of Gawthorpe Hall, in the County of Lancaster, at Smithils and Gawthorpe, from September 1582 to October 1621, ed. John Harland. (Chetham Society Publication, Vols. 43 and 46.) London, 1858. (Both volumes contain glossaries with descriptions of the weights and measures found on this estate.)

Silegrave Silegrave, Henry of. A Chronicle of English History from the Earliest Period to A.D. 1274, ed. C. Hook. London, 1849. (Printed for the Caxton Society and published from a Cottonian MS.)

Simmonds Simmonds, P. L. The Commercial Dictionary of Trade Products, Manufacturing and Technical Terms: with a Definition of the Moneys, Weights, and Measures of All Countries Reduced to the British Standard. London, 1883.

Skene 1 Skene, Sir John. De verborum Significatione. London, 1597. (Some valuable metrological entries.)

_____ **2** _____. Regiam majestatem, the Auld Lawes and Constitutions of Scotland Faithfullie Collected. London, 1609.

Skilling Skilling, Thomas. The Farmer's Ready Reckoner or Glasnevin Agricultural Tables. Dublin, 1848.

Skinner Skinner, Frederick George. Weights and Measures: Their Ancient Origins and Their Development in Great Britain up to AD 1855. London, 1967.

Slade Slade, C. F. The Leicestershire Survey: c.A.D. 1130. Leicester, 1956. (Preface by Frank M. Stenton; No. 7 in the Department of English Local History Occasional Papers, ed. H. P. R. Finberg.)

Southampton 1 The Local Port Book of Southampton for 1435-36, ed. Brian Foster. Southampton, 1963. (This and the port book for 1439-40 provide numerous examples of the types of capacity and quantity measures which were used for certain products. Occasionally there are descriptions of the measures themselves.)

_____ **2** The Local Port Book of Southampton for 1439-40, ed. Henry S. Cobb. Southampton, 1961.

Speed Speed, William. Tables for Ascertaining the Weight of Cattle, Calves, Sheep, and Hogs, by Measure. London, 1847.

Spelman Spelman, Henry. Glossarium Archaiologicum. London, 1664.

Stat. The Whole Volume of Statutes at Large, which at Anie Time Heeretofore Have Beene Extant in Print. London, 1587. (This is one of the earliest collections of the statutes. It checks out well against later editions and is notable primarily because of its unusual spellings and metrological constructions.)

Stat. Charles A Collection of the Statutes Made in the Reigns of King Charles the I and King Charles the II, ed. Tho. Manby. London, 1687.

Stat. Irel. The Statutes of Ireland, Beginning the Third Yere of K. Edward the Second, and Continuing untill the End of the Parliament, Begunne in the Eleventh Yeare of the Reign of Our Most Gratious Soveraigne Lord King James. Dublin, 1621.

Steele Steele, John. The Hay and Straw Measurer. London, 1882.

Sternberg Sternberg, Thomas. The Dialect and Folk-Lore of Northamptonshire. London, 1851. (Sternberg has entries for 8 measures.)

Stevens Stevens, A. B. Arithmetic of Pharmacy. 4th ed. New York, 1920. (One of the few books containing metrological information specifically for druggists.)

Stevin Stevin, Simon. Disme: the Art of Tenths, or Decimall Arithmetike, trans R. Norton. London, 1608.

Stonor The Stonor Letters and Papers: 1290-1483, ed. Charles Lethbridge Kingsford. (Camden Society Publication, Third Series, Vols. 29 and 30.) London, 1919.

Strachan Strachan, James. A New Set of Tables for Computing the Weight of Cattle by Measurement. London, 1849. (1843 edition used also.)

Stratton Stratton, Samuel W. Report to the International High Commission Relative to the Use of the Metric System in Export Trade. (64th Congress, 1st Session, Senate Document No. 241.) Washington, 1916.

Sussex Domesday Book in Relation to the County of Sussex, ed. William Douglas Parish. Sussex, 1886.

Swinfield A Roll of the Household Expenses of Richard de Swinfield, Bishop of Hereford during Part of the Years 1289 and 1290, ed. Rev. John Webb. (Camden Society Publication, Vol. 59.) London, 1854. (The glossary contains descriptions of several weights and measures.)

Swinton Swinton, John. A Proposal for Uniformity of Weights and Measures in Scotland, By Execution of the Laws Now in Force. Edinburgh, 1779. (Pages 23-130 contain extremely valuable tables of the weights and measures common in the various Scottish shires.)

Swithun A Consuetudinary of the Fourteenth Century for the Refectory of the House of S. Swithun in Winchester, ed. George William Kitchin. London, 1886.

Tait Tait, J. "Hides and Virgates at Battle Abbey," in English Historical Review, 18 (1903), 705-08.

Tap 1 Tap, John. The Path-Way to Knowledge; Containing the Whole Art of Arithmeticke, Both in Numbers and Fractions. London, 1613.

_____ **2** _____. The Seamans Kalender, or, An Ephemerides of the Sunne, Moone, and Certaine of the Most Notable Fixed Starres. 9th edition. London, 1625.

Tarbé Tarbé, M. Nouveau manuel complet des poids et mesures. Paris, 1845. (Mainly important for French metrology, but some English materials are included.)

Taylor Taylor, Isaac. "The Ploughland and the Plough," in Domesday

Studies, ed. P. Edward Dove. Vol. 1, pp. 143-88. London, 1888.

Third Rep. "Third Report of Standards Commission, February 1, 1870." Metric System Pamphlets, 1 (1878), No. 2.

Thomson Thomson, John. New and Correct Tables Shewing, Both in Scots and Sterling Money, the Price of Any Quantity of Grain. Edinburgh, 1761.

Thorpe Thorpe, John. Custumale Roffense, from the Original Manuscript in the Archives of the Dean and Chapter of Rochester. London, 1788.

Thor. Rogers 1 Rogers, James E. Thorold. A History of Agriculture and Prices in England. Vol. 1. Oxford, 1866. (Rogers describes 34 weights and measures and examines the standards upon which they were based. His descriptions, however, are not detailed.)

_____ **2** _____. Six Centuries of Work and Wages. New York, 1884. (Rogers discusses Arabic numbers, the hide, and the gallon.)

Thurston Thurston, Robert Henry. Conversion Tables of Metric and British or United States Weights and Measures. New York, 1883.

Topham Topham, John. Wardrobe Account of the 28th Year of King Edward the First. London, 1789.

Tower An Exact Abridgement of the Records in the Tower of London, ed. Sir Robert Cotton. London, 1657. (This collection contains few references to weights and measures, but occasionally there are summaries of the principal statutes and ordinances that dealt with them.)

Townsend Townsend, John R. "Metric Versus English Systems," in Systems of Units: National and International Aspects, ed. Carl F. Kayan. (Publication No. 57 of the American Association for the Advancement of

Science.) Washington, 1959.

Tracts Old and Scarce Tracts on Money, ed. J. R. McCulloch. London, 1933. (Some metrological information.)

Trade Board of Trade. Report of the Committee on Weights & Measures Legislation. London, 1951. (Presented by the President of the Board of Trade to Parliament by command of King George VI.)

Triulzi Triulzi, Antonio Maria. Bilancio dei pesi e misure di tutte le piazze mercantili dell´ Europa. 5th edition. Venice, 1803.

T. Robinson Robinson, Thomas. The Common Law of Kent: The Customs of Gavelkind. London, 1897.

Tusser Tusser, Thomas. Fiue Hundreth Pointes of Good Husbandrie. (English Dialect Society.) London, 1878.

Val. Leigh Leigh, Valentine. Moste Profitable and Commendable Science of Surueying of Landes, Tenementes, and Hereditamentes. London, 1577.

Vaughan Vaughan, Rice. A Discourse of Coin and Coinage. London, 1675.

Vinogradoff Vinogradoff, P. "Sulung and Hide," in English Historical Review. 19 (1904), 282-86.

Violet Violet, Thomas. The Advancement of Merchandize. London, 1651.

Wagstaff Wagstaff, W. H. The Metric System of Weights and Measures Compared with the Imperial System. London, 1896.

Warburton Warburton, Rev. W. Edward III. London, 1924. (There are short definitions of the sack and pack of wool.)

Warren Warren, Charles. The Ancient Cubit and Our Weights and Measures. London, 1903.

Waterston Waterston, William. A Manual of Commerce. Edinburgh, 1840. (Some interregional comparisons.)

W. C. Dickinson Dickinson, William Croft. Scotland from the Earliest Times to 1603. Edinburgh, 1962.

Wedgwood Wedgwood, Hensleigh. A Dictionary of English Etymology. London, 1878. (Wedgwood gives etymological derivations for 53 units.)

Weights "Weights, Measures and Coins." House of Commons Accounts and Papers, 58 (1864), 1-35.

Weinbaum Weinbaum, Martin. "London unter Eduard I und II." Vierteljahrschrift für Sozial und Wirtschaftsgeschichte, Suppl. 29 (1933). (There is information on the Assisa Panis and the clove.)

Wellingborough Wellingborough Manorial Accounts: A.D. 1258-1323: from the Account Rolls of Crowland Abbey, ed. Frances M. Page. Northamptonshire, 1936.

Welsh The Welsh Port Books (1550-1603), ed. Edward Arthur Lewis. (Cymmrodorion Record Series, Vol. 12.) London, 1927. (Similar in format to the Southampton Port Books.)

W. Hardwicke Hardwicke, W. W. "Currency and Weights and Measures: the Case for Reform." Empire Review, December 1915, 503-510.

Wheeler Wheeler, John. A Treatise of Commerce, Wherin Are Shewed the Commodies Arising By a Wel Ordered, and Rvled Trade. Middleburgh, 1601. (Reprinted in 1931 by the New York University Press, and edited by George Burton Hotchkiss.)

Wigan Wigan, Eleazar. Practical Arithmetick. London, 1695.

Wil. Airy Airy, Wilfrid. "On the Origin of the British Measures of Capacity, Weight and Length." Minutes of Proceedings of the Institution of Civil Engineers, 175 (1909), 164-76. (Airy discusses the origins of the pint, foot, and avoirdupois pound. The appendix includes some drawings of Roman and Egyptian measures together with several extracts from the laws of William I concerning standardization.)

Wilkinson Wilkinson, Robert Oliver. The Druggist´s Price-Book, or a Catalogue of the Drugs, Chemicals, & Perfumery, Generally Sold by Chemists & Druggists, with the Doses and Old Names Annexed. 2nd edition. London, 1832.

Willis Willis, Browne. The History and Antiquities of the Town, Hundred, and Deanry of Buckingham. London, 1755. (Willis discusses the larger superficial measures such as the virgate, bovate, plowland, and hide.)

Wills Wills and Inventories from the Registers of the Commissary of Bury St. Edmund´s and the Archdeacon of Sudbury, ed. Samuel Tymms. (Camden Society Publication, Vol. 49.) London, 1850.

Winchester Documents Relating to the Foundation of the Chapter of Winchester: A.D. 1541-1547, ed. George Williams Kitchin and Francis Thomas Madge. London, 1889.

Winter Winter, George. A Compendious System of Husbandry. London, 1797. (Some interesting metrological comments.)

Wise Wise, Charles. The Compotus of the Manor of Kettering for A.D. 1292. Kettering, 1899.

W. Miller Miller, W. H. "On the Construction of the New Imperial

Standard Pound, and Its Copies of Platinum; and on the Comparison of the Imperial Standard Pound with the Kilogramme des Archives." Philosophical Transactions, 146 (1856), 753-946. (Miller's article contains information on the tower, troy, mercantile, and avoirdupois pounds.)

Wood Wood, William. A Survey of Trade. London, 1718.

Worcester Registrum sive liber irrotularius et consuetudinarius prioratus Beatæ Mariæ Wigorniensis, ed. William Hale. (Camden Society Publication, Vol. 91.) London, 1865.

Worlidge Worlidge, John. Systeme agriculturæ; the Mystery of Husbandry Discovered. London, 1669.

Wurtele Wurtele, Arthur. Tables for Reducing English, Old French and Metrical Measures. Montreal, 1861.

Yates Yates, James. "Narrative of the Origin and Formation of the International Association for Obtaining a Uniform Decimal System of Measures, Weights and Coins." Metric System Pamphlets, 1 (1856), No. 11.

Year Bk. Year Books of Edward IV: 10 Edward IV and 49 Henry VI, ed. N. Neilson. (Selden Society Publication, Vol. 47.) London, 1931. (This volume contains occasional references to infractions of metrological regulations and to punishments imposed for violations of statutory provisions.)

York Mem. 1 York Memorandum Book: Part I (1376-1419). (Surtees Society Publication, Vol. 120.) Durham, 1912. (There are numerous references to weights and measures in addition to a table defining many linear and superficial units. Several documents are concerned with the proper sealing of weights and measures and with the scope of the mayor's jurisdiction in supervising regular assays.)

_____ **2** York Memorandum Book: Part II (1388-1493). (Surtees Society Publication, Vol. 125.) Durham, 1915. (The glossary has definitions for the selion and the butt of land.)

York Mer. The York Mercers and Merchant Adventurers: 1356-1917. (Surtees Society Publication, Vol. 129.) Durham, 1918. (There are several documents that discuss the purchase of new weights and measures to replace the old and defective standards. The glossary contains information on the keel, fatt, and last.)

Young Young, William. The History of Dulwich College. 2 vols. Edinburgh, 1889. (Volume 2 contains some interesting variant spellings of weights and measures in a list of pharmaceutical recipes and in the diary of Edward Allen.)